Lecture Notes in Computer Science 13348

More information about this series at https://link.springer.com/bookseries/558

Reneta P. Barneva · Valentin E. Brimkov ·
Giorgio Nordo (Eds.)

Combinatorial Image Analysis

21st International Workshop, IWCIA 2022
Messina, Italy, July 13–15, 2022
Proceedings

Springer

Editors
Reneta P. Barneva
State University of New York at Fredonia
Fredonia, NY, USA

Giorgio Nordo
University of Messina
Messina, Italy

Valentin E. Brimkov
SUNY Buffalo State
Buffalo, NY, USA

Institute of Mathematics and Informatics
Bulgarian Academy of Sciences
Sofia, Bulgaria

ISSN 0302-9743 ISSN 1611-3349 (electronic)
Lecture Notes in Computer Science
ISBN 978-3-031-23611-2 ISBN 978-3-031-23612-9 (eBook)
https://doi.org/10.1007/978-3-031-23612-9

This Springer imprint is published by the registered company Springer Nature Switzerland AG
The registered company address is: Gewerbestrasse 11, 6330 Cham, Switzerland

Preface

This volume contains the proceedings of the 21st International Workshop on Combinatorial Image Analysis (IWCIA 2022) organized by the MIFT Department (Mathematics, Computer Science, Physics, and Earth Sciences) of the University of Messina, Italy, and held during July 13–15, 2022.

Image analysis provides theoretical foundations and methods for solving real life problems arising in various areas of human practice, such as medicine, robotics, defense, and security. Since typically the input data to be processed are discrete, the "combinatorial" approach to image analysis is a natural one and therefore its applicability is expanding. Combinatorial image analysis often provides advantages in terms of efficiency and accuracy over the more traditional approaches based on continuous models that require numerical computation.

The IWCIA workshop series provides a forum for researchers throughout the world to present cutting-edge results in combinatorial image analysis, to discuss recent advances and new challenges in this research area, and to promote interaction with researchers from other countries. IWCIA had successful prior meetings in Paris (France) 1991, Ube (Japan) 1992, Washington DC (USA) 1994, Lyon (France) 1995, Hiroshima (Japan) 1997, Madras (India) 1999, Caen (France) 2000, Philadelphia, PA (USA) 2001, Palermo (Italy) 2003, Auckland (New Zealand) 2004, Berlin (Germany) 2006, Buffalo, NY (USA) 2008, Playa del Carmen (Mexico) 2009, Madrid (Spain) 2011, Austin, TX (USA) 2012, Brno (Czech Republic) 2014, Kolkata (India) 2015, Plovdiv (Bulgaria) 2017, Porto (Portugal) 2018, and Novi Sad (Serbia) 2020. The workshop organized by the University of Messina retained and enriched the international spirit of these previous editions. The IWCIA 2022 Program Committee was comprised of renowned experts coming from 19 different countries in Asia, Europe, and North and South America.

The workshop received 24 submissions from authors based in seven different countries. Each submitted paper was sent to at least three reviewers for a double-blind review. Easychair provided a convenient platform for smoothly carrying out the rigorous review process. The most important selection criterion for acceptance or rejection of a paper was the overall score received. Other criteria included relevance to the workshop topics, correctness, originality, mathematical depth, clarity, and presentation quality. We believe that as a result, only papers of high quality have been accepted for publication in this volume, comprising the 19 papers presented at IWCIA 2022.

Excellent keynote talks were given by our invited speakers. Bhargab Bhattacharya from the Indian Institute of Technology Kharagpur spoke about digital geometry applications in medical diagnostics and biochemistry. Benedek Nagy from the Eastern Mediterranean University gave a talk on the advances and challenges of the non-traditional 2D grids in combinatorial imaging. Jessica Zhang from Carnegie Mellon University presented machine learning enhanced simulation and PDE-constrained optimization for material transport control in neurons.

The contributed papers included in this volume are grouped into four sections. The first one consists of one invited talk. The second section contains seven papers devoted to digital geometry and topology. The third consists of six papers discussing picture languages. The last section, containing six papers, is devoted to various applications. We believe that many of these papers would be of interest to a broader audience, including researchers in scientific areas such as computer vision, shape modeling, pattern analysis and recognition, and computer graphics.

Many individuals and organizations contributed to the success of IWCIA 2022. The editors are indebted to IWCIA's Steering Committee for endorsing the candidacy of Messina for the 21st edition of the workshop. We wish to thank everybody who submitted their work to IWCIA 2022. We are grateful to all participants and especially to the contributors of this volume. Our most sincere thanks go to the IWCIA 2022 Program Committee whose cooperation in carrying out high-quality reviews was essential in establishing a strong scientific program. We express our sincere gratitude to the keynote speakers, Bhargab B. Bhattacharya, Benedek Nagy, and Jessica Zhang, for the excellent talks and overall contribution to the workshop program.

We are indebted to the organizations and individuals who supported IWCIA 2022 – to the Mayor of Reggio Calabria, Paolo Brunetti, to the Metropolitan Mayor of Reggio Calabria, Carmelo Versace, to the Delegate of the Rector of the University of Messina, Filippo Grasso, to the MIFT Department Research Delegate, Valentina Venuti, and to the University of Messina, the Metropolitan City of Reggio Calabria, the city of Messina, Comune di Reggio Calabria, Bronzi di Riace 50th Years Committee, and Calabria Formazione.

The success of the workshop would not have been possible without the hard work of the Organizing Committee. We are grateful to the host organization – the MIFT Department of the University of Messina – for its support. Finally, we wish to thank the team at Springer for the efficient and kind cooperation in the timely production of this book.

October 2022

Reneta P. Barneva
Valentin E. Brimkov
Giorgio Nordo

Organization

The 21st International Workshop on Combinatorial Image Analysis, IWCIA 2022, was organized by the MIFT Department (Mathematics, Computer Science, Physics and Earth Sciences) of the University of Messina and was held during July 13–15, 2022.

General Chair

Giorgio Nordo	University of Messina, Italy

Program Chairs

Reneta P. Barneva	SUNY Fredonia, USA
Maddalena Bonanzinga	University of Messina, Italy
Valentin E. Brimkov	SUNY Buffalo State, USA
Mario De Salvo	University of Messina, Italy
Saeid Jafari	Topositus, Denmark

Steering Committee

Gabor Herman	CUNY Graduate Center, USA
Valentin E. Brimkov	SUNY Buffalo State, USA
Tibor Lukić	University of Novi Sad, Serbia
Renato M. Natal Jorge	University of Porto, Portugal
Joao Manuel R. S. Tavares	University of Porto, Portugal

Program Committee

Eric Andres	Université de Poitiers, France
Buda Bajić	University of Novi Sad, Serbia
Péter Balaźs	University of Szeged, Hungary
George Bebis	University of Nevada at Reno, USA
Partha Bhowmick	Indian Institute of Technology Kharagpur, India
Arindam Biswas	Indian Institute of Engineering Science and Technology, Shibpur, India
Boris Brimkov	Slippery Rock University, USA
Alfred M. Bruckstein	Technion, Israel
Li Chen	University of the District of Columbia, USA
Lidija Čomić	University of Novi Sad, Serbia

Mousumi Dutt	St. Thomas College of Engineering and Technology, India
Fabien Feschet	Université d'Auvergne, France
Leila De Floriani	University of Maryland, USA
Chiou-Shann Fuh	National Taiwan University, Taiwan
Atsushi Imiya	Chiba University, Japan
Krassimira Ivanova	Bulgarian Academy of Sciences, Bulgaria
Kamen Kanev	Shizuoka University, Japan
Kostadin Koroutchev	Universidad Autónoma de Madrid, Spain
Walter G. Kropatsch	TU Wien, Austria
Jerome Liang	SUNY Stony Brook, USA
Tibor Lukić	University of Novi Sad, Serbia
Benedek Nagy	Eastern Mediterranean University, North Cyprus
Kálmán Palágyi	University of Szeged, Hungary
Meenakshi Paramasivan	University of Trier, Gemany
Hemerson Pistori	Dom Bosco Catholic University, Brazil
Konrad Polthier	Freie Universitaet Berlin, Germany
Paolo Remagnino	Kingston University London, UK
Nikolay Sirakov	Texas A&M University – Commerce, USA
Josef Slapal	Technical University of Brno, Czech Republic
Ivan Štajduhar	University of Rijeka, Croatia
K. G. Subramanian	Madras Christian College, India
João Manuel R. S.	Tavares University of Porto, Portugal
D. G. Thomas	Madras Christian College, India
László Varga	University of Szeged, Hungary
Petra Wiederhold	CINVESTAV-IPN, Mexico
Jinhui Xu	SUNY Buffalo, USA

Invited Speakers

Bhargab B. Bhattacharya	Indian Institute of Technology Kharagpur, India
Benedek Nagy	Eastern Mediterranean University, North Cyprus
Jessica Zhang	Carnegie Mellon University, USA

Organizing Committee

Bhimraj Basumatary	Bodoland University, India
Maddalena Bonanzinga	University of Messina, Italy
Mario De Salvo	University of Messina, Italy
Farkhanda Afzal	National University of Sciences and Technology, Pakistan
Saeid Jafari	Topositus, Denmark

Arif Mehmood — University of Science and Technology Bannu, Pakistan
Giorgio Nordo — University of Messina, Italy

Additional Reviewers

Henriette-Sophie Lipschuetz
Ulrich Reitebuch
Eric Zimmermann

Contents

Invited Paper

Non-traditional 2D Grids in Combinatorial Imaging – Advances and Challenges

Benedek Nagy(✉)

Department of Mathematics, Faculty of Arts and Sciences,
Eastern Mediterranean University, Famagusta, North Cyprus, Mersin-10, Turkey
nbenedek.inf@gmail.com

Abstract. On the one hand, the digital image processing and many other digital applications are mostly based on the square grid. On the other hand, there are two other regular grids, the hexagonal and the triangular grids. Moreover, there are eight semi-regular grids based on more than one type of tiles. These non-traditional grids and their dual grids have various advantages over the square grid, e.g., on some of them no topological paradoxes occur. Most of them have more symmetries, i.e., more directions of symmetry axes and also a smaller angle rotation may transform most of these grids into themselves. However, since most of these grids are not point lattices, we need to face some challenges to work with them; they may define various digital geometries. We show how a good coordinate system can be characterized, what type of digital distances are studied, tomography and distance transform. Other grid transformations, including translations and rotations with some of their interesting properties are mentioned. Mathematical morphology and cell complexes are also shown. The advantages and challenges are overviewed by various examples on the triangular grid, as a characteristic example for a non-traditional grid.

Keywords: Digital geometry · Regular and semi-regular grids · Dual grids · Coordinate systems · Digital distance · Discrete tomography · Digital topology · Topological paradoxes · Transformations · Mathematical morphology · Cell complexes

1 Introduction – Why (Not) the Traditional Square Grid?

In this paper, we give an overview about the usage of non-traditional grids in combinatorial image processing and in digital geometry. In contrast, the most used grid is the square grid, and it is referred to as the traditional grid in this context. It is "traditional" as most people including researchers met with this grid already in elementary school, the Cartesian coordinate frame (restricting the used coordinates to integer values) fits very well to this grid, and also both

R. P. Barneva et al. (Eds.): IWCIA 2022, LNCS 13348, pp. 3–27, 2023.
https://doi.org/10.1007/978-3-031-23612-9_1

hardware and software industry use it almost as if it were the only possibility to have a 2D digital platform. Unfortunately, there are people who identify the digital plane by $\mathbb{Z}^2$.

On the other hand, there are various "problems" with this grid that could make the life of researchers and software engineers hard. As many times people, including students, researchers, reviewers, are asking why do you use this or that grid, I would say that they should ask, why do they use solely the square/rectangular grid, even if it is not the ideal choice. Both nature [102] and ancient people used various other structures (Fig. 1), as our world is not square/rectangular grid based.

Fig. 1. Various tilings on the floor in the ancient city of Salamis.

On the one hand, the geometry of every "digital plane/world" differs from the Euclidean geometry [44,45], as e.g., in any discrete space based on a grid, the neighborhood relation plays an essential role. In contrast, there are no neighbor points in the Euclidean plane, but for any $\varepsilon > 0$, there are continuum many points that have distance less than ε from any point of the plane. On the other hand, the square grid is not the best choice to "digitize" the world/plane. The following facts are well-known. The square grid has the following topological paradoxes: There are lines that go through on each other without a common tile, e.g., think about the two diagonal lines of a usual chessboard: the white and black diagonals connect opposite corners, but there is no common tile. In this way the inner and outer parts of a closed curve can also be connected (violating the Jordan curve theorem of Euclidean geometry). In contrast, on the hexagonal grid similar paradoxes do not occur (see, e.g., [19]). Also, there are various traditional hardware, i.e., kitchenware product, where circular objects are arranged according to the square grid, see, e.g., muffin-trays (Fig. 2) or the most usual egg-trays. It is probably by the lack of knowledge of the designers/producers to put these into square grid based structure. Already the bees know that the hexagonal grid would be the most efficient way to build minimal perimeter for maximal-size storages (for more precise mathematical description and proof see [35]) and also it gives the ideal way to pack same-size circular objects [23].

Fig. 2. Usual arrangements for muffins/cupcakes.

We should also mention here, last in this section, but not with least importance, that the digital distances based on the sole usage of any of the two type of motions on the square grid results digital distance functions that has $\sqrt{2}$ factor difference for points with the same Euclidean distance [103]. The digital disks defined by distances based on any of the sole neighborhood relation are squares. (In this way, as one may interpret, the unsolvable ancient problem to square a circle is finally resolved, but, in fact, rather it shows how rough are the approximations of the Euclidean distance – L_2 metric – by these digital distances – equivalent to L_1 and L_∞, respectively.) To reduce this error on the square grid, first, octagonal distances, i.e., distances based on neighborhood sequences were introduced and studied [26–28,66,78,103,111,112] with the approximation of the Euclidean distance made by their help [25,32]. The digital discs by these distances, i.e., the set of pixels that have at most a given distance (radius) from a pixel (called center) are octagons with these distances, hence the name. However, these distances may easily violate the triangular inequality, and thus, the metricity condition, too. (To obtain a metric distance a sufficient and necessary condition has been published in [66,72], we discuss it with similar issues in Sect. 4). Chamfer (or also called weighted) distances [15] are always metric, based on the two types of neighborhood, the digital disks are, again, octagons (based on the two type of neighborhood) [16]. These distances have also been expanded by using larger motions, e.g., knight movements or even larger neighborhood [16,20] with various weights. Further techniques to have digital distances to approximate the Euclidean distance are also investigated as t-cost distances [29], their combination with chamfer distances [61,62] and the combination of weighted and neighborhood sequence based distances [108]. The weight-sequences [97–99] allows to have a digital distance that has perfect Euclidean distance of the points of the perimeter of a square from its midpoint. Another technique to have digital (path-based) distances that have small rotational dependency is to use a non-traditional grid, and this leads us to one of the main topics of this paper. In the next section we give a brief visit to some non-traditional grids.

2 Regular, Semi-regular Grids and Their Duals

Let us fix some terminology first. When we are talking about grids, we think about tessellations of the plane based on one or more geometric shapes, referred to as tiles, without overlaps and gaps in a periodic manner (such that they are periodic with two independent directions of the plane). In fact, we use periodic representations of infinite planar graphs with polygon regions put in a side-by-side manner. The grid-points or vertices of a grid are those points where some gridlines meet. The regions of the planar graph, the tiles, as we mainly refer to imaging, are also called pixels.

There are three regular grids in the plane and eight semi-regular grids [22,34]. All these grids are also referred to as Archimedean, they are built up by regular polygons in an isogonal way (the grid-points of the tiling are identical, isometric transformations may map any of them to any other). Already Kepler studied and described them. The dual (by planar duality) of a grid is obtained by inverting the roles of the pixels and gridpoints (corners): by putting a point to the midpoint of each tile/pixel and connecting those which points representing side-neighbor pixels, the dual grid is obtained. The first observation is that the square grid is self-dual, thus the same, Cartesian coordinate system fits also to the dual. In fact, one can decide which representation has more meaning, to use and address the pixels or the grid points. In image processing usually the pixel-based representation is preferred, but in other disciplines, e.g., building and simulating communication networks, the dual approach is more adequate. The second observation is that the other two regular grids, the hexagonal and the triangular grids are in the duality relation.

Tables 1 and 2 show the names and the patterns of the non-traditional grids including also the three regular grids, the eight semi-regular grids and their dual tilings. Each semi-regular grid has the property that their grid points are indistinguishable. One of their official naming method is based on that, by listing

Table 1. Various grids with some of their properties. The three regular grids (on the left), and three of the eight semi-regular grids (on the right), and their dual grids are below.

Name	Square	Hexagonal	Tri-hexagonal	Truncated hexagonal	Truncated trihexagonal
Notation	(4,4,4,4)	(6,6,6)	(3,6,3,6)	(3,12,12)	(4,6,12)
Tiles	1: square	1: hexagon	2: triangles and hexagon	2: triangles and dodecagon	3: squares, hexagon, and dodecagon
Neighbors	8 (4)	6 (6)	12 (6); 6 (3)	12 (12); 3 (3)	12(12);6(6);4(4)
Rotation angle	90	60	60	60	60
Symmetry axes angle	45	30	30	30	30
Grid Dualgrid	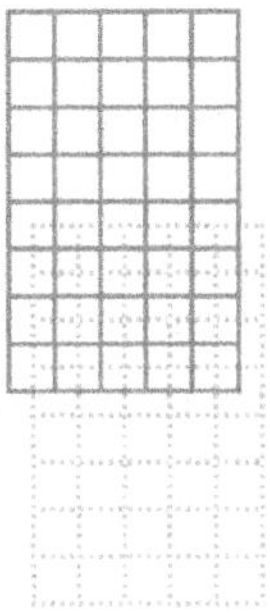	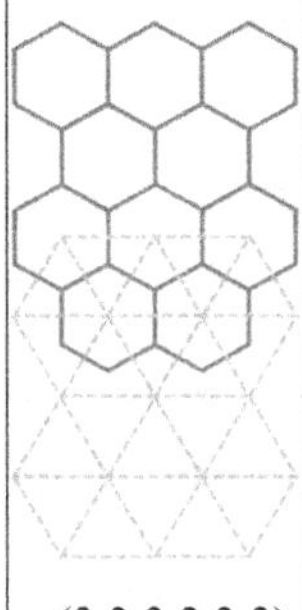 (3,3,3,3,3,3)	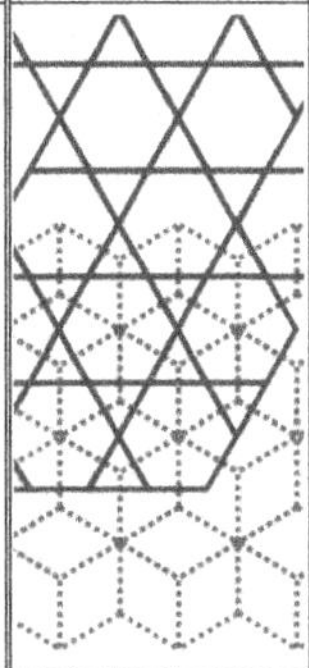	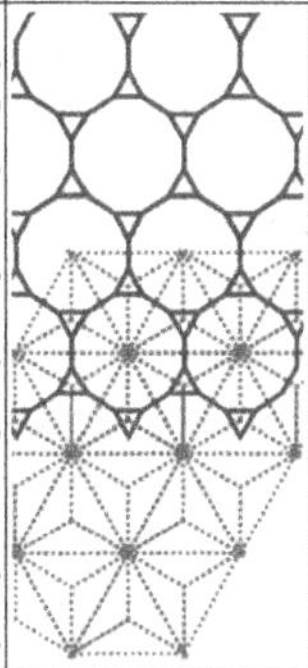	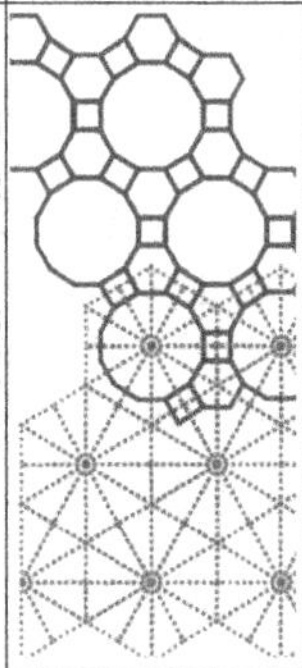
Dual name	Square	Triangular	Rhombille	Kisdeltille	Kisrhombille
Tiles		2: triangles	3: rhombuses	6: triangles	12: triangles
Neighbors		12 (3)	10 (4)	21 (3)	16 (3)

the regular polygons in the order they are next to each other at a common vertex (see the row "Notation" for a possible way to write this information), moreover the types of the polygons of the tiles are also listed (we wrote a polygon in singular form if each of their occurrences are identical up to translations and used plural form if the polygon appears in more than one orientations). Based on the property that the gridpoints of the semi-regular tilings are identical, the tiles of their dual grids are indistinguishable, i.e., the dual grids built up by a sole tile (but usually with various orientations), this information is given in the row "Tiles" under the name of the dual grid. Actually, for this reason, to have same size pixels, we may recommend to use the dual grids of the semi-regular grids in image processing, pictures on them could have better properties from this point of view than on the semi-regular grids (on which sometimes there are large differences on the size of the various polygons). Further, the table shows the smallest rotation angle that transforms the grid into itself (it is 90° for the square grid), the smallest angle between two symmetry axes of different directions (that is 45° for the square grid). This information applies for both the grid and its dual in the given column. The values highlighted by purple color show smaller values than the actual values for the square grid, thus one may easily see that many of the non-traditional grids overperform the square grid, i.e., usually smaller angles are enough for rotations, and there are symmetry axes in more directions than on the square grid. The table gives also the number of different types of tiles (for semi-regular grids) and the number of different orientations (for the triangular and the dual of the semi-regular grids). As one may see that only the square and hexagonal grids are point lattices, i.e., they are discrete subgroups of the 2-dimensional Euclidean space, meaning that any of their pixel has exactly the same role, every vector connecting the midpoints of any two pixels translate grid into itself. None of the other grids have a similar property, and thus, the triangular grid may share various properties with the dual grids of the semi-regular ones. For this reason, in some of the other sections, we concentrate on the triangular grid. The table also show the number of pixels that share at least one corner with a given pixel, we call those tiles *adjacent* to the original tile. This number is 8 for the square grid. As we can see, in most of the non-traditional grids, except the hexagonal, the Khalimsky grid (i.e., truncated quadrille or truncated square tiling), the Cairo pattern (also called 4-fold pentille) and the iso(4-)pentille (that is also called prismatic pentagonal) tiling, the non-traditional grids have larger extended neighborhood of pixels which may either give more flexibility to play with various types of neighborhood or provide more isometric directions of the plane to move.

Almost all grids have alternative names: the triangular grid is also called isometric grid, the hexagonal grid has an alternative name honeycomb grid. Actually, these alternative names may fit better if instead of the pixels, the gridpoints are used and the edges of the grid give the moving directions between them. However, as we already mentioned, in image processing, in computer graphics and in imaging applications, usually, the tiles are used as pixels. The tri-hexagonal grid is also called hexadeltille tiling. The truncated hexagonal tiling is also called truncated hextille. The truncated trihexagonal is also called truncated hexadeltille tiling. The rhombitrihexagonal is also called rhombihexadeltille tiling. The snub hexagonal tiling is also called snub hextille. The snub square tiling is also called snub quadrille tiling. The isosnub quadrille tiling is also called elongated triangular tiling. Their dual grids also have alternative names, e.g., the kisdeltille tiling is also called triakis triangular grid; the tetrille tiling is also called deltoidal trihexagonal grid; and the 6-fold pentille tiling is also called floret pentagonal tiling.

In the rows entitled neighbor, the number of tiles sharing at least one point on their boundary are counted and in brackets the number of neighbors with which a full edge (side) is shared. That latter is actually the number of sides of the given tile. At the hexagonal, the truncated hexagonal, the truncated trihexagonal and the Khalimsky grids, green color shows that the number of all neighbors of each tile is the same as the number of side-neighbors and in this way, the topological paradox mentioned in the previous section does not occur in these grids. Actually,

Table 2. Various grids with some of their properties (cont.) Five of the eight semi-regular grids and their dual grids are below. The '–' sign in the row Symmetry means that in fact the given semi-regular grid and its dual are not axial symmetric, instead there are two variants of both these grids, and the variants are axial mirror images of each-other.

Name	Rhombitri-hexagonal	Snub-hexagonal	Snub-square	Isosnub quadrille	Khalimsky
Notation	(3,4,6,4)	(3,3,3,3,6)	(3,3,4,3,4)	(3,3,3,4,4)	(4,8,8)
Tiles	3: triangles, squares and hexagon	2: triangles and hexagon	2: triangles and squares	2: triangles and square	2: square and octagon
Neighbors	12(6);8(4);6(3)	18 (6); 9 (3)	12 (4); 9 (3)	12 (4); 9 (3)	8 (8); 4 (4)
Rotation angle	60	60	90	180	90
Symmetry axes angle	30	–	90	90	45
Grid Dualgrid					
Dual name	Tetrille	6-Fold Pentille	Cairo pattern	Iso(4-)Pentille	Kisquadrille
Tiles	6: deltoids	6: pentagons	4: pentagons	2: pentagons	4: triangles
Neighbors	9 (4)	8 (5)	7 (5)	7 (5)	14 (3)

this property gives the importance of the Khalimsky grid in various applications [42]. Furthermore, the Khalimsky grid shares the symmetric properties with the square grid, and thus, it is relatively easy to convert images and algorithms from the traditional square grid to this grid.

Based on the displayed properties, we may conclude that non-traditional grids provide nicer underlying structures with more symmetries and better properties (e.g., larger and more flexible neighborhood structure) for various applications than the square grid (see e.g. [86]).

To use them in, e.g., image processing and other computer oriented disciplines, some mathematical background, the first steps of digital geometry should be given by coordinate systems that are elegant and easy to use.

3 Coordinate Systems

As grids can be seen as graphs, they can be bipartite and non-bipartite. In bipartite grids, based on steps/moves on side neighbors, every cycle has an even length, while at non-bipartite grids there are paths between the same two tiles such that the difference of their lengths is odd [46]. For bipartite grids one may find a coordinate frame such that between two side neighbor tiles exactly one of the coordinates is changing and it is changing by ± 1. However, for a grid that is not bipartite, it is impossible to have such a coordinate system with integer triplets that is changing exactly one of the coordinate by ± 1 in each move to a side neighbor tile (pixel). One may check that some of the grids have a bipartite structure, e.g., the square grid, the triangular and the trihexagonal grids and the kisquadrille tiling. Other grids, e.g., the hexagonal, the rhombille, the 6-fold pentille and the isosnub quadrille and the Khalimsky grids are not bipartite.

Thus, in some cases, e.g., in the hexagonal grid, to preserve also the symmetry of the grid, extra coordinate(s) can be introduced. As Her [37,38] proposed, zero-sum integer triplets can efficiently be used for the hexagonal grid (Fig. 3). In this way, by a move to a neighbor two of the coordinates are changing, one of them is increased, the other is decreased by 1.

Actually, the triangular grid is a bipartite grid, and it is easy to see it by the orientations of its tiles. The symmetry of the grid suggests to use coordinate triplets [63,68,107] with sum 0 and 1 according to the two types of pixels. The 0-sum integer triplets are used to address the even, the 1-sum integer triplets to address the odd trixels (triangle pixels). As every even pixel has three odd side neighbors, and vice versa, the names even and odd are very apt for these tiles. A part of the grid with coordinates is shown in Fig. 4.

The formal, mathematical description of the neighbor relations is also nice and reflects the symmetry caught by the 3-valued coordinate systems.

In the hexagonal grid two hexagonal pixels (hexels) are neighbors if and only if one of their coordinate values are common, and the difference on the other two coordinates are $+1$ and -1, respectively. The three types of neighborhood on the triangular grid was already mentioned and used in the 1970's [30] in relation to image processing. They can be written as: the triangular pixels (trixels) $p(\hat{p}_1, \hat{p}_2, \hat{p}_3)$ and $q(\hat{q}_1, \hat{q}_2, \hat{q}_3)$ are m-neighbors (with $m \in \{1, 2, 3\}$) if and only if

- $|\hat{p}_i - \hat{q}_i| \leq 1$ for each $i \in \{1, 2, 3\}$, and
- $\sum_i |\hat{p}_i - \hat{q}_i| = m$.

Observe that these conditions are pretty much the same for the two types of neighborhood on the square grid (where $m \in \{1, 2\}$ the number of changing coordinates of a cityblock or a chessboard move). We note that in [36], to avoid some of the difficulties, the moves for the 1-neighbors are called 'half moves', while moves to 2-neighbors were called 'full moves', and 3-neighbors were not used.

In fact, based on the previously described coordinate systems, the hexagonal and triangular grids can be seen as those subspaces of the cubic grid which are built up by the points of one and two parallel oblique planes. In this way, the

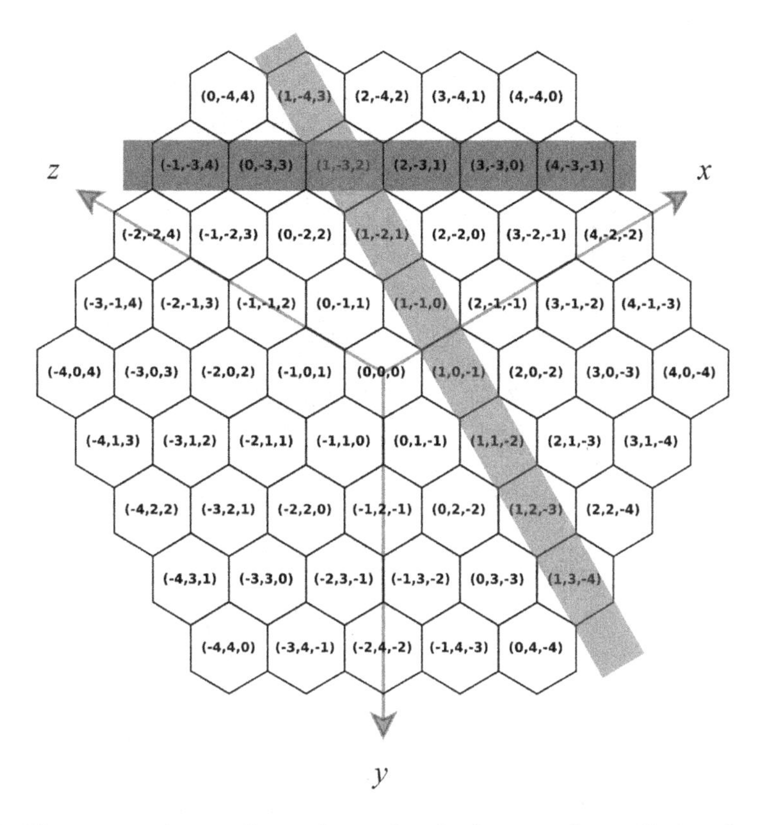

Fig. 3. The symmetric coordinate frame for the hexagonal tessellation. Lanes are obtained by fixing a value of a coordinate (e.g., $x = 1$ at the pixels with red color and $y = -3$ with blue color).(Color figure online)

trihexagonal grid can be seen as a grid built up by three parallel oblique planes [67,69], and thus, integer coordinate triplets can efficiently be used with sum $-1, 0$ and 1 to address the tiles of this grid.

Some of the dual semi-regular grids have also been addressed, see, e.g., for a general method to assign a coordinate system for periodic tessellations in [106]. Digital geometry of various grids are shown based on various coordinate systems, e.g., on the truncated hexagonal [51], snub-square [17], Khalimsky [48,49], rhombille, tetrille tilings [106], on the Cairo pattern [47], and on the kisquadrille tiling [21,106], just to mention a few possible solutions.

To show why it is important to find a good coordinate frame that reflects the symmetries of the grid, we show discrete tomography examples.

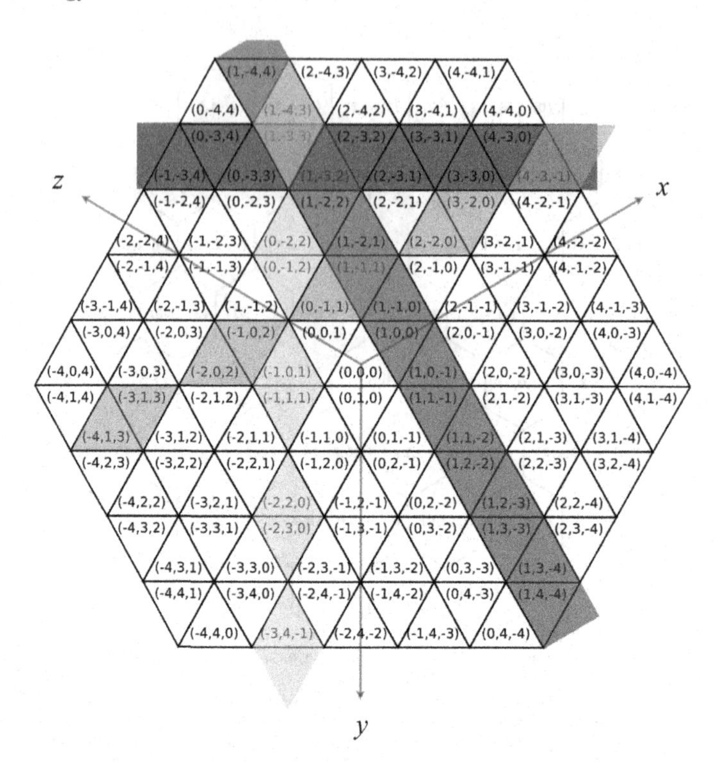

Fig. 4. The symmetric coordinate frame for the triangular tessellation. Lanes are obtained by fixing a value of a coordinate in both grids (e.g., $x = 1$ at the pixels with red color and $y = -3$ with blue color), while diamond chains are also shown (e.g., at the pixels with yellow color $x - z = -2$ and with green color $y - z = -2$). (Color figure online)

3.1 Discrete Tomography

In discrete (and especially, in binary) tomography, the sum of the values of some special subsets of the pixels are given (projections), and the task is to find (reconstruct) an image that fits (maybe perfectly or with a relatively small error) to these values.

The original problem on the square grid was based on two orthogonal projections, i.e., by row and column sums [33,104]. Observe that these projections are summing up the pixel values for pixels with either a fixed first coordinate (column) or a fixed second coordinate (row).

On the hexagonal grid usually three directions are considered based on the natural structure (symmetry) of the grid [55–57]. Actually, the *lanes* of the hexagonal grids are sets of hexels where one of the coordinates is fixed, e.g., the first coordinate is 1; or, another example is when the second coordinate is -2 (see also Fig. 3). In this way, all the three lane directions are analogous to the rows/columns of the square grid. Since there is only type of neighborhood on the hexagonal grid, and by the three lane directions through a hexel all its six neighbors are already taken into account, it is very opt to use these three natural grid directions in tomography.

On the square grid (based on the other type of neighborhood), diagonal projections can also be considered to decrease the number of possible solutions. These directions can be described by either fixing the sum or the difference of the two coordinates.

Now, we show how elegant is the coordinate system for the triangular grid. The main directions of the grid are by lanes, i.e., by the set of trixels with a fixed given coordinate. These lanes, also by their mathematical description, are closely analogous to the rows and columns of the square grid also on the triangular grid. The first paper on binary tomography on the triangular grid used these three directions and based on a genetic algorithm [60]. Another approach was based on Ryser's algorithm, in which for two directions perfect projection values were obtained and the error was minimized for the third direction by using switching pairs (for the first two directions, not to destroy those errorless values) [95].

Moreover, since the perpendicular lines to the lanes are not lanes on the triangular grid, but in fact, they are exactly the bisectors of the angles of two lanes, they play similar roles as the diagonal lines on the square grid. These *diamond-chain* directions are described by fixing the difference of two of the coordinates (see also Fig. 4). Binary tomography based on these three alternative directions were studied in [93], while all the six directions were used in [94] in a memetic approach. Some types of ghosts (switching patterns) causing multiple solutions for projections by the three lane-directions were also presented in [94].

Finally, in this section, we recall a phenomenon that is not on the square grid. By measuring the lengths of the projection lines inside the trixels, with various tricks, the number of even and odd trixels of the image can also be differentiated and computed separately, in this way supporting (a better) solution of the tomographic problem (we used energy minimization methods in [90–92]).

4 Digital Distances

The second step of creating digital geometry on a grid is to compute path based, digital distances between any pair of pixels. Digital distance on the hexagonal grid based on the usual neighborhood is determined in [54] with two coordinates, and in [65] by the symmetric coordinate system obtaining a more symmetric formula: for two hexels $p(\hat{p}_1, \hat{p}_2, \hat{p}_3)$ and $q(\hat{q}_1, \hat{q}_2, \hat{q}_3)$, their path-based distance, i.e., the number of steps to neighbor hexels to reach p from q or vice versa is

$$\max_{i \in \{1,2,3\}} |\hat{p}_i - \hat{q}_i| = \frac{\sum_{i=1}^{3} |\hat{p}_i - \hat{q}_i|}{2}.$$

Further, in this section, we concentrate on the triangular grid. First, we present a formula for distances based on the 1-neighbors (analogously to the cityblock distance of the square grid) [65]. Let $p(\hat{p}_1, \hat{p}_2, \hat{p}_3)$ and $q(\hat{q}_1, \hat{q}_2, \hat{q}_3)$ be two trixels; their (1)-distance is $\sum_{i=1}^{3} |\hat{p}_i - \hat{q}_i|$.

To complement it, now a formula for the distance based on adjacent trixels, i.e., allowing to use any of the three types of neighborhood in every step of the path (analogously to the chessboard distance of the square grid) is shown [88]. The (3)-distance of p and q is $\max_{i \in \{1,2,3\}} |\hat{p}_i - \hat{q}_i|$.

As we can see, the symmetric coordinate systems allow us to use very similar distance formulae than the ones on the square grid.

Further in this section, we briefly recall neighborhood sequence based and weighted distances on the triangular grid with some of their interesting properties. These types of distances are already mentioned in the introduction for the square grid, here we recall their definitions on the triangular grid (they can be defined analogously on the other grids).

A *neighborhood sequence* is an infinite sequence $B = (b(i))_{i=1}^{\infty}$ of possible neighbor relations $b(i) \in \{1, 2, 3\}$ (for all $i \geq 0$). A path $p = p_0, p_1, \ldots, p_n = q$ of adjacent trixels is a B-path if p_{i-1} and p_i are at most $b(i)$-neighbors (we call such moves $b(i)$-steps) for each $1 \leq i \leq n$. Obviously, for any two trixels p and q there are various B-paths for any neighborhood sequence B, however some (at least one) of them has the minimal length, i.e., the minimal number of steps/moves. The number of steps n of such a minimal path defines the B-distance of p and q. These distances were introduced in [63,64] and further studied in various other papers. If the sequence B is periodic, we may abbreviate its writing giving only the first period. Observe that the (1)-distance and the (3)-distance are special, actually, extremal cases of the neighborhood sequence based distances. A wide variety of distances can be defined and studied between these two distances.

We present some examples to highlight the interesting properties of these distances.

Example 1. Let $B_1 = (3, 1)$. Further let $p = (0, 0, 0)$, $q = (1, 1, -2)$, $r = (1, 1, -1)$ and $s = (2, 1, -2)$ be trixels. Then, p, r, q is a B_1-path with a 3-step and a 1-step, and it is the shortest, yielding that the distance from p to q is 2. On the other hand, from q to p a shortest path is $q = (1, 1, -2), (0, 1, -1), (0, 1, 0), p = (0, 0, 0)$, thus this distance is 3. Observe that the first 3-step of this path is also a 2-step, i.e., it is not to a 3-neighbor.

Now, from p to r the distance is 1, as a 3-step is enough to reach r from p. From r to s, the distance is 1: a 3-step suffices (in fact a 2-step would also be enough). Now, from p to s, a shortest path is $p = (0, 0, 0), r = (1, 1, -1), (1, 1, -2)$, $s = (2, 1, -2)$, thus this distance is 3. (The second step must be a 1-step according to B_1.) Thus, the B_1-distance fulfills neither the symmetry nor the triangular inequality.

Let $B_2 = (1, 3, 3, 3, 3, 3, ...)$ having only one value 1, then B_2 does not fulfill symmetry, but fulfills the triangular inequality.

Let $B_3 = (2, 1)$, then the triangular inequality does not hold in general, but this distance is symmetric.

The $(1, 1, 2)$-distance and the $(1, 2, 3, 2, 2, 2, ...)$-distance are metric.

A formula to compute B-distance on the triangular grid is provided in [71,75] (in such a way that B-paths of the cubic grids are restricted to use only 0-sum and 1-sum triplets). Another interesting phenomenon is that there are neighborhood sequences that define the same distance function. (This is impossible on the square grid, for any two neighborhood sequences B_1 and B_2 that are not identical we can find two pixels such that their B_1- and B_2-distances differ.) On the triangular grid two consecutive 3-steps are equivalent to a 3-step and then a 2-step. This gives equivalent classes of neighborhood sequences, and allow to define the smallest element of each class, namely the *minimal equivalent neighborhood sequence* (see [64] where it is shown how to obtain such sequence).

A necessary and sufficient condition for B to define a metrical distance is as follows [64,76]. B defines a metric if both of the following conditions hold:

- if B contains an element 3, then let ℓ be the smallest value such that $b(\ell) = 3$, further
  - the sum of the elements before the first occurrence of 3, i.e. $\sum_{i=1}^{\ell-1} b(i)$ is even, and
  - after the occurrence of 3, only 2s and 3s may occur, i.e., $b(i) > 1$ if $i > \ell$.
- $\sum_{k=1}^{i} b(k) \leq \sum_{k=j+1}^{j+i} b(k)$ for all pairs i, j with $i + j < \ell$ (or generally for all $i, j \in \mathbb{N}$ if 3 does not occur in B).

Actually, the first part of the condition is taking care about not to violate the symmetry, while the last condition (which actually regulates the distance to fulfill the triangular inequality) is very similar to the condition of metricity on the square grid [72]. This latter can be written formally as follows: for all j, k $\sum_{k=1}^{i+1} b(k) \leq \sum_{k=j+1}^{j+i} b(k)$.

The digital disks based on a digital distance, as we have already mentioned, contains the set of pixels that are at most the given distance (called radius) from its centre pixel. The digital disks by neighborhood sequences are dodecagons (maybe degenerated to enneagon, hexagon or triangle, they are fully characterized in [70,74]). Approximation of the Euclidean distance and Euclidean disks are studied in [77,96]. The approximation results, not surprisingly, better than the best approximations on the square grid (as a dodecagon is much closer to the circle than an octagon). Moreover, in [96], the idea to use the intersection of two digital disks was mentioned and it was shown that by this method the approximation can be even better than without this trick. To have two disks with the same radii such that none of them contains the other is again such a phenomenon that do not occur on the square grid, and it made possible on the triangular grid to use the intersection of two disks as a more round object than the disks themselves.

Moreover, a digital disk as a set of trixels may have more than one radius depending on the used neighborhood sequence, i.e., by radius 2 with the sequence $(1, 1, \ldots)$ and by radius 1 with the sequence $(2, \ldots)$ the same disk is obtained. This is again such a property which do not occur on the square grid. Some of the mentioned interesting properties were studied and used in various communication scenarios in [83].

Now, we turn to the other type of popular digital distance functions generalizing the (1)- and (3)-distances in another way. The *weighted/chamfer distances* are also defined on the triangular grid based on three positive weights $\alpha, \beta, \gamma \in \mathbb{R}^+$ assigned to the three types of neighborhood relations [81]. As usual, the natural condition $0 < \alpha \leq \beta \leq \gamma$ may be applied. A path $p = p_0, p_1, \ldots, p_n = q$ of adjacent trixels has its cost/weight

$$\alpha \cdot |\{i|(p_i - 1, p_i) \text{ are 1-neighbors}\}| + \beta \cdot |\{i|(p_i - 1, p_i) \text{ are 2-neighbors}\}| \\ + \gamma \cdot |\{i|(p_i - 1, p_i) \text{ are 3-neighbors}\}|.$$

Again, usually for any two trixels p and q there are various paths with various costs, however some (at least one) of them has the minimal cost, and this minimal cost defines the (α, β, γ)-weighted distance of p and q. It is known that the chamfer distances defines metric in any grid structure.

We highlight here one of the main advantages of the digital disks defined by chamfer distances on the triangular grid over the similar approaches on the square grid. On the one hand, there are digital disks which are polygons with many corners and sides, e.g., a 63-gon is obtained by $\alpha = 8, \beta = 15, \gamma = 18$ and $radius = 723$. On the other hand, the roundness of the digital disk with data $\alpha = 29, \beta = 56, \gamma = 68, radius = 892$ measured by its non-compactness (also called isoperimetric) ratio, $\kappa = \frac{(perimeter)^2}{area}$ is 12.628 [59] (this value is the smallest for the Euclidean circle, and it is exactly $4\pi \approx 12.566$, while it is 16 for the square). To compare these results to the analogous results on the square grid, we recall that by two weights for the usual two types of neighbors on the square grid, the chamfer balls are octagons. On the square grid by 3 weights (including knight moves, and in this way, based on 5×5 size neighborhood), the chamfer disks are 16-gons, further, by 7×7 size neighborhood (with 5 weights), they are 32-gons [20]. As we have recalled, we have obtained better results (using only the 3 natural neighborhood) on the triangular grid.

Since, the shortest path is usually not unique, a related combinatorial topic is the *path-counting* which can be used to analyse images [103], but plays also a significant role in networking to find routing/rerouting strategies. The number of shortest paths were computed on the square grid by neighborhood sequences [24, 85] and by chamfer distances [6], as well. There are also results on the hexagonal grid, by the distance based on the sole neighborhood [43, 88] and on the triangular grid by paths based on 1- or 2-neighborhood [31], 3-neighborhood [88] and also for shortest weighted paths [89] for some of the chamfer distances. In some cases, the results can easily be written by the binomial coefficients, however in other cases, the obtained formulae are more complex and, e.g., the Fibonacci numbers may also appear.

As our main aim is to consider image processing and analysis oriented applications as well, we recall the concept of *distance transform* [16]. In a binary image, the pixels may have two values, and thus we may differentiate them as object and background pixels. The distance transform computes the distance map, i.e., it assigns to each object pixel its distance to the closest background point. Usually, path based distances are very opt to compute distance transform, as incremental algorithms can efficiently be used to generate the distance map of an object.

We should mention that various digital distances, e.g., simple step based and/or weighted distances have already been introduced and computed, for instance, on the trihexagonal grid in [50], on the truncated hexagonal tiling in [51], on the Cairo pattern in [47], and on the Khalimsky grid in [48,49]. In this latter, digital disks were also characterized, with their holes and islands. Already on the square grid non-convex balls can be obtained when the weights of the diagonal moves is smaller than the weight of the cityblock moves. It is an interesting phenomenon that holes and islands may also appear in the semi-regular grids due to their various types of tiles. (After computing their necessary and sufficient conditions, one may decide if for the given application point of view they are advantageous or disadvantageous, and choose the weights for the distance according to his or her decision.) In the grids which are not point lattices, in general, the distance cannot be translation invariant (and norm). However, this leads us to the next section, to investigate how images can be transformed on non-traditional grids.

5 Transformations

With the distance transform in the previous section we have already mentioned transformations. Among many possible transformations, in this section we concentrate on digitized isometric transformations of the grid, i.e., mapping a picture in the grid as if the analogue of an isometric transformation is done.

In the Euclidean plane, the isometric transformations, the translations, rotations and various reflections (mirroring) are always bijective. This is not the usual case of similar transformations of the digital plane and, therefore, of the digital images. In fact, there are usually very few rotation angles that may map the grid into itself. And it is a real practical problem, as we are making images by smart phones such that the horizontal and vertical directions may not precisely match, but when we want to print out the picture or we want to show on a tv screen or on a projector, the image should be shown in a rotated way having the horizontal and vertical directions matching. However, as it is well-known, digitized transformations, especially, rotations are usually degrading the quality of the image [41], moreover, iterated use of rotations usually leads to lose more and more information till all information has been lost.

For the triangular grid, all isometric transformations that map the grid into itself were shown in [73,79]. On the other hand, when the transformation does not map the grid into itself, by using a nontraditional grid, usually similar problems arisen as with similar transformations on the square grid. However, as some of the non-traditional grids are more symmetric, i.e., they are mapped to themselves by smaller rotation angles, or with other words, by more angles (of the interval $[0°, 360°]$), one may expect better average performance by using a non-traditional grid (somewhat similarly as digital disks are more round on a non-traditional grid, as we have discussed in the previous section). Rotations with arbitrary angles on the triangular grid were also studied recently by various approaches, e.g., by neighborhood motion maps (where a pattern by a pixel and its neighbors are studied under the given transformation) [10–12] and by shear based method [8].

On the other hand, discretized rotations have a relatively large literature on the square grid, see, e.g., [7,18,109]. Rotations on the hexagonal grid has also been investigated with various approaches, e.g., by neighborhood motion maps representing the hexels by Eisenstein integers of the complex plane [100,101] and shear based bijective rotations in [9]. Comparisons of rotations on the three regular grids are presented in [13,14] showing that non-traditional grids really have a good potential also from this point of view: in various experiments (with various rotation centers) each integer degree angle of the interval $[0°, 360°]$ is studied, rotations are classified to three classes:

- where the neighborhood motion map are not bijective, i.e., there are pixels that mapped to the same pixel by the rotation;
- where it is bijective, but the result is not digitally continuous, i.e., there are pixels which were neighbors, but they are mapped not to neighbor pixels;
- where the neighborhood motion map shows bijectivity and also the neighborhood is kept (in this small pattern).

On the other hand, we should also mention translations. While on the point lattices, i.e., on the square and hexagonal grids, all (digitized) translations are bijections, it is not the case on the other grids. Whenever a translation vector is taken that does not map the grid into itself (and if the grid is not a point lattice, then there are grid vectors also having this property), one may face some interesting phenomena. Here again, we restrict ourself to the case of the triangular grid, but similar things happen also on the other grids with the non-lattice property. On the triangular grid there are three types of translations [4,5]:

- non-bijective translations which map two neighbor triangles to the same triangle tile (see Fig. 5, left).
- strongly bijective translations: they are bijective and digitally continuous keeping the neighbor pixels as neighbors. These translations map each type of tile to the same type as the original tile (see Fig. 5, middle).
- weakly bijective translations: they are bijective, but digitally not continuous, i.e., some neighbor pixels are mapped to non-neighbor pixels. These translations map each pixel to the opposite type of pixel (see Fig. 5, right).

We should mention here that to work with transformations that may map some gridpoints (i.e., midpoints of pixels) not to gridpoints, as the general case of the digitized isometric and many other transformations, continuous extensions of the symmetric coordinate systems could be very helpful. They play a

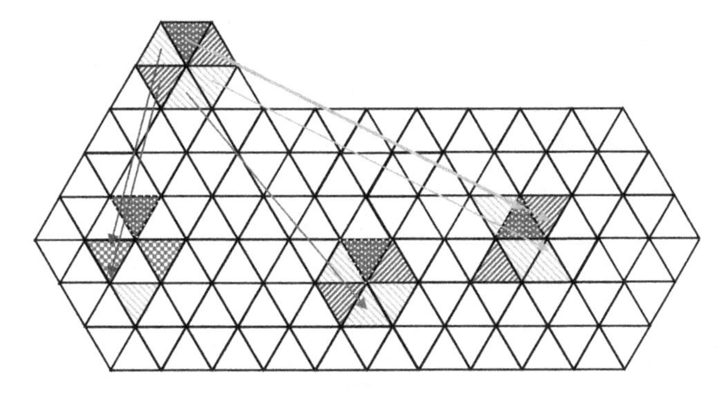

Fig. 5. Three types of translations on the triangular grid: non-bijective (left), strongly bijective (middle) and weakly bijective (i.e., bijective but not continuous, right).

similar role, as, at the case of square grid, we may use $\mathbb{R}^2$ in the computation instead of $\mathbb{Z}^2$, and we may do a digitization, a rounding operation, in the end just before displaying the result. The continuous extension of the symmetric hexagonal coordinate system uses all 0-sum triplets of $\mathbb{R}^3$. While continuous extension of the triangular coordinate system is introduced and studied in [3,87], in this system at least one of the coordinated is always an integer, and the sum of the three values is always in the interval $[-1, 1]$.

5.1 Mathematical Morphology

As we have already seen, the translations are not trivial on non-traditional grids. However, some important operators on images, including operators of mathematical morphology are usually defined based on local translations. Since we have more than one type of pixels, their neighborhood structures are different, and thus it is an interesting challenge to define morphological operators on non-traditional grids. In mathematical morphology an object image is given, and we may perform the operations by the help of another image-like object, by the so-called structural element.

On the one hand, if the structural element is restricted to the neighbor trixels, cellular automata may perform dilation and erosion based on the number of object pixels in the neighborhood (e.g., in [105] 1-neighborhood was used).

Let us consider a more general approach allowing a larger set of possible structural elements, as usual also in point lattices. The simplest approach is when only those vectors are allowed to use for translations, i.e., in the structural elements, which transform the grid into itself [2]. Another approach allows to define the structural element as a pair, and in this way to provide the set of vectors for the even and odd trixels independently [1]. As there is no free lunch, we need to pay the fee that the grid is not a point lattice, we need to release some of the usual properties of the morphological operators, i.e., the role of the picture and structural elements cannot be interchanged without any additional condition, and thus the dilation is not generally commutative. In other approaches we may force the commutative property of dilation, but we may loose the adjunction relation of dilation and erosion. Some possible solutions to define morphological operators on the triangular grid were already coined (e.g., in [84]) and details are coming in forthcoming papers.

5.2 Thinning and Abstract Cell Complexes

Finally, we have arrived back to one of our initial points, namely to the topological paradoxes and to the coordinate systems. As we have stated, the topological paradoxes occur in various grids (including the square grid). One way to avoid these paradoxes, as we have already proposed, to shift the underlying grid structure to an appropriate tessellations where all neighbor tiles share a full side. There is also another method, namely the use of topology based on abstract cell complexes. In these complexes, not only the tiles, but the edges (sides) and also the vertices of the grid are parts of the objects and images, they also have values, see, e.g., [110].

To make all these things in an elegant way, topological (also called combinatorial) coordinate systems can be used that are addressing not only the largest dimensional cells of the structure, but all smaller dimensional cells. To see more about on the square grid, see, e.g., [52,53]. The self-duality of the square grid allows to use, e.g., even-valued integer pairs to address the tiles, odd-valued integer pairs to address the gridpoints (the corners), and pairs with an odd and an even value to address the edges of the grid, where the position of the even value determines also the direction of the edge. Combinatorial coordinate system for the hexagonal and the triangular grids are presented in [80,82], to use only integer coordinates the original coordinate system has also been rescaled (similarly as it is rescaled in the case of the square grid), some details are recalled in Table 3. By the help of these coordinate systems, it is easy to perform various

Table 3. Some properties of topological/combinatorial coordinate system for the hexagonal and triangular grids.

Grid	Triangular	Hexagonal
Pixel	3 odd coordinates −1-sum for even +1-sum for odd	3 even coordinates 0-sum
Side/Edge	2 odd, 1 even coordinate, 0-sum the even coordinate position determines the direction	
Vertex	3 even coordinates with 0-sum	3 odd coordinates ±1 sum depending on the corner type (Y, λ)

operations on cell complexes, e.g., collapsing, cuts, thinning and skeletonization. Thinning and skeletonization algorithms are important techniques used in various applications, e.g., in character recognition. It is of high importance to keep the topology of the original object. We note here that there are also various algorithms based on objects represented solely by their pixels, some of those techniques have also been implemented on the hexagonal and triangular grids [39,40].

6 Final Comment

Finally, we recall that various image processing algorithms were considered on the hexagonal grid in [58] (as it is the simplest non-traditional grid and it is a point lattice). With this paper, we encourage the members of the image processing communities to consider also other non-traditional grids which may show better performance from various points of view.

Acknowledgements. The author thanks the work of many of his collaborators, including his PhD students and discussions with various members of the community.

References

1. Abdalla, M., Nagy, B.: Dilation and erosion on the triangular tessellation: an independent approach. IEEE Access **6**, 23108–23119 (2018). https://doi.org/10.1109/ACCESS.2018.2827566
2. Abdalla, M., Nagy, B.: Mathematical morphology on the triangular grid: the strict approach. SIAM J. Imaging Sci. **13**, 1367–1385 (2020). https://doi.org/10.1137/19M128017X
3. Abuhmaidan, K., Aldwairi, M., Nagy, B.: Vector arithmetic in the triangular grid. Entropy **23**(3), paper 373 (2021). https://doi.org/10.3390/e23030373
4. Abuhmaidan, K., Nagy, B.: Non-bijective translations on the triangular plane. In: 16th World Symposium on Applied Machine Intelligence and Informatics (SAMI 2018, Kosice, Slovakia), pp. 183–188. IEEE (2018). https://doi.org/10.1109/SAMI.2018.8324836
5. Abuhmaidan, K., Nagy, B.: Bijective, non-bijective and semi-bijective translations on the triangular plane. Mathematics **8**(1), paper 29 (2020). https://doi.org/10.3390/math8010029
6. Alzboon, L., Khassawneh, B., Nagy, B.: Counting the number of shortest chamfer paths in the square grid. Acta Polytechnica Hungarica **17**(4), 67–87 (2020). https://doi.org/10.12700/APH.17.4.2020.4.4
7. Andres, E.: Discrete circles, and discrete rotations. Ph.D. thesis, Universite Louis Pasteur, France (1992)
8. Andres, E., Largeteau-Skapin, G., Zrour, R.: Shear based bijective digital rotation in triangular grids. HAL report, hal-01900149 (2018/2022)
9. Andres, E., Largeteau-Skapin, G., Zrour, R.: Shear based bijective digital rotation in hexagonal grids. In: Lindblad, J., Malmberg, F., Sladoje, N. (eds.) DGMM 2021. LNCS, vol. 12708, pp. 217–228. Springer, Cham (2021). https://doi.org/10.1007/978-3-030-76657-3_15
10. Avkan, A., Nagy, B., Saadetoğlu, M.: Digitized rotations of closest neighborhood on the triangular grid. In: Barneva, R.P., Brimkov, V.E., Tavares, J.M.R.S. (eds.) IWCIA 2018. LNCS, vol. 11255, pp. 53–67. Springer, Cham (2018). https://doi.org/10.1007/978-3-030-05288-1_5

11. Avkan, A., Nagy, B., Saadetoğlu, M.: On the angles of change of the neighborhood motion maps on the triangular grid. In: 11th International Symposium on Image and Signal Processing and Analysis (ISPA 2019), pp. 76–81. IEEE (2019). https://doi.org/10.1109/ISPA.2019.8868526
12. Avkan, A., Nagy, B., Saadetoğlu, M.: Digitized rotations of 12 neighbors on the triangular grid. Ann. Math. Artif. Intell. **88**(8), 833–857 (2020). https://doi.org/10.1007/s10472-019-09688-w
13. Avkan, A., Nagy, B., Saadetoğlu, M.: A comparison of digitized rotations of neighborhood motion maps of closest neighbors on 2D regular grids. Signal Image Video Process. **16**(2), 505–513 (2022). https://doi.org/10.1007/s11760-021-01993-4
14. Avkan, A., Nagy, B., Saadetoglu, M.: A comparison of 2D regular grids based on digital continuity of rotations. In: ISAIM 2022 (abstract)
15. Borgefors, G.: Chamfering: a fast method for obtaining approximations of the Euclidean distance in N dimensions. In: 3rd Scandinavian Conference on Image Analysis, Copenhagen, Denmark, pp. 250–255 (1983)
16. Borgefors, G.: Distance transformations in digital images. Comput. Vis. Graph. Image Process. **34**(3), 344–371 (1986)
17. Borgefors, G.: A semiregular image grid. J. Vis. Commun. Image Represent. **1**(2), 127–136 (1990)
18. Breuils, S., Kenmochi, Y., Sugimoto, A.: Visiting bijective digitized reflections and rotations using geometric algebra. In: Lindblad, J., Malmberg, F., Sladoje, N. (eds.) DGMM 2021. LNCS, vol. 12708, pp. 242–254. Springer, Cham (2021). https://doi.org/10.1007/978-3-030-76657-3_17
19. Brimkov, W.E., Barneva, R.P.: Analytical honeycomb geometry for raster and volume graphics. Comput. J. **48**(2), 180–199 (2005)
20. Butt, M.A., Maragos, P.: Optimum design of chamfer distance transforms. IEEE Trans. Image Process. **7**(10), 1477–1484 (1998)
21. Comic, L.: A combinatorial coordinate system for the vertices in the octagonal $C_4C_8(S)$ grid. In: 12th International Symposium on Image and Signal Processing and Analysis (ISPA 2021), pp. 235–240. IEEE (2021)
22. Conway, J.H., Burgiel, H., Goodman-Strauss, C.: The Symmetries of Things. AK Peters (2008)
23. Conway, J.H., Sloane, N.J.A.: Sphere Packings, Lattices and Groups. Grundlehren der mathematischen Wissenschaften, vol. 290. Springer, New York (1993). https://doi.org/10.1007/978-1-4757-6568-7
24. Das, P.P.: Counting minimal paths in digital geometry. Pattern Recognit. Lett. **12**, 595–603 (1991). https://doi.org/10.1016/0167-8655(91)90013-C
25. Das, P.P.: Best simple octagonal distances in digital geometry. J. Approx. Theory **68**, 155–174 (1992)
26. Das, P.P., Chakrabarti, P.P., Chatterji, B.N.: Generalised distances in digital geometry. Inf. Sci. **42**, 51–67 (1987)
27. Das, P.P., Chakrabarti, P.P., Chatterji, B.N.: Distance functions in digital geometry. Inf. Sci. **42**, 113–136 (1987)
28. Das, P.P., Chatterji, B.N.: Octagonal distances for digital pictures. Inf. Sci. **50**, 123–150 (1990)
29. Das, P.P., Mukherjee, J., Chatterji, B.N.: The t-cost distance in digital geometry. Inf. Sci. **59**(1–2), 1–20 (1992)
30. Deutsch, E.S.: Thinning algorithms on rectangular, hexagonal and triangular arrays. Commun. ACM **15**(3), 827–837 (1972)

31. Dutt, M., Biswas, A., Nagy, B.: Number of shortest paths in triangular grid for 1- and 2-neighborhoods. In: Barneva, R.P., Bhattacharya, B.B., Brimkov, V.E. (eds.) IWCIA 2015. LNCS, vol. 9448, pp. 115–124. Springer, Cham (2015). https://doi.org/10.1007/978-3-319-26145-4_9
32. Farkas, J., Baják, Sz., Nagy, B.: Notes on approximating the Euclidean circle in square grids. Pure Math. Appl. PU.M.A. **17**, 309–322 (2006)
33. Gale, D.: A theorem on flows in networks. Pac. J. Math. **7**(2), 1073–1082 (1957)
34. Grünbaum, B., Shephard, G.C.: Tilings by regular polygons. Math. Mag. **50**(5), 227–247 (1977)
35. Hales, T.: The honeycomb conjecture. Discret. Comput. Geom. **25**, 1–22 (2001)
36. Hartman, N.P., Tanimoto, S.L.: A hexagonal pyramid data structure for image processing. IEEE Trans. Syst. Man Cybern. **14**(2), 247–256 (1984)
37. Her, I.: A symmetrical coordinate frame on the hexagonal grid for computer graphics and vision. ASME J. Mech. Des. **115**(3), 447–449 (1993)
38. Her, I.: Geometric transformations on the hexagonal grid. IEEE Trans. Image Proc. **4**, 1213–1221 (1995)
39. Kardos, P., Palágyi, K.: Topology preservation on the triangular grid. Ann. Math. Artif. Intell. **75**(1), 53–68 (2015)
40. Kardos, P., Palágyi, K.: On topology preservation of mixed operators in triangular, square, and hexagonal grids. Discret. Appl. Math. **216**, 441–448 (2017)
41. Kaufman, A.: Voxels as a computational representation of geometry. Presented at the SIGGRAPH 1999/Course 29, Los Angeles Convention Center, Los Angeles, CA, USA, 8–13 August 1999, pp. 14–58 (1999)
42. Khalimsky, E.D., Kopperman, R., Meyer, P.R.: Computer graphics and connected topologies on finite ordered sets. Topol. Appl. **36** (1990)
43. Khassawneh, B., Nagy, B.: Polynomial and multinomial coefficients in terms of number of shortest paths. C. R. Acad. Bulgare Sci. **75**(4), 495–503 (2022). https://doi.org/10.7546/CRABS.2022.04.03
44. Kiselman, C.O.: Elements of Digital Geometry, Mathematical Morphology, and Discrete Optimization. World Scientific, Singapore (2022)
45. Klette, R., Rosenfeld, A.: Digital Geometry - Geometric Methods for Digital Picture Analysis. Morgan Kaufmann, Elsevier Science B.V. (2004)
46. Kovács, G., Nagy, B., Stomfai, G., Turgay, N.D., Vizvári, B.: On chamfer distances on the square and body-centered cubic grids: an operational research approach. Math. Probl. Eng. **2021**, 9, Article ID 5582034 (2021). https://doi.org/10.1155/2021/5582034
47. Kovács, G., Nagy, B., Turgay, N.D.: Distance on the Cairo pattern. Pattern Recogn. Lett. **145**, 141–146 (2021). https://doi.org/10.1016/j.patrec.2021.02.002
48. Kovács, G., Nagy, B., Vizvári, B.: On weighted distances on the Khalimsky grid. In: Normand, N., Guédon, J., Autrusseau, F. (eds.) DGCI 2016. LNCS, vol. 9647, pp. 372–384. Springer, Cham (2016). https://doi.org/10.1007/978-3-319-32360-2_29
49. Kovács, G., Nagy, B., Vizvári, B.: Weighted distances and digital disks on the Khalimsky grid. J. Math. Imaging Vis. **59**(1), 2–22 (2017). https://doi.org/10.1007/s10851-016-0701-5
50. Kovács, G., Nagy, B., Vizvári, B.: Weighted distances on the trihexagonal grid. In: Kropatsch, W.G., Artner, N.M., Janusch, I. (eds.) DGCI 2017. LNCS, vol. 10502, pp. 82–93. Springer, Cham (2017). https://doi.org/10.1007/978-3-319-66272-5_8
51. Kovács, G., Nagy, B., Vizvári, B.: Weighted distances on the truncated hexagonal grid. Pattern Recogn. Lett. **152**, 26–33 (2021). https://doi.org/10.1016/j.patrec.2021.09.015

52. Kovalevsky, V.: Algorithms in digital geometry based on cellular topology. In: Klette, R., Žunić, J. (eds.) IWCIA 2004. LNCS, vol. 3322, pp. 366–393. Springer, Heidelberg (2004). https://doi.org/10.1007/978-3-540-30503-3_27
53. Kovalevsky, V.A.: Geometry of Locally Finite Spaces (Computer Agreeable Topology and Algorithms for Computer Imagery), editing house Dr. Bärbel Kovalevski, Berlin (2008)
54. Luczak, E., Rosenfeld, A.: Distance on a hexagonal grid. IEEE Trans. Comput. **5**, 532–533 (1976)
55. Lukić, T., Nagy, B.: Regularized binary tomography on the hexagonal grid. Physica Scripta **94**, paper 025201, 9 p. (2019). https://doi.org/10.1088/1402-4896/aafbcb
56. Matej, S., Herman, G.T., Vardi, A.: Binary tomography on the hexagonal grid using Gibbs priors. Int. J. Imaging Syst. Technol. **9**, 126–131 (1998)
57. Matej, S., Vardi, A., Herman, G.T., Vardi, E.: Binary tomography using Gibbs priors. In: Herman, G.T., Kuba, A. (eds.) Discrete Tomography: Foundations, Algorithms and Applications, chap. 8, pp. 191–212. Birkhäuser, Boston (1999)
58. Middleton, L., Sivaswamy, J.: Hexagonal Image Processing: A Practical Approach. Springer, London (2005)
59. Mir-Mohammad-Sadeghi, H., Nagy, B.: On the chamfer polygons on the triangular grid. In: Brimkov, V.E., Barneva, R.P. (eds.) IWCIA 2017. LNCS, vol. 10256, pp. 53–65. Springer, Cham (2017). https://doi.org/10.1007/978-3-319-59108-7_5
60. Moisi, E., Nagy, B.: Discrete tomography on the triangular grid: a memetic approach. In: 7th IEEE International Symposium on Image and Signal Processing and Analysis (ISPA 2011), Dubrovnik, Croatia, pp. 579–584. IEEE (2011)
61. Mukherjee, J.: Linear combination of weighted t-cost and chamfering weighted distances. Pattern Recogn. Lett. **40**, 72–79 (2014)
62. Mukhopadhyay, J.: Approximation of Euclidean Metric by Digital Distances. Springer, Heidelberg (2020)
63. Nagy, B.: Finding shortest path with neighborhood sequences in triangular grids. In: Proceedings of ITI-ISPA 2001: 2nd IEEE R8-EURASIP International Symposium on Image and Signal Processing and Analysis, Pula, Croatia, pp. 55–60. IEEE (2001)
64. Nagy, B.: Metrics based on neighbourhood sequences in triangular grids. Pure Math. Appl. **13**, 259–274 (2002)
65. Nagy, B.: Shortest path in triangular grids with neighbourhood sequences. J. Comput. and Inf. Tech. **11**, 111–122 (2003)
66. Nagy, B.: Distance functions based on neighbourhood sequences. Publicationes Mathematicae Debrecen **63**, 483–493 (2003)
67. Nagy, B.: A family of triangular grids in digital geometry. In: 3rd International Symposium on Image and Signal Processing and Analysis (ISPA 2003), Rome, Italy, pp. 101–106. IEEE (2003)
68. Nagy, B.: A symmetric coordinate frame for hexagonal networks. In: Theoretical Computer Science - Information Society (ACM Conference), Ljubljana, Slovenia, pp. 193–196 (2004)
69. Nagy, B.: Generalized triangular grids in digital geometry. Acta Mathematica Academiae Paedagogicae Nyiregyháziensis **20**, 63–78 (2004)
70. Nagy, B.: Characterization of digital circles in triangular grid. Pattern Recogn. Lett. **25**(11), 1231–1242 (2004). https://doi.org/10.1016/j.patrec.2004.04.001
71. Nagy, B.: Calculating distance with neighborhood sequences in the hexagonal grid. In: Klette, R., Žunić, J. (eds.) IWCIA 2004. LNCS, vol. 3322, pp. 98–109. Springer, Heidelberg (2004). https://doi.org/10.1007/978-3-540-30503-3_8

72. Nagy, B.: Metric and non-metric distances on $\mathbb{Z}^n$ by generalized neighbourhood sequences. In: 4th International Symposium on Image and Signal Processing and Analysis (ISPA 2005), Zagreb, Croatia, pp. 215–220. IEEE (2005)
73. Nagy, B.: Transformations of the triangular grid. In: Third Hungarian Conference on Computer Graphics and Geometry (GRAFGEO), Budapest, Hungary, pp. 155–162 (2005)
74. Nagy, B.: Geometry of neighborhood sequences in hexagonal grid. In: Kuba, A., Nyúl, L.G., Palágyi, K. (eds.) DGCI 2006. LNCS, vol. 4245, pp. 53–64. Springer, Heidelberg (2006). https://doi.org/10.1007/11907350_5
75. Nagy, B.: Distances with neighbourhood sequences in cubic and triangular grids. Pattern Recogn. Lett. **28**, 99–109 (2007). https://doi.org/10.1016/j.patrec.2006.06.007
76. Nagy, B.: Nonmetrical distances on the hexagonal grid using neighborhood sequences. Pattern Recogn. Image Anal. **17**, 183–190 (2007)
77. Nagy, B.: Optimal neighborhood sequences on the hexagonal grid. In: 5th International Symposium on Image and Signal Processing and Analysis, (ISPA 2007), Istanbul, Turkey, pp. 310–315. IEEE (2007)
78. Nagy, B.: Distance with generalized neighbourhood sequences in nD and ∞D. Discret. Appl. Math. **156**(12), 2344–2351 (2008). https://doi.org/10.1016/j.dam.2007.10.017
79. Nagy, B.: Isometric transformations of the dual of the hexagonal lattice. In: Proceedings of the 6th International Symposium on Image and Signal Processing and Analysis, pp. 432–437. IEEE (2009)
80. Nagy, B.: Cellular topology on the triangular grid. In: Barneva, R.P., Brimkov, V.E., Aggarwal, J.K. (eds.) IWCIA 2012. LNCS, vol. 7655, pp. 143–153. Springer, Heidelberg (2012). https://doi.org/10.1007/978-3-642-34732-0_11
81. Nagy, B.: Weighted distances on a triangular grid. In: Barneva, R.P., Brimkov, V.E., Šlapal, J. (eds.) IWCIA 2014. LNCS, vol. 8466, pp. 37–50. Springer, Cham (2014). https://doi.org/10.1007/978-3-319-07148-0_5
82. Nagy, B.: Cellular topology and topological coordinate systems on the hexagonal and on the triangular grids. Ann. Math. Artif. Intell. **75**(1-2), 117–134 (2015). https://doi.org/10.1007/s10472-014-9404-z
83. Nagy, B.: Application of neighborhood sequences in communication of hexagonal networks. Discret. Appl. Math. **216**, 424–440 (2017). https://doi.org/10.1016/j.dam.2015.10.034
84. Nagy, B.: Binary morphology on the triangular grid. In: Workshop on Digital Topology and Mathematical Morphology on the Occasion of the Retirement of Gilles Bertand, ESIEE Paris (2019). (Preconference Workshop of DGCI 2019)
85. Nagy, B.: On the number of shortest paths by neighborhood sequences on the square grid. Miskolc Math. Notes **21**, 285–301 (2020). https://doi.org/10.18514/MMN.2020.2790
86. Nagy, B.: Diagrams based on the hexagonal and triangular grids. Acta Polytechnica Hungarica **19**(4), 27–42 (2022)
87. Nagy, B., Abuhmaidan, K.: A continuous coordinate system for the plane by triangular symmetry. Symmetry **11**(2), 17, Article no. 191 (2019). https://doi.org/10.3390/sym11020191
88. Nagy, B., Akkeleş, A.: Trajectories and traces on non-traditional regular tessellations of the plane. In: Brimkov, V.E., Barneva, R.P. (eds.) IWCIA 2017. LNCS, vol. 10256, pp. 16–29. Springer, Cham (2017). https://doi.org/10.1007/978-3-319-59108-7_2

89. Nagy, B., Khassawneh, B.: On the number of shortest weighted paths in a triangular grid. Mathematics **8**(1), paper 118 (2020). https://doi.org/10.3390/math8010118
90. Nagy, B., Lukić, T.: Dense projection tomography on the triangular tiling. Fund. Inform. **145**, 125–141 (2016). https://doi.org/10.3233/FI-2016-1350
91. Nagy, B., Lukić, T.: Binary tomography on triangular grid involving hexagonal grid approach. In: Barneva, R.P., Brimkov, V.E., Tavares, J.M.R.S. (eds.) IWCIA 2018. LNCS, vol. 11255, pp. 68–81. Springer, Cham (2018). https://doi.org/10.1007/978-3-030-05288-1_6
92. Nagy, B., Lukić, T.: Binary tomography on the isometric tessellation involving pixel shape orientation. IET Image Proc. **14**(1), 25–30 (2020). https://doi.org/10.1049/iet-ipr.2019.0099
93. Nagy, B., Moisi, E.V.: Binary tomography on the triangular grid with 3 alternative directions - a genetic approach. In: 22nd International Conference on Pattern Recognition (ICPR 2014), Stockholm, Sweden, pp. 1079–1084. IEEE Computer Society (2014). https://doi.org/10.1109/ICPR.2014.195
94. Nagy, B., Moisi, E.V.: Memetic algorithms for reconstruction of binary images on triangular grids with 3 and 6 projections. Appl. Soft Comput. **52**, 549–565 (2017). https://doi.org/10.1016/j.asoc.2016.10.014
95. Nagy, B., Moisi, E.V., Cretu, V.I.: Discrete tomography on the triangular grid based on Ryser's results. In: 8th International Symposium on Image and Signal Processing and Analysis (ISPA 2013), Trieste, Italy, pp. 794–799. IEEE (2013). https://doi.org/10.1109/ISPA.2013.6703846
96. Nagy, B., Strand, R.: Approximating Euclidean circles by neighbourhood sequences in a hexagonal grid. Theoret. Comput. Sci. **412**, 1364–1377 (2011). https://doi.org/10.1016/j.tcs.2010.10.028
97. Nagy, B., Strand, R., Normand, N.: A weight sequence distance function. In: Hendriks, C.L.L., Borgefors, G., Strand, R. (eds.) ISMM 2013. LNCS, vol. 7883, pp. 292–301. Springer, Heidelberg (2013). https://doi.org/10.1007/978-3-642-38294-9_25
98. Nagy, B., Strand, R., Normand, N.: Distance functions based on multiple types of weighted steps combined with neighborhood sequences. J. Math. Imaging Vis. **60**, 1209–1219 (2018). https://doi.org/10.1007/s10851-018-0805-1
99. Nagy, B., Strand, R., Normand, N.: Distance transform based on weight sequences. In: Couprie, M., Cousty, J., Kenmochi, Y., Mustafa, N. (eds.) DGCI 2019. LNCS, vol. 11414, pp. 62–74. Springer, Cham (2019). https://doi.org/10.1007/978-3-030-14085-4_6
100. Pluta, K., Romon, P., Kenmochi, Y., Passat, N.: Honeycomb geometry: rigid motions on the hexagonal grid. In: Kropatsch, W.G., Artner, N.M., Janusch, I. (eds.) DGCI 2017. LNCS, vol. 10502, pp. 33–45. Springer, Cham (2017). https://doi.org/10.1007/978-3-319-66272-5_4
101. Pluta, K., Roussillon, T., Coeurjolly, D., Romon, P., Kenmochi, Y., Ostromoukhov, V.: Characterization of bijective digitized rotations on the hexagonal grid. J. Math. Imaging Vis. **60**(5), 707–716 (2018)
102. Radványi, A.G.: On the rectangular grid representation of general CNN networks. Int. J. Circuit Theory Appl. **30**(2–3), 181–193 (2002)
103. Rosenfeld, A., Pfaltz, J.L.: Distance functions on digital pictures. Pattern Recogn. **1**, 33–61 (1968)
104. Ryser, H.J.: Combinatorial properties of matrices of zeros and ones. Can. J. Math. **9**, 371–377 (1957)

105. Saadat, M.R., Nagy, B.: Cellular automata approach to mathematical morphology in the triangular grid. Acta Polytechnica Hungarica (J. Appl. Sci.) **15**(6), 45–62 (2018)
106. Saadat, M., Nagy, B.: Digital geometry on the dual of some semi-regular tessellations. In: Lindblad, J., Malmberg, F., Sladoje, N. (eds.) DGMM 2021. LNCS, vol. 12708, pp. 283–295. Springer, Cham (2021). https://doi.org/10.1007/978-3-030-76657-3_20
107. Stojmenovic, I.: Honeycomb networks: topological properties and communication algorithms. IEEE Trans. Parallel Distrib. Syst. **8**, 1036–1042 (1997)
108. Strand, R., Nagy, B.: A weighted neighbourhood sequence distance function with three local steps. In: 7th International Symposium on Image and Signal Processing and Analysis (ISPA 2011), Dubrovnik, Croatia, pp. 564–568. IEEE (2011)
109. Thibault, Y., Kenmochi, Y., Sugimoto, A.: Computing admissible rotation angles from rotated digital images. In: Brimkov, V.E., Barneva, R.P., Hauptman, H.A. (eds.) IWCIA 2008. LNCS, vol. 4958, pp. 99–111. Springer, Heidelberg (2008). https://doi.org/10.1007/978-3-540-78275-9_9
110. Wiederhold, P., Morales, S.: Thinning on quadratic, triangular, and hexagonal cell complexes. In: Brimkov, V.E., Barneva, R.P., Hauptman, H.A. (eds.) IWCIA 2008. LNCS, vol. 4958, pp. 13–25. Springer, Heidelberg (2008). https://doi.org/10.1007/978-3-540-78275-9_2
111. Yamashita, M., Honda, N.: Distance functions defined by variable neighborhood sequences. Pattern Recogn. **17**, 509–513 (1984)
112. Yamashita, M., Ibaraki, T.: Distances defined by neighborhood sequences. Pattern Recogn. **19**, 237–246 (1986)

Digital Geometry and Topology

Rectangularization of Digital Objects and Its Relation with Straight Skeletons

Anukul Maity[1], Mousumi Dutt[2(✉)], and Arindam Biswas[3]

[1] Narula Institute of Technology, Kolkata, India
[2] St. Thomas' College of Engineering and Technology, Kolkata, India
duttmousumi@gmail.com
[3] Indian Institute of Engineering Science and Technology, Howrah, Shibpur, India

Abstract. The rectangular partitioning of a digital object, A (without holes) is presented here. The partitioning is obtained in such a way that the set of connected output rectangles are related to the straight skeleton of the corresponding digital object. The given digital object, A is imposed on background grid of size, g (say) and its inner isothetic cover, P is obtained which is the maximum area orthogonal polygon inside the digital object. The combinatorial rules are formulated to apply those on P to partition it into a set of rectangles such that it is related to the straight skeleton of P. The partitioning algorithm discussed here runs in $O(n/g \log n/g)$ where n being the number of pixels on the periphery of digital object and g being the grid size. The experimental result shows the efficiency of the algorithm.

Keywords: Rectangular decomposition · Straight skeleton · Inner isotheic cover · Minimal partition · Orthogonal polygon · Shape analysis

1 Introduction

The polygonal decomposition is two types: covering problem and partitioning problem, where the input polygon may be hole-free or with holes. In this work, the partitioning is performed on non self-intersecting and hole-free orthogonal polygons. Decomposition of hole-free orthogonal polygon is a crucial and challenging task in many areas of scientific analysis due to its cognitive content. It is heavily used in computer vision and VLSI layout design [21]. The typical applications of shape description, shape analysis, and shape matching include art, architecture, robotic vision, cell biology, satellite imagery, neuron morphology, psycholinguistics, qualitative reasoning, image processing [14], pattern recognition [2], chip manufacturing [20], etc.

There are many research works on polygon decomposition in different domains, of which some of the problems are NP-hard [24]. In [7,8,19,25], it is shown that the decomposition of polygon with holes into minimum number of components is NP-hard. An optimal solution of polygon decomposition for

R. P. Barneva et al. (Eds.): IWCIA 2022, LNCS 13348, pp. 31–45, 2023.
https://doi.org/10.1007/978-3-031-23612-9_2

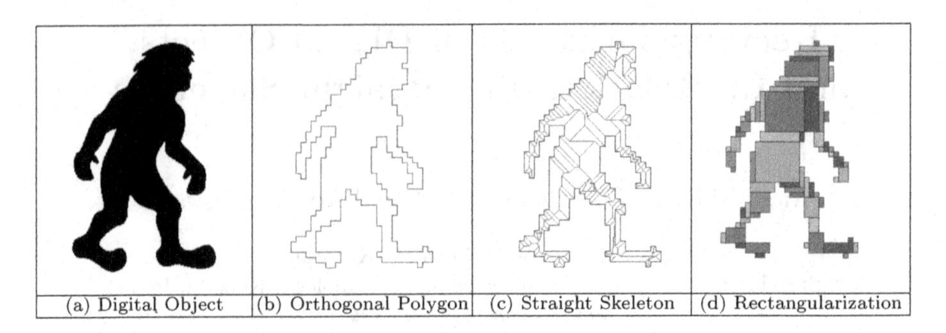

Fig. 1. (a) The 8-connected digital object, A, (b) Inner isothetic cover, P, for $g = 4$, (c) The straight skeleton, $S(P)$, (d) Rectangularization, R.

the polygons without holes into minimum number of convex pieces is proposed in [8] whereas the corresponding covering problem is NP-hard [9]. The decomposition of polygons with holes allowing Steiner points is NP-hard [18]. As the convex decomposition into minimum number of components is NP-hard, approximate decomposition in 3D domain is presented in [17]. The rectangular polygon decomposition are discussed in literature [11,26]. In [8,13,15,23], the polygon partitioning into minimum number of convex components are given. Optimum rectangular partitioning of simple polygon by minimising the stabbing number for the problem of finding a spanning tree or a triangulation is NP-hard [1,3,4,12]. In this paper, the rectangularization is related to the corresponding straight skeleton. The shape skeleton influences the perception of shape structure and have important consequence in visual processing and human vision [22].

Here, in this paper a 8-connected digital object, A is taken as input (Fig. 1(a)). A is imposed on background grid of size g and its inner isothetic cover, P [5,6] is obtained (Fig. 1(b)) which is an orthogonal polygon. The combinatorial rules are applied on P to partition it into rectangles such that the set of rectangles are related to the straight skeleton P. The straight skeleton, $S(P)$ and its partitioning into rectangles (R) are shown in Fig. 1(c) and Fig. 1(d) respectively for the digital object shown in Fig. 1(a). This paper is organized as follows. The definitions are stated in Sect. 2. Section 3 explains the combinatorial rules of partitioning an orthogonal polygon. In Sect. 4, the procedure of partitioning orthogonal polygons into rectangles is explained along with the algorithm, time complexity analysis, and demonstration The experimental results and its analysis are depicted in Sect. 5. Section 6 presents concluding remarks.

2 Definitions

Definition 1 *(Digital object): A digital object A is defined as k-connected ($k = 4$ or 8) subset of $\mathbb{Z}^2$ [16].*

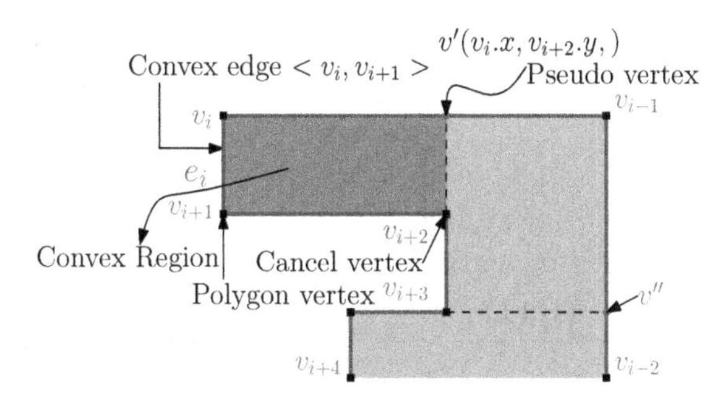

Fig. 2. Convex edge $e_i = \langle v_i, v_{i+1} \rangle$ is represented by two consecutive type-1 vertices along with different types of vertices (polygon vertex, v, internal vertex, u, pseudo vertex, v'). Solid line indicates the edge of polygon and dotted line indicates common edge of two adjacent rectangles. The convex region is marked by lightcyan color. (Color figure online)

In this paper, 8-connected digital objects are used whose background grid is 4-connected.

Definition 2 *(Digital grid): The digital grid is defined as $\mathcal{G} = (\mathcal{H}, \mathcal{V})$, where $\mathcal{H}$ is the set of horizontal grid lines and $\mathcal{V}$ is the set of vertical grid lines which are equi-distant from each other. The grid size (g) is defined as the distance between two consecutive horizontal or two consecutive vertical grid lines and intersection of a horizontal and a vertical grid line is called grid point. The edge between two consecutive grid points is called grid edge. A unit grid block (UGB) is the smallest unit square block in grid consisting of four grid points and four grid edges.*

Definition 3 *(Orthogonal polygon): An orthogonal polygon P is a polygon whose edges are axes parallel.*

Here, in this paper simple orthogonal polygon (isothetic polygon) is considered, i.e., the orthogonal polygon without holes and without self intersecting edges. The two consecutive edges of an orthogonal polygon either intersects at 90° or 270°. The vertex with 90° angle is termed as type 1 vertex and the vertex with 270° angle is termed as type 3 vertex. The type 3 vertex is also sometimes termed as *Reflex Vertex*. The inner isothetic cover (which is an orthogonal polygon) of 8-connected digital object is obtained here using the algorithm in [5,6]. To obtain more (less) tightly fitted inner isothetic cover of digital objects grid size needs to be decreased (increased).

Definition 4 *(Convex edge and Convex region): An edge $\langle v_i, v_{i+1} \rangle$ of P is called a convex edge if and only if two consecutive vertices have the same vertex type 1 (90°) and the rectangular area R_{CR} corresponding to the convex edge is called convex region as shown in Fig. 2.*

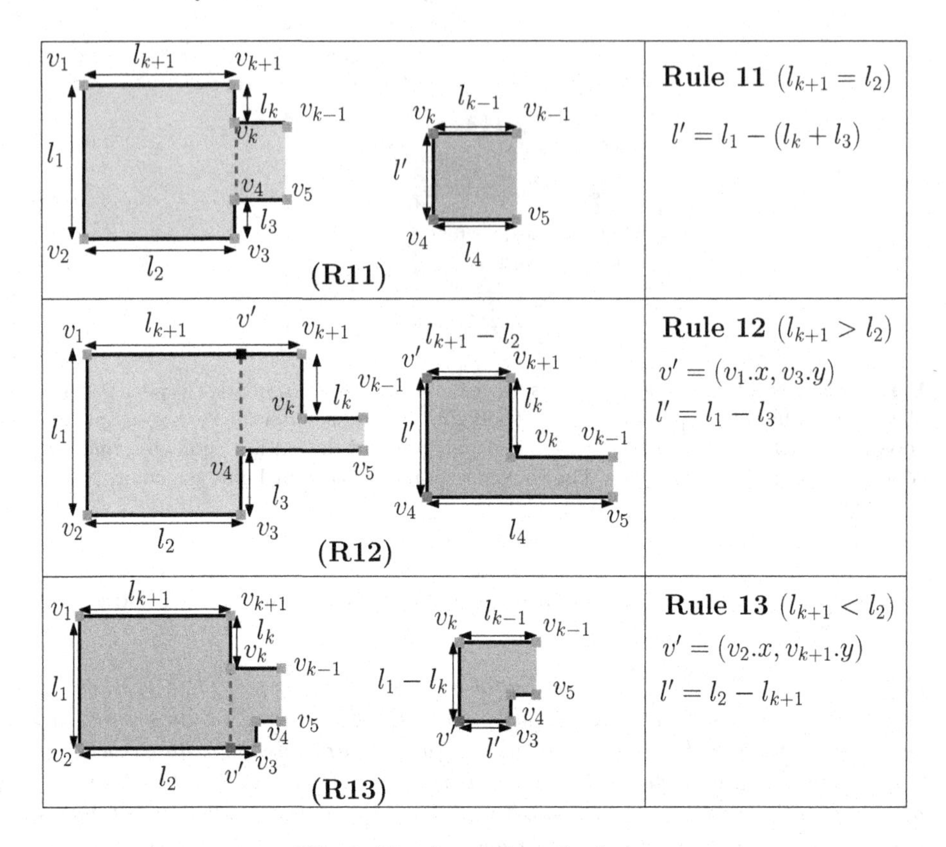

Fig. 3. The description of rule 1.

The conventional definition of the *straight skeleton*, $S(P)$, of a polygon as stated in [10] is as follows. Imagine shrinking δP via a parallel translation of all edges at the same speed inward, with each vertex following the angle bisector. Reflex vertices also travel on angle bisectors, which implies that the incident edge grows in length at that endpoint. The shrinking continues until one of two events occurs.

- An edge shrinks to zero length. This is exactly the event we saw with the medial axis of convex polygons and, just as in that circumstance, the process continues with the new vertex tracking the bisector of the neighboring edges.
- A reflex vertex collides with an edge. At this point, the original polygon is "pinched off", creating two new polygons. The shrinking process then continues on the two polygons independently.

3 Rules for Partitioning into Rectangles

The inner isothetic cover, P is traversed anticlockwise from top-left corner. While traversing anticlockwise from the vertex v_i to v_{i+1}, the direction of traversal of v_{i+1} is d_{i+1}, which is obtained from the previous direction $d_{i+1} = (d_i + t_i) \bmod 4$ ($d_i \in \langle 0, 1, 2, 3 \rangle$, denoting the direction towards right, top, left, or bottom respectively). Let l_i be the length of edge $e_i(v_i, v_{i+1})$. At each step while the rule is applied and desired rectangle is identified, the resulting rectangle is temporarily discarded and P is updated. The rules are explained as follows.

Rule 1: This rule is applied when there is a sequence of vertices of types $\langle 3, 1, 1, 1, 1, 3 \rangle$ as shown in Fig. 3. The rectangle for the convex region is detected as marked by blue dashed line and accordingly the rectangle is discarded from the polygon. Let l_k be the length of edge associated with vertex v_k. Depending on the length of l_{k+1} and l_2 three cases are there as shown in Fig. 3. The three cases are as follows.

Rule R11 ($l_{k+1} = l_2$)**:** This rule is applied when $l_{k+1} = l_2$. Let the rectangle be consisted of the sequence of vertices $\langle v_1, v_2, v_3, v_{k+1} \rangle$. The rectangle associated with convex region are determined and thereby the following vertices v_1, v_2, v_3, and v_{k+1} are removed to update the polygon and l_k is modified to $l_1 - (l_3 + l_k)$.

Rule R12 ($l_{k+1} > l_2$)**:** This rule is applied to remove the convexity where l_{k+1} is greater than l_2. After removing the resulting rectangle, the length of the corresponding edges are updated from l_{k+1} and l_3 to $l_{k+1} - l_2$ and $l' = l_1 - l_3$ respectively.

Rule R13 ($l_{k+1} < l_2$)**:** This is another variation of **Rule R12**.

Rule 2: This rule is applied when there is a sequence of vertices of types $\langle 3, 1, 1, 1, 3 \rangle$ as shown in Fig. 4. Depending on the length of l_{k+1} and l_2, there are three cases as discussed in the following. In each cases, the resulting rectangle is marked by blue dashed line.

Rule R21 ($l_{k+1} = l_2$)**:** This rule is applied when the length of l_2 and l_{k+1} are equal. The sequence of vertices of the resulting rectangle is $\langle v_1, v_2, v_3, v_{k+1} \rangle$. The vertices v_1, v_2, v_3, and v_{k+1} are removed and the length of the corresponding edge is updated from l_3 to $l_1 - l_k + l_3$.

Rule R22 ($l_{k+1} < l_2$)**:** This rule is applied when l_{k+1} is less than l_2. A pseudo vertex (edge point) v' is determined where the coordinate of v' is $(v_3.x, v_{k+1}.y)$. The vertices v_1, v_2, and v_{k+1} are removed. The length of the corresponding edges are updated from l_k and l_2 to $l_1 - l_k$ and $l' = l_2 - l_{k+1}$ respectively.

Rule R23 ($l_{k+1} > l_2$)**:** This rule is applied when l_{k+1} is greater than l_2. The length of the corresponding edges are updated from l_k and l_2 to $l' = l_1 - l_k$ and $l'' = l_{k+1} - l_2$ respectively. The vertices v_1, v_2, and v_{k+1} are removed.

Rule 3: This rule is applicable when there is a sequence of vertices of type $\langle 3, 1, 1, 3 \rangle$ as shown in Fig. 5. There are three cases as discussed below. In each cases the resulting rectangle is marked by blue dashed line.

Rule R31 ($l_{k+1} = l_2$)**:** This rule is applied when the length l_2 and l_{k+1} are equal. The vertices v_2, v_3, v_4, and v_5 are removed on detecting the resulting rectangle. The length of corresponding edge is modified from l_1 to $l_1 + l_3 + l_5$.

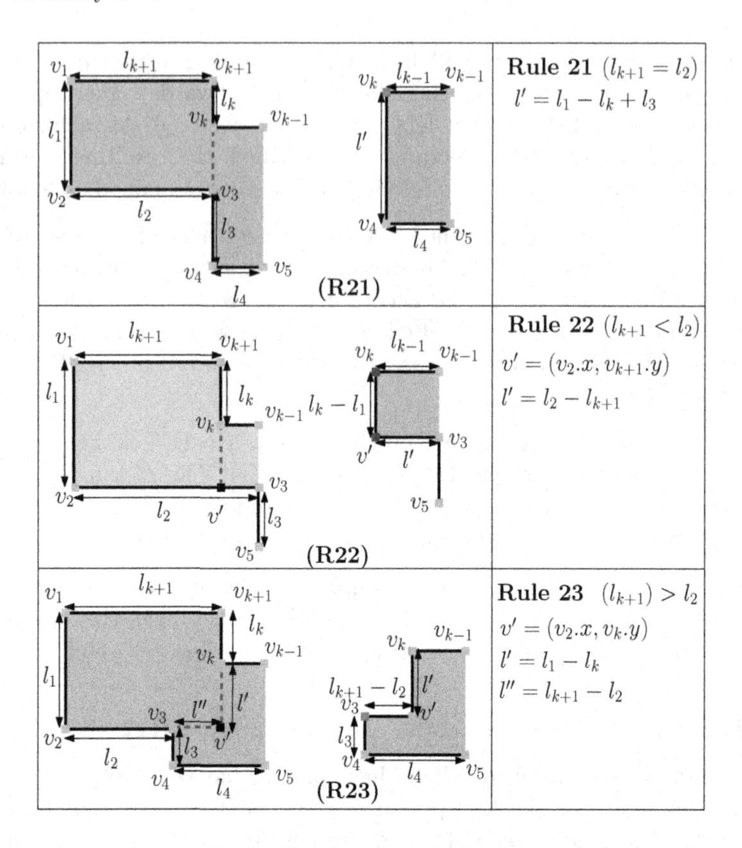

Fig. 4. The description of rule 2.

Rule R32 $(l_{k+1} < l_2)$: This rule is applied when $l_{k+1} < l_2$. The vertices v_3, v_4, and v_5 are removed and length of corresponding edges are modified from l_5 and l_2 to $l' = l_2 - l_4$ and $l_3 + l_5$ respectively.

Rule R33 $(l_{k+1} > l_2)$: This rule is applied when $l_{k+1} > l_2$. The vertices v_2, v_3, and v_4 are removed and the length of corresponding edges modified from l_1 and l_4 to $l_1 + l_3$ and $l' = l_4 - l_2$ respectively.

Rule 4: This rule is applicable when there is a sequence of vertices of more than four type 1 vertices as shown in Fig. 6. The sequence of type 1 vertices ends with a type 3 vertex (v_7 in Fig. 6). From that type 3 vertex two rectangles are considered R_1 and R_2 as shown in lightblue and lightgreen color respectively in Fig. 6. There are two cases based on the areas of R_1 and R_2 as discussed below.

Rule R41: When the area of R_1 is less than R_2, the corresponding result is shown in Fig. 6. The vertices v_5, v_6, and v_7 are deleted and v_5' is inserted.

Rule R42: When the area of R_1 is greater than R_2, the corresponding result is shown Fig. 6. The vertices v_4, v_5, and v_6 are deleted and v' is inserted.

On application of Rule 4, the polygon will be reduced and number of consecutive Type 1 vertices will be reduced such that again rules can be applied to

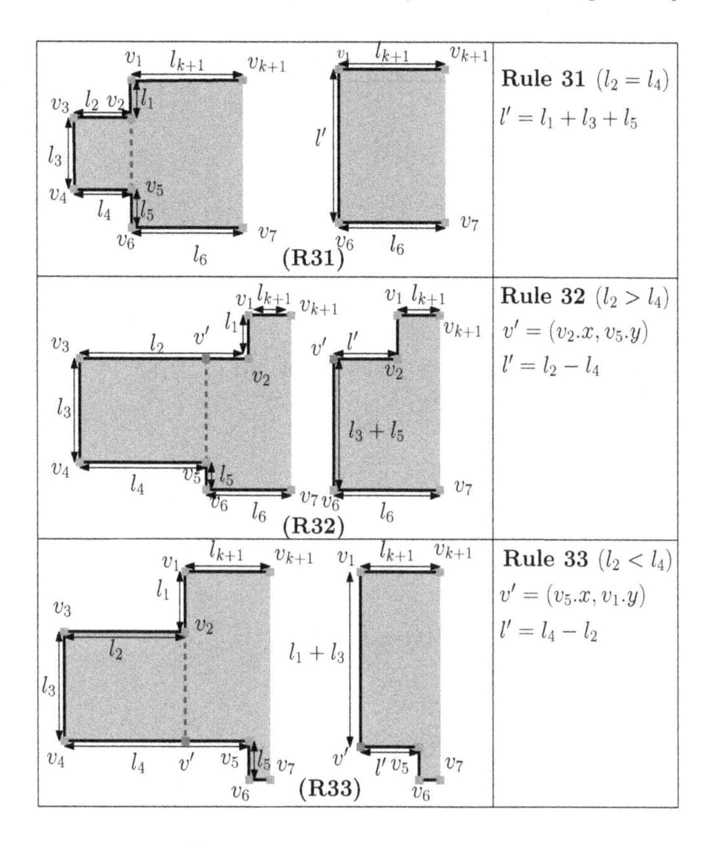

Fig. 5. The description of rule 3.

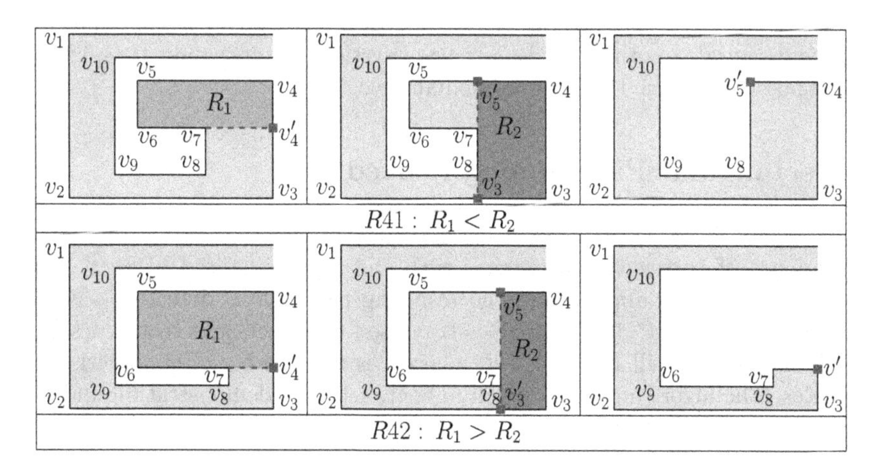

Fig. 6. The description of rule 4.

Algorithm 1: RECTANGULAR-POLYGON-PARTITION

```
    Input: A, g
    Output: R
 1  L, L_x, L_y ← FIND-IIC(A); R ← {φ}
 2  L ← INITPOLY(L)
 3  while L ≠ {φ} do
 4      if v_i → t = 1 & v_{i+1} → t = 1 then
 5          if v_{i-1} → Type = 3, & v_{i+2} → Type = 3 then
 6              R_i ← APPLY-RULE3(L, v_i, v_{i+1});
 7          else
 8              val ← FIND-SEQUENCE-TYPE1(v_i, v_{i+1});
 9              if val = 3 then
10                  R_i ← APPLY-RULE2(L, v_i, v_{i+1});
11              else if val = 4 then
12                  R_i ← APPLY-RULE1(L, v_i, v_{i+1});
13              else if val > 4 then
14                  R_i ← APPLY-RULE4(L, v_i, v_{i+1});
15              else
16                  R_i ← FIND-RECT(v_{i-1}, v_i, v_{i+1}, v_{i+1});
17      R ← R ∪ R_i
18      UPDATE-POLY(P)
19  return R
```

solve the rectangularization. Since the number of Type 1 vertices is four more than the number of Type 3 vertices in an orthogonal polygon, existence of convex region is essential. In other words, sequence vertices of types only $\langle 1, 3, 1, 3, \ldots \rangle$ are not possible. Hence the rules are exhaustive.

4 Rectangular Partitioning Procedure

The orthogonal polygon, P is traversed anticlockwise from top left corner. When the sequence of vertex types matches with the rules as stated in Sect. 3, the corresponding rule is applied and the resulting rectangle is detected (say, R_i). R_i is discarded from P and again P is traversed to detect next rectangles. This procedure continues till $P = \sum_{i=1}^{k} R_i$ where k is the number of non-overlapping rectangles. The algorithm is discussed in Sect. 4.1. The demonstration and time complexity are presented in Sect. 4.2 and Sect. 4.3 respectively.

4.1 Algorithm

The Algorithm 1 RECTANGULAR-POLYGON-PARTITION is used to find the rectangles of orthogonal polygon (inner isothetic cover that tightly inscribes the 8-connected digital object imposed on 4-connected background grid of size g).

Procedure InitPoly($L = \langle v_1, v_2, v_3, \ldots, v_n \rangle$)

```
1 n ← length(L);
2 for i → 1 to n do
3     v_i → l = distance(v_i);
4     v_i → s = 1;
5     if (d_{i+1} − d_i) mod 4 == 1 then
6         v_i → t = 1;
7     else
8         v_i → t = 3;
```

The digital object, A and the grid size, g are the input of the algorithm and it returns the set of rectangles, R as the output. The partitioning is obtained in such a way that the set of connected output rectangles are related to the straight skeleton. The linear list, L, is used to store the sequence of vertices of P generated by invoking the procedure FIND-IIC as stated in [5,6] (Step 1). The lexicographically sorted lists L_x and L_y are also obtained which are required to find vertices along a vertical line or horizontal line. L_x is lexicographically sorted list w.r.t. x as primary key and y as secondary key and L_y is lexicographically sorted list w.r.t. y as primary key and x as secondary key. L_x and L_y contain vertices and edge points (grid points on the edge of the polygon). The vertices of P contain vertex type, t, length of edge $(v_i v_{i+1})$, l, and vertex status, s ($s_i \in \{1, 2, 3\}$, where 1, 2, and 3 indicate original vertex, pseudo vertex and cancel vertex respectively) using INITPOLY procedure (Step 2). The rules stated in Sect. 3 are applied in Steps 3–18. The loop is executed until all the vertices in L is traversed (Step 3). When there are two consecutive type 1 vertices (Step 4), the rules are checked in Steps 5–16. When there is a sequence of vertices of types $\langle 3, 1, 1, 3 \rangle$, Rule 3 is applied (Steps 5–6) by calling the procedure APPLY-RULE3. Otherwise, the total number of consecutive type 1 vertices is counted and stored in val in Step 8 using the procedure FIND-SEQUENCE-TYPE1. If there are three consecutive type 1 vertices $\langle 3, 1, 1, 1, 3 \rangle$, i.e., $val = 3$, Rule 2 is applied by calling the procedure APPLY-RULE2 (Steps 9–10). If there are four consecutive type 1 vertices $\langle 3, 1, 1, 1, 1, 3 \rangle$, i.e., $val = 4$, Rule 1 is applied by calling the procedure APPLY-RULE1 (Steps 11–12). Otherwise, Rule 4 is applied (Steps 13–14) when $val > 4$ (by the procedure APPLY-RULE4). When there are only four vertices of type 1 or the polygon is reduced to a rectangle, the last rectangle is identified by using the procedure FIND-RECT (Steps 15–16). On application of rules, the rectangle is detected and added to the set R (Step 17) and the polygon is updated by calling the procedure UPDATE-POLY (Step 18). The set of partitioned rectangles, R is returned in Step 19.

4.2 Demonstration

The polygon P is defined as an ordered sequence of vertices $\langle 1, 2, 3, \ldots, 16 \rangle$ (Fig. 7(a)). The vertices 3 and 4 are two consecutive type 1 vertices and Rule 3

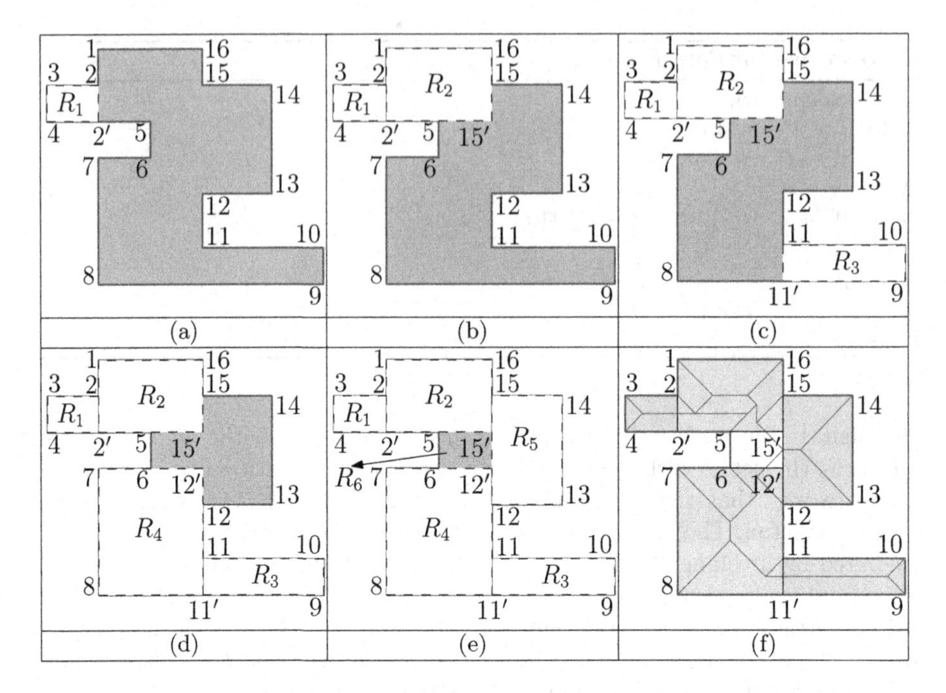

Fig. 7. Demonstration of partitioning into rectangles and corresponding straight skeleton.

is applied as the sequence of vertices of types $\langle 3, 1, 1, 3 \rangle$ is there. R_1 is obtained as shown in Fig. 7(a). The polygon is updated and the vertex $2'$ is added and the vertices 2, 3, and 4 are deleted. Rule 2 is applied on the sequence of vertices $\langle 15, 16, 1, 2', 5 \rangle$ of types $\langle 3, 1, 1, 1, 3 \rangle$ and R_2 is obtained as shown in Fig. 7(b). The vertices 1, $2'$ and 16 are deleted and $15'$ is added. Next, consecutive type 1 vertices are obtained for the vertices $\langle 6, 7, 8, 9, 10, 11 \rangle$ with types $\langle 3, 1, 1, 1, 1, 3 \rangle$. Rule 1 is applied and the rectangle R_3 is obtained as shown in Fig. 7(c). The vertices 9, 10 and 11 are deleted and $11'$ is added. Now, Rule 2 is applied on the vertices $\langle 6, 7, 8, 11', 12 \rangle$ with types $\langle 3, 1, 1, 1, 3 \rangle$ as shown in Fig. 7(d). The rectangle R_4 is obtained and the vertices $12'$ are added and the vertices 7, 8, and $11'$ are deleted. Rule 1 is applied on the sequence of vertices $\langle 12', 12, 13, 14, 15, 15' \rangle$ with types $\langle 3, 1, 1, 1, 1, 3 \rangle$. The rectangle R_5 is obtained as shown in Fig. 7(e). The vertices 12, 13, 14, and 15 are deleted. Now the remaining part of the polygon is a rectangle identified by the procedure FIND-RECT. The algorithm stops and R_6 is enlisted as shown in Fig. 7(e). The corresponding straight skeleton is shown in Fig. 7(f) which has a relation with the set of rectangles as partitioned.

4.3 Time Complexity Analysis

The digital object is imposed on background grid of size g and the number of pixels on the periphery of the digital object is n. The inner isothetic cover, P of

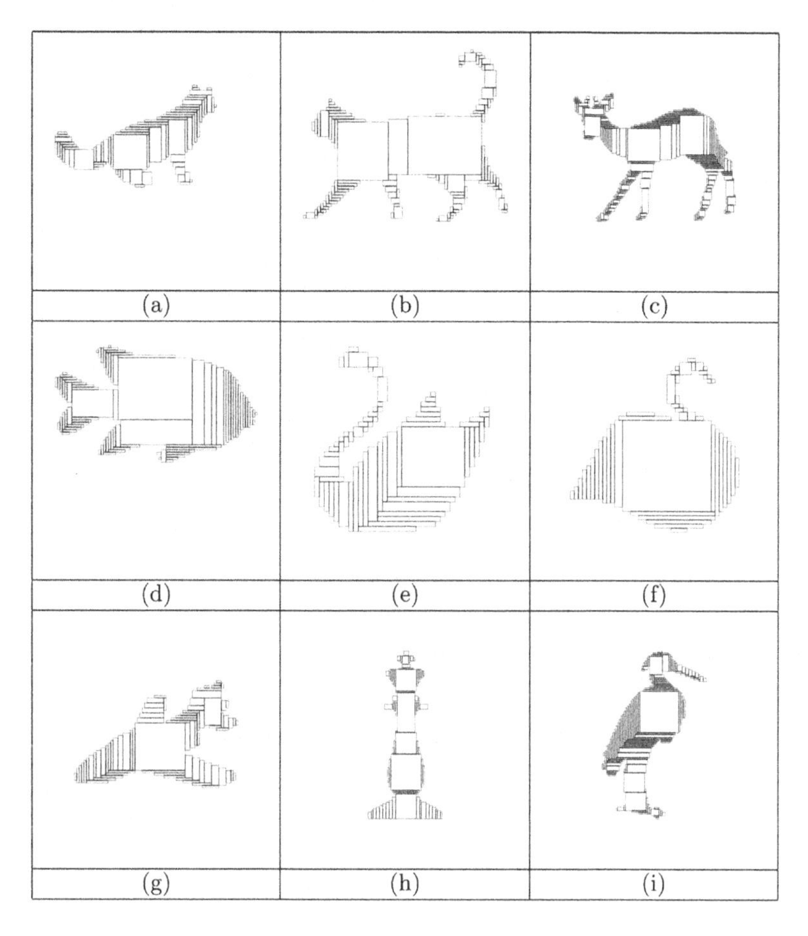

Fig. 8. Experimental results of partitioning into rectangles for a set of 8-connected digital objects.

the digital object is obtained in $O(n/g)$ time and lexicographically ordered lists, L_x and L_y, are constructed in $O(\log n/g)$ time. P is partitioned into rectangles using the combinatorial rules. To apply the rules, there is a need to search for vertices and requires to add and/ or delete from vertex list, L. To find the opposite vertex (horizontally or vertically), L_x and L_y are traversed in $O(\log(n/g))$. The addition and deletion of vertices from L needs linear time w.r.t. the total number of vertices in L. The polygon P is traversed once in anti-clockwise manner from top-left corner once to find the rectangularization by applying the rules which needs $O((n/g) + \log(n/g)) = O(n/g \log n/g)$ time.

5 Experimental Results

The proposed combinatorial algorithm is implemented using C programming language in Ubuntu 14.04 environment. The algorithm is tested on various types of

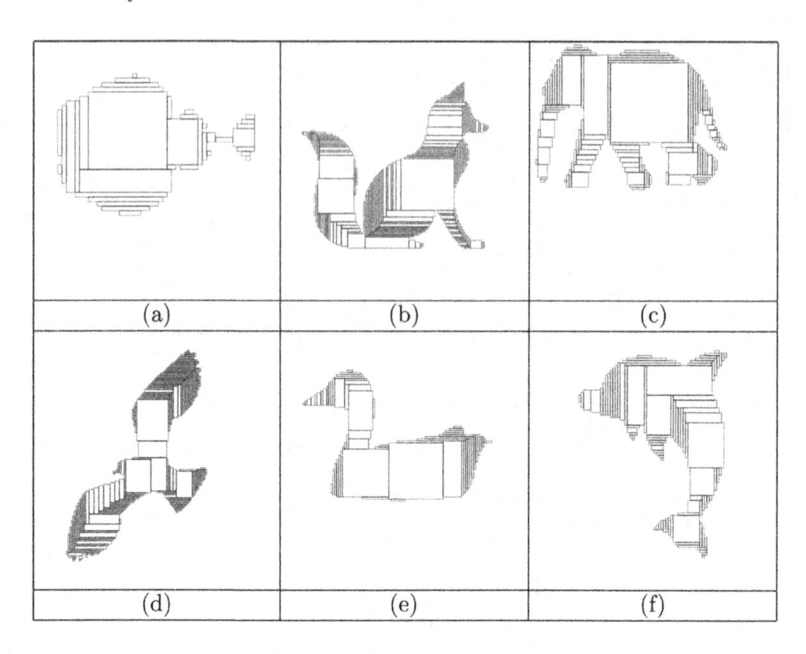

Fig. 9. Experimental results of partitioning into rectangles for another set of 8-connected digital objects.

Table 1. The data of experimental results shown in Fig. 8 and Fig. 9.

Digital object	# Vertices of P	Area of P	#Rectangles	Maximum area Rectangle	CPU Time (ms)
Fig. 8(a)	162	39816	72	7056	4.949
Fig. 8(b)	230	75900	108	25480	6.426
Fig. 8(c)	158	79600	69	14000	4.324
Fig. 8(d)	110	95226	36	37026	2.633
Fig. 8(e)	146	89712	68	22464	6.209
Fig. 8(f)	130	90500	49	48400	3.294
Fig. 8(g)	172	50292	77	14364	6.133
Fig. 8(h)	166	29328	65	6688	3.942
Fig. 8(i)	258	38608	117	9600	5.732
Fig. 9(a)	202	87552	83	27900	7.802
Fig. 9(b)	350	97984	160	17920	8.987
Fig. 9(c)	256	104076	108	36864	5.868
Fig. 9(d)	344	61088	149	8000	8.556
Fig. 9(e)	150	64675	62	18224	3.063
Fig. 9(f)	202	72396	90	11664	6.21

complex digital objects (8-connected) as shown in Fig. 8 and Fig. 9. All the operations are performed in the integer domain. The experimental results demonstrate the correctness of the proposed algorithm. The number of resulting rectangles, the number of vertices, the number of convex regions, the area and the perimeter of resulting rectangles are important features for shape analysis. Some of the data are shown in Table 1. The way the rectangularization occurs here has a relation with the corresponding straight skeleton w.r.t. the shape of the digital objects. A tree can be formed by the centers of the partitioned rectangles, which has a similarity with the straight skeleton tree. The advantage is that the rectangularization tree has less storage compared to the straight skeleton tree. If the number of rectangles with less area is more and they are consecutive, it implies complexity on that part of the digital object. For simpler digital objects, the number of rectangles are less and rectangles are comparatively larger. For symmetric objects, w.r.t. the line of symmetry, the number of rectangles are same and correspondingly their area also. The complexity of the shape of digital objects can be determined from these data. The symmetry of digital objects can also be obtained from these data along with the degree of symmetry. All these features are useful for defining the shape signature of the digital objects. With some more features, these data will be useful for shape classification.

6 Conclusion

The polygon partitioning into minimum number of rectangles is NP-hard. Here, the rectangularization is performed in such a way that straight skeleton has a relation with it. Thus the resulting set of rectangles are useful shape signature of digital objects. The rectangularization algorithm presented here applies combinatorial rules and runs in $O(n/g \log n/g)$ where n being the number of pixels on the periphery of digital object and g being the grid size on which the digital object is imposed. The experimental results are also presented. The problem has applications mainly in shape analysis. The set of rectangles represent the underlying structure of digital objects and preserves the geometrical and topological information of orthogonal polygons. This approach can be extended to orthogonal polyhedron as the future scope of this work.

References

1. Abam, M.A., Aronov, B., De Berg, M., Khosravi, A.: Approximation algorithms for computing partitions with minimum stabbing number of rectilinear and simple polygons. In: Proceedings of the Twenty-Seventh Annual Symposium On Computational Geometry, SOCG 2011, pp. 407–416. Association for Computing Machinery, New York (2011)
2. Avis, D., Toussaint, G.T.: An efficient algorithm for decomposing a polygon into star-shaped polygons. Pattern Recogn. **13**(6), 395–398 (1981)
3. Berg, M.D., Cheong, O., Kreveld, M.V., Overmars, M.: Computational Geometry: Algorithms and Applications, 3rd edn. Springer-Verlag, Heidelberg (2008)
4. Berg, M.D., Kreveld, M.V.: Rectilinear decompositions with low stabbing number. Inf. Process. Lett. **52**(4), 215–221 (1994)

5. Biswas, A., Bhowmick, P., Bhattacharya, B.B.: TIPS: on finding a tight Isothetic polygonal shape covering a 2D object. In: Kalviainen, H., Parkkinen, J., Kaarna, A. (eds.) SCIA 2005. LNCS, vol. 3540, pp. 930–939. Springer, Heidelberg (2005). https://doi.org/10.1007/11499145_94
6. Biswas, A., Bhowmick, P., Bhattacharya, B.B.: Construction of Isothetic covers of a digital object: a combinatorial approach. J. Vis. Commun. Image Represent. **21**(4), 295–310 (2010)
7. Chazelle, B.: A theorem on polygon cutting with applications. In: Proceedings of 23rd Annual Symposium on Foundations of Computer Science, pp. 339–349. IEEE Computer Society (1982)
8. Chazelle, B., Dobkin, D.: Decomposing a polygon into its convex parts. In: Proceedings of the Eleventh Annual ACM Symposium on Theory of Computing, pp. 38–48. Association for Computing Machinery, New York (1979)
9. Culberson, J.C., Reckhow, R.A.: Covering polygons is hard. In: Proceedings of 29th Annual Symposium on Foundations of Computer Science, pp. 38–48. IEEE Computer Society (1988)
10. Devadoss, S.L., Rourke, J.O.: Discrete and Computational Geometry. Princeton University Press, Princeton (2011)
11. Dutt, M., Biswas, A., Bhowmick, P.: Approximate partitioning of 2D objects into orthogonally convex components. Comput. Vis. Image Underst. **117**(4), 326–341 (2013)
12. Fekete, S., Lübbecke, M., Meijer, H.: Minimizing the stabbing number of matchings, trees, and triangulations. Discret. Comput. Geom. **40**, 595–621 (2008). https://doi.org/10.1007/s00454-008-9114-6
13. Garey, M., Johnson, D., Preparata, F., Tarjan, R.: Triangulating a simple polygon. Inf. Process. Lett. **7**(4), 175–179 (1978)
14. Gourley, K., Green, D.: A polygon-to-rectangle conversion algorithm. IEEE Comput. Graph. Appl. **3**(1), 31–36 (1983)
15. Keil, J.M.: Decomposing a polygon into simpler components. SIAM J. Comput. **14**(4), 799–817 (1985)
16. Klette, R., Rosenfeld, A.: Digital Geometry: Geometric Methods for Digital Picture Analysis. Morgan Kaufmann, San Francisco (2004)
17. Lien, J.M., Amato, N.M.: Approximate convex decomposition of polygons. Comput. Geom. Theory Appl. **35**(1), 100–123 (2006)
18. Lingas, A.: The power of non-rectilinear holes. In: Nielsen, M., Schmidt, E.M. (eds.) Automata, Languages and Programming. Lecture Notes in Computer Science, vol. 140, pp. 369–383. Lecture Notes in Computer Science (LNCS), Springer, Cham (1982). https://doi.org/10.1007/BFb0012784
19. Lingas, A., Soltan, V.: Minimum convex partition of a polygon with holes by cuts in given directions. In: Asano, T., Igarashi, Y., Nagamochi, H., Miyano, S., Suri, S. (eds.) Algorithms and Computation. Lecture Notes in Computer Science, pp. 315–325. Springer, Heidelberg (1996). https://doi.org/10.1007/BFb0009508
20. Liu, C., et al.: An effective chemical mechanical polishing fill insertion approach. ACM Trans. Des. Autom. Electron. Syst. **21**(3), 1–21 (2016)
21. Lopez, M.A., Mehta, D.P.: Efficient decomposition of polygons into L-shapes with application to VLSI layouts. ACM Trans. Des. Autom. Electron. Syst. **1**(3), 371–395 (1996)
22. Lowet, A.S., Firestone, C., Scholl, B.J.: Seeing structure: shape skeletons modulate perceived similarity. Atten. Percept. Psychophys. **80**(5), 1278–1289 (2018). https://doi.org/10.3758/s13414-017-1457-8

23. Lubiw, A.: Decomposing polygonal regions into convex quadrilaterals. In: Proceedings of the First Annual Symposium on Computational Geometry, pp. 97–106. Association for Computing Machinery, New York (1985)
24. Rourke, J.O., Supowit, K.J.: Some NP-hard polygon decomposition problems. IEEE Trans. Inf. Theory **29**(2), 181–190 (1983)
25. Schachter, B.: Decomposition of polygons into convex sets. IEEE Trans. Comput. **27**(11), 1078–1082 (1978)
26. Suk, T., Höschl, C., Flusser, J.: Decomposition of binary images- a survey and comparison. Pattern Recogn. **45**(12), 4279–4291 (2012)

On the Number of 0-Tandems in Simple nD Digital 0-Connected Curves

Lidija Čomić(✉)

Faculty of Technical Sciences, University of Novi Sad, Novi Sad, Serbia
comic@uns.ac.rs

Abstract. A 0-tandem is a configuration of two voxels (n-cells) sharing exactly one vertex. We propose a formula connecting the number of 0-tandems in a simple nD digital open or closed 0-connected curve γ with the number of cells in γ. Our formula generalizes the formula by Brimkov et al. (2006) from closed to open curves, and the formula by Maimone and Nordo (2015) from open curves in 3D to open or closed curves in nD. We also propose an alternative formula valid for 3D curves.

Keywords: Digital topology · Digital curves · Tandems · Gaps · Critical configurations

1 Introduction

Topological analysis of images and shapes is an active research field with many applications. One important topological property of a digital object O is its well-composedness or manifoldness [2,17,18]. The numerical characterization of this property, the number of critical configurations (tandems or gaps) in the object O, has been widely investigated in the literature. Establishing the relations between this descriptor and the number of cells of different types in the decomposition of the digital object O into a cell complex Q (the set of voxels in O plus all of their faces) enables a better understanding and a deeper insight into the topological structure of the object O and the complex Q. Tandems and gaps play an important role in ray-casting based rendering of digital curves and surfaces [10,25], and in particular of digital planes [1,12,13].

Many different formulas have been proposed expressing the number of $(n-2)$-tandems ($(n-2)$-gaps) in an nD digital object O through the number of cells of various types in Q, or some other topological descriptors of O, like its Euler characteristic [3,6–8,11,19,20]. For simple digital curves, formulas expressing the number of the k-cells through the number of n-cells and j-tandems of a closed curve in nD [3,8], $k \leq j \leq n$, or expressing the number of 0-tandems through the number of cells of an open curve in 3D [21,23], have been proposed as well.

We propose a formula expressing the number of 0-tandems in a simple open or closed 0-connected curve γ in nD through the number of cells in γ, thus generalizing the results by Brimkov et al. [3,8] from closed to open curves, and

R. P. Barneva et al. (Eds.): IWCIA 2022, LNCS 13348, pp. 46–55, 2023.
https://doi.org/10.1007/978-3-031-23612-9_3

the results by Maimone and Nordo [21,23] from open curves in 3D to open or closed curves in nD. We also propose an alternative formula valid for simple open or closed curves in 3D.

2 Preliminaries

We introduce some basic notions on the cubic grid [14,15], on tandems and gaps in digital objects in this grid [3,6,12,19], and on simple digital curves [5,14,21].

2.1 The Cubic Grid

Definition 1. *The nD cubic grid is a set of closed unit axis-aligned n-cubes (n-cells or voxels) centered at points in $\mathbb{Z}^n$. The naturally associated (cubical) cell complex is composed of all voxels, and all their k-faces (k-cells), $0 \leq k \leq n-1$.*

Each k-face is a k-cube in the space with $n-k$ Cartesian coordinates fixed to half-integer values [16]. Due to the regular structure of the grid, the number of j-cells incident with a given k-cell is constant.

Lemma 1. *Each voxel has $2^{n-k}\binom{n}{k}$ k-faces, $0 \leq k \leq n-1$. Each k-cell is incident with 2^{n-k} voxels, $0 \leq k \leq n-1$.*

Different types of adjacency relation are defined between the voxels in the grid, depending on their intersection.

Definition 2. *Two voxels are k-adjacent if they share a common k-cell. They are strictly k-adjacent if they are k-adjacent but not $(k+1)$-adjacent.*

2.2 Tandems and Gaps in Digital Objects

Definition 3. *An nD* digital object *O is a finite set of voxels in the nD cubic grid.*

The voxels in O are called black (object) voxels. The voxels in the complement O^c of O are called white (background).

Definition 4. *The cubical complex Q associated with an object O consists of all the voxels in O and all their k-faces, $0 \leq k \leq n-1$.*

The number of k-faces (k-cells) in Q is denoted by c_k.

Definition 5. *A $2^{n-k} \times 1^k$ block of voxels is called a $2^{n-k}1^k$-block, $0 \leq k \leq n-1$. The $2^{n-k}1^k$-block $B_k(e)$ centered at a k-cell e consists of 2^{n-k} voxels incident with e.*

Definition 6. *A pair of strictly k-adjacent voxels through a k-cell e is called a k-tandem over e, $0 \leq k \leq n-1$.*

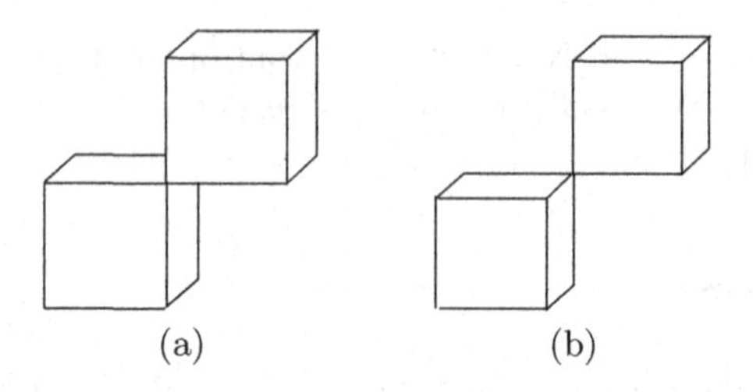

Fig. 1. Two strictly (a) 1-adjacent and (b) 0-adjacent voxels in 3D forming a 1-tandem (and 1-gap) and 0-tandem, respectively.

Definition 7. *An object O has a k-tandem at a k-cell e if $B_k(e) \cap O$ is a k-tandem over e.*

Definition 8. *An object O has a k-gap at a k-cell e if $B_k(e) \setminus O$ is a k-tandem over e.*

Thus, tandems and gaps are dual to each other: a k-tandem of an object O is a k-gap of the complement of O. A k-tandem is determined by a pair of strictly k-adjacent black voxels through a k-cell e, the other voxels incident with e being white. Similarly, a k-gap is determined by a pair of strictly k-adjacent white voxels in the $2^{n-k}1^k$-block centered at a k-cell e.

In 2D, a 0-tandem (and a 0-gap) occurs at a vertex incident with two strictly 0-adjacent black pixels (and to two white ones). In 3D, a 1-tandem (and a 1-gap) occurs at the edge incident with exactly two strictly 1-adjacent black voxels (and to two white ones), see Fig. 1 (a). A 0-tandem occurs at a vertex incident with two strictly 0-adjacent black voxels (and to six white ones), see Fig. 1 (b). A 0-gap occurs at a vertex incident with two strictly 0-adjacent white voxels (and to six black ones). The numbers of k-tandems and k-gaps in an object are denoted by t_k and g_k, respectively.

Definition 9. *A boundary (free) k-cell in Q, $0 \leq k \leq n-1$, is a k-cell incident both to a voxel in O and a voxel in O^c. An interior cell is incident to voxels in O only.*

A k-cell e in Q is interior if the $2^{n-k}1^k$-block $B_k(e)$ centered at e is contained in O. Otherwise, e is a boundary k-cell [7,20]. The number of interior and boundary k-cells in Q is denoted by c'_k and c^*_k, respectively, and

$$c_k = c'_k + c^*_k.$$

Definition 10. *A totally boundary k-cell in Q, $0 \leq k \leq n-1$, is a k-cell incident with exactly one voxel in O. A non totally boundary cell belongs to the shared face of at least one tandem in dimension $j \geq k$.*

The numbers of totally boundary and non totally boundary k-cells in Q are denoted by n_k^{tb} and c_k^{ntb}, respectively, and

$$c_k = c_k^{tb} + c_k^{ntb}.$$

2.3 Digital Curves

Definition 11. [5,21,23] *For an adjacency relation k, a simple closed k-connected digital curve γ of length m is a set $\{v_0, v_1, ..., v_m = v_0\}$ of voxels such that any voxel v_i in γ is k-adjacent in γ only to v_{i-1} and v_{i+1} (modulo m), and v_{i-1} and v_{i+1} are not k-adjacent to each other.*

Definition 12. [5] *A simple open k-connected curve is a k-connected proper subset of a simple closed k-connected curve.*

If the open curve contains at least two voxels, then it contains exactly two voxels, called end voxels, that have exactly one neighbor in the curve. All other voxels have exactly two such neighbors. Specially, if the open curve contains only one voxel, this voxel has no neighbors in γ.

3 Related Work

We review relevant work related to the number of 0-gaps in 2D, as well as some relations between the number of tandems and the numbers of cells of different types for digital curves.

3.1 2D Objects

Two equivalent [22] formulas have been proposed for the number g_0 of 0-gaps (i.e., the number t_0 of 0-tandems) in a 2D digital object O. One [6] states that

$$g_0 = t_0 = c_0 - 2(c_2 + c^0 - h^1) + c_0',$$

where c^0 is the number of 0-components (maximal connected components of black pixels with respect to 0-adjacency), and h^1 is the number of 1-holes (maximal finite connected components of white pixels with respect to 1-adjacency).

An alternative formula for the number of gaps in 2D [4,7,9] expresses g_0 in terms of boundary cells in Q as

$$g_0 = t_0 = c_1^* - c_0^* = c_1 - c_1' - c_0 + c_0'.$$

This relation has been obtained independently in the context of calculating the number of holes in a 2D binary image [24]. The number of 0-gaps in 2D has been related to the dimension of a digital object O [22].

3.2 Digital Curves

Two sets of formulas have been proposed relating the numbers of cells and the numbers of tandems in a simple digital 0-connected curve γ. The first one [3,8], proposed for closed 0-connected curves in nD, gives a connection between the numbers c_k of k-cells, $0 \leq k \leq n-1$ and t_j of j-tandems $k \leq j \leq n-1$ and c_n of n-cells as

$$c_k = 2^{n-k}\binom{n}{k}c_n - \sum_{i=0}^{n-k-1} 2^i \binom{k+i}{k} t_{i+k}. \tag{1}$$

The second one [21,23], proposed for open 0-connected curves in 3D, expresses the number t_0 of 0-tandems in γ through the numbers c_k of k-cells in γ, $0 \leq k \leq 3$ as

$$t_0 = \sum_{i=0}^{3}(-1)^{i+1}2^i c_i = -c_0 + 2c_1 - 4c_2 + 8c_3. \tag{2}$$

The proof of Formula 1 is based on distinguishing the cells in γ as belonging or not to some tandem in γ, i.e., on counting the (non) totally boundary cells in γ. The proof of Formula 2 relies on some advanced combinatorial notions.

4 0-Tandems in nD Digital Curves

Let us denote by A the sum $\sum_{i=0}^{n}(-1)^{i+1}2^i c_i$. We will show that $t_0(\gamma) = A$ for both open and closed 0-connected simple curves in nD, thus extending the results of [3,8] to open curves and of [21,23] to closed ones (and to arbitrary dimensions).

Proposition 1. *Let γ be a simple 0-connected digital open curve in nD with c_i i-cells, $0 \leq i \leq n$. Then*

$$t_0(\gamma) = \sum_{i=0}^{n}(-1)^{i+1}2^i c_i.$$

Proof. The proof is by induction on the length (the number of voxels) m of γ.

1. For $m = 1$, γ consists of one voxel, it has no 0-tandems and

$$\begin{aligned} A &= \sum_{i=0}^{n}(-1)^{i+1}2^i c_i \\ &= \sum_{i=0}^{n}(-1)^{i+1}2^i 2^{n-i}\tbinom{n}{i} \\ &= -2^n \sum_{i=0}^{n}(-1)^i \tbinom{n}{i} \\ &= -2^n(1-1)^n \\ &= t_0(\gamma). \end{aligned}$$

2. Let $t_0(\gamma) = A$ for each curve of length m, $m \in \mathbb{N}$, and let $\gamma = \{v_1, v_2, ..., v_{m+1}\}$ be a curve of length $m+1$. Let δ be the curve obtained from γ by removing its end voxel v_{m+1}, and let d_i be the number of i-cells in δ. Then, by inductive hypothesis, $t_0(\delta) = \sum_{i=0}^{n}(-1)^{i+1}2^i d_i$. We distinguish between two cases

(a) If v_{m+1} and v_m are strictly 0-adjacent, then $t_0(\gamma) = t_0(\delta) + 1$, and

$$\begin{aligned} A &= \sum_{i=0}^{n} (-1)^{i+1} 2^i c_i \\ &= \sum_{i=0}^{n} (-1)^{i+1} 2^i d_i + \sum_{i=0}^{n} (-1)^{i+1} 2^i 2^{n-i} \binom{n}{i} + 1 \\ &= t_0(\delta) + 0 + 1 \\ &= t_0(\gamma). \end{aligned}$$

(b) If v_{m+1} and v_m are strictly k-adjacent, $1 \le k \le n-1$, then $t_0(\gamma) = t_0(\delta)$, and

$$\begin{aligned} A &= \sum_{i=0}^{n} (-1)^{i+1} 2^i c_i \\ &= \sum_{i=0}^{n} (-1)^{i+1} 2^i d_i + \sum_{i=0}^{n} (-1)^{i+1} 2^i 2^{n-i} \binom{n}{i} - \sum_{i=0}^{k} (-1)^{i+1} 2^i 2^{k-i} \binom{k}{i} \\ &= t_0(\delta) + 0 - 0 \\ &= t_0(\gamma). \end{aligned}$$

Proposition 2. *Let γ be a 0-connected digital closed curve in nD with c_i i-cells, $0 \le i \le n$. Then*

$$t_0(\gamma) = \sum_{i=0}^{n} (-1)^{i+1} 2^i c_i.$$

Proof. Let $\delta = \{v_1, v_2, , , , v_{m-1}\}$ be the curve obtained from $\gamma = \{v_0, v_1, v_2, ..., v_m = v_0\}$ by removing the voxel $v_m = v_0$, and let d_i be the number of i-cells in δ. Let v_m be strictly k-adjacent to v_{m-1} and strictly j-adjacent to v_1, $k, j \in \{0, 1, ..., n-1\}$. By Proposition 1, $t_0(\delta) = \sum_{i=0}^{n} (-1)^{i+1} 2^i d_i$. We distinguish between three cases

1. If $k = j = 0$, then $t_0(\gamma) = t_0(\delta) + 2$, and

$$\begin{aligned} A &= \sum_{i=0}^{n} (-1)^{i+1} 2^i c_i \\ &= \sum_{i=0}^{n} (-1)^{i+1} 2^i d_i + \sum_{i=0}^{n} (-1)^{i+1} 2^i 2^{n-i} \binom{n}{i} + 2 \\ &= t_0(\delta) + 0 + 2 \\ &= t_0(\gamma). \end{aligned}$$

2. If $k = 0$, $j > 0$ (and similarly for $k > 0$, $j = 0$), then $t_0(\gamma) = t_0(\delta) + 1$, and

$$\begin{aligned} A &= \sum_{i=0}^{n} (-1)^{i+1} 2^i c_i \\ &= \sum_{i=0}^{n} (-1)^{i+1} 2^i d_i + \left(\sum_{i=0}^{n} (-1)^{i+1} 2^i 2^{n-i} \binom{n}{i} - 1 \right) - \sum_{i=0}^{k} (-1)^{i+1} 2^i 2^{k-i} \binom{k}{i} \\ &= t_0(\delta) + 1 - 0 \\ &= t_0(\gamma). \end{aligned}$$

3. If $k, j > 0$, then $t_0(\gamma) = t_0(\delta)$, and

$$\begin{aligned} A &= \sum_{i=0}^{n}(-1)^{i+1}2^i c_i \\ &= \sum_{i=0}^{n}(-1)^{i+1}2^i d_i + \sum_{i=0}^{n}(-1)^{i+1}2^i 2^{n-i}\binom{n}{i} - \\ &\quad \sum_{i=0}^{k}(-1)^{i+1}2^i 2^{k-i}\binom{k}{i} - \sum_{i=0}^{j}(-1)^{i+1}2^i 2^{j-i}\binom{j}{i} \\ &= t_0(\delta) + 0 - 0 - 0 \\ &= t_0(\gamma). \end{aligned}$$

5 0-Tandems in 3D Digital Curves

We give an alternative formula for the number of 0-tandems in simple open or closed 0-connected digital curves in 3D.

Proposition 3. *Let γ be an open 0-connected digital curve in 3D with c_3 cubes (voxels), c_2 faces, c_2^* boundary faces and c_1 edges. Then*

$$t_0(\gamma) = c_3 - c_2 - c_2^* + c_1 - 1.$$

Proof. The proof is by induction on the length m of γ. Let $A = c_3 - c_2 - c_2^* + c_1 - 1$. We will show that $A = t_0(\gamma)$.

1. For $m = 1$, γ consists of one voxel, it has no 0-tandems and

$$A = 1 - 6 - 6 + 12 - 1 = 0 = t_0(\gamma)$$

2. Let $A = t_0(\gamma)$ for each open curve of length m, $m \in \mathbb{N}$, and let $\gamma = \{v_1, v_2, ..., v_{m+1}\}$ be a curve of length $m + 1$. Let δ be the curve obtained from γ by removing its end voxel v_{m+1}, and let d_3, d_2, d_2^* and d_1 be the number of voxels, faces, boundary faces and edges in δ. Then, by inductive hypothesis, $t_0(\delta) = B$ where $B = d_3 - d_2 - d_2^* + d_1$. We distinguish between three cases
   (a) If v_{m+1} and v_m are strictly 0-adjacent, then $t_0(\gamma) = t_0(\delta) + 1$, and

$$\begin{aligned} A &= c_3 - c_2 - c_2^* + c_1 - 1 \\ &= (d_3 + 1) - (d_2 + 6) - (d_2^* + 6) + (d_1 + 12) - 1 \\ &= d_3 - d_2 - d_2^* + d_1 \\ &= t_0(\delta) + 1 \\ &= t_0(\gamma). \end{aligned}$$

   (b) If v_{m+1} and v_m are strictly 1-adjacent, then $t_0(\gamma) = t_0(\delta)$, and

$$\begin{aligned} A &= c_3 - c_2 - c_2^* + c_1 - 1 \\ &= (d_3 + 1) - (d_2 + 6) - (d_2^* + 6) + (d_1 + 11) - 1 \\ &= d_3 - d_2 - d_2^* + d_1 - 1 \\ &= t_0(\delta) \\ &= t_0(\gamma). \end{aligned}$$

(c) If v_{m+1} and v_m are strictly 2-adjacent, then $t_0(\gamma) = t_0(\delta)$, and

$$\begin{aligned} A &= c_3 - c_2 - c_2^* + c_1 - 1 \\ &= (d_3 + 1) - (d_2 + 5) - (d_2^* + 4) + (d_1 + 8) - 1 \\ &= d_3 - d_2 - d_2^* + d_1 - 1 \\ &= t_0(\delta) \\ &= t_0(\gamma). \end{aligned}$$

Intuitively, for each maximal 2-connected component of γ (parallel to one of the coordinate axes) the number of voxels is greater by one than the number of inner faces, and each voxel belongs to exactly one such component. Thus, $c_3 - c_2'$ is equal to the number of 2-components of γ. Let us denote this number by X. At each 1-tandem, two such 2-components merge, creating 1-components. Thus, the number Y of 1-components of γ is equal to $X - t_1$. Since $t_1 = g_1 = 2c_2^* - c_1^* = 2c_2^* - c_1$ [6], we have that $Y = c_3 - c_2' - t_1 = c_3 - c_2' - 2c_2^* + c_1 = c_3 - c_2 - c_2^* + c_1$. At each 0-tandem, 1-components merge to produce the (connected) curve γ, which has one 0-component. Thus, $1 = Y - t_0$, i.e., $t_0(\gamma) = c_3 - c_2 - c_2^* + c_1 - 1$.

Proposition 4. *Let γ be a closed 0-connected digital curve in 3D with c_3 cubes (voxels), c_2 faces, c_2^* boundary faces and c_1 edges. Then*

$$t_0(\gamma) = c_3 - c_2 - c_2^* + c_1.$$

Proof. Let $\delta = \{v_1, v_2, ..., v_{m-1}\}$ be the curve obtained from $\gamma = \{v_0, v_1, v_2, ..., v_m = v_0\}$ by removing the voxel $v_m = v_0$, and let d_i be the number of i-cells in δ. Let r and s be the cells shared by the voxel v_m and the voxels v_1 and v_{m-1}, respectively. The curve δ is 0-connected, because γ is. Thus, the cells r and s are incident with two opposite faces f and g of v_m. Let $C = c_3 - c_2 - c_2^* + c_1$, $D = d_3 - d_2 - d_2^* + d_1 = t_0(\delta) + 1$ and $\Delta = D - C$.

The cells of the voxel v_m which are not incident with either f or g contribute four faces (all four are boundary faces) and four edges to Δ (with the appropriate sign), i.e., they add -3 to Δ. If v_m and v_1 are face-adjacent, then f adds 1 to Δ (the face f is a boundary face in δ but not in γ). If v_m and v_1 are edge-adjacent, then f adds 1 to Δ (the face f and its three non-shared edges are not present in δ and f is a boundary face in γ). If v_m and v_1 are vertex-adjacent, then f adds 2 to Δ (it contributes one face which is a boundary face, together with its four edges). It also creates a 0-tandem in γ which was not present in δ. Thus, f adds $1 + t_f$ to Δ, where t_f is the number of tandems created by f ($t_f = 1$ if v_m and v_1 are 0-adjacent (if they form a 0-tandem), and $t_f = 0$ otherwise). Similar considerations and notations apply for the face g, and $t_0(\gamma) = t_0(\delta) + t_f + t_g$. Then

$$\begin{aligned} C &= D - \Delta \\ &= (t_0(\delta) + 1) - \Delta \\ &= (t_0(\delta) + 1) - 3 + (1 + t_f) + (1 + t_g) \\ &= t_0(\delta) + t_f + t_g \\ &= t_0(\gamma). \end{aligned}$$

The intuitive reasoning is similar as above, except that the final tandem does not merge two different components of γ, but connects two endpoints of γ to produce a closed curve.

Acknowledgement. This work has been partially supported by the Ministry of Education, Science and Technological Development of the Republic of Serbia through the project no. 451-03-68/2020-14/200156.

References

1. Andres, E., Acharya, R., Sibata, C.H.: Discrete analytical hyperplanes. CVGIP: Graph. Model Image Process. **59**(5), 302–309 (1997)
2. Boutry, N., Géraud, T., Najman, L.: A Tutorial on Well-Composedness. J. Math. Imaging Vis. **60**(3), 443–478 (2018). https://doi.org/10.1007/s10851-017-0769-6
3. Brimkov, V.E.: Formulas for the number of $(n-2)$-gaps of binary objects in arbitrary dimension. Discret. Appl. Math. **157**(3), 452–463 (2009)
4. Brimkov, V.E., Barneva, R.P.: Linear time constant-working space algorithm for computing the genus of a digital object. In: Bebis, G., et al. (eds.) ISVC 2008. LNCS, vol. 5358, pp. 669–677. Springer, Heidelberg (2008). https://doi.org/10.1007/978-3-540-89639-5_64
5. Brimkov, V.E., Klette, R.: Curves, hypersurfaces, and good pairs of adjacency relations. In: Klette, R., Žunić, J. (eds.) IWCIA 2004. LNCS, vol. 3322, pp. 276–290. Springer, Heidelberg (2004). https://doi.org/10.1007/978-3-540-30503-3_21
6. Brimkov, V.E., Maimone, A., Nordo, G.: An explicit formula for the number of tunnels in digital objects. CoRR abs/cs/0505084 (2005). http://arxiv.org/abs/cs/0505084
7. Brimkov, V.E., Maimone, A., Nordo, G.: Counting gaps in binary pictures. In: Reulke, R., Eckardt, U., Flach, B., Knauer, U., Polthier, K. (eds.) IWCIA 2006. LNCS, vol. 4040, pp. 16–24. Springer, Heidelberg (2006). https://doi.org/10.1007/11774938_2
8. Brimkov, V.E., Moroni, D., Barneva, R.: Combinatorial relations for digital pictures. In: Kuba, A., Nyúl, L.G., Palágyi, K. (eds.) DGCI 2006. LNCS, vol. 4245, pp. 189–198. Springer, Heidelberg (2006). https://doi.org/10.1007/11907350_16
9. Brimkov, V.E., Nordo, G., Barneva, R.P., Maimone, A.: Genus and dimension of digital images and their time- and space-efficient computation. Int. J. Shape Model. **14**(2), 147–168 (2008)
10. Cohen-Or, D., Kaufman, A.E.: 3D line voxelization and connectivity control. IEEE Comput. Graph. Appl. **17**(6), 80–87 (1997)
11. Čomić, L.: On gaps in digital objects. In: Barneva, R.P., Brimkov, V.E., Tavares, J.M.R.S. (eds.) IWCIA 2018. LNCS, vol. 11255, pp. 3–16. Springer, Cham (2018). https://doi.org/10.1007/978-3-030-05288-1_1
12. Françon, J., Schramm, J.-M., Tajine, M.: Recognizing arithmetic straight lines and planes. In: Miguet, S., Montanvert, A., Ubéda, S. (eds.) DGCI 1996. LNCS, vol. 1176, pp. 139–150. Springer, Heidelberg (1996). https://doi.org/10.1007/3-540-62005-2_12
13. Kenmochi, Y., Imiya, A.: Combinatorial topologies for discrete planes. In: Nyström, I., Sanniti di Baja, G., Svensson, S. (eds.) DGCI 2003. LNCS, vol. 2886, pp. 144–153. Springer, Heidelberg (2003). https://doi.org/10.1007/978-3-540-39966-7_13

14. Klette, R., Rosenfeld, A.: Digital Geometry Geometric: Methods for Digital Picture Analysis. Morgan Kaufmann Publishers, San Francisco, Amsterdam (2004)
15. Kong, T.Y., Rosenfeld, A.: Digital topology: introduction and survey. Comput. Vis. Graph. Image Process. **48**(3), 357–393 (1989)
16. Kovalevsky, V.A.: Geometry of Locally Finite Spaces (Computer Agreeable Topology and Algorithms for Computer Imagery). Editing House Dr. Bärbel Kovalevski, Berlin (2008)
17. Latecki, L.J.: 3D well-composed pictures. CVGIP: Graph. Model Image Process. **59**(3), 164–172 (1997)
18. Latecki, L.J., Eckhardt, U., Rosenfeld, A.: Well-composed sets. Comput. Vis. Image Underst. **61**(1), 70–83 (1995)
19. Maimone, A., Nordo, G.: On 1-gaps in 3D digital objects. Filomat **22**(3), 85–91 (2011)
20. Maimone, A., Nordo, G.: A formula for the number of $(n-2)$-gaps in digital n-objects. Filomat **27**(4), 547–557 (2013)
21. Maimone, A., Nordo, G.: 0-gaps on 3D digital curves. Appl. Math. Math. Phys. **1**, 119–128 (2015)
22. Maimone, A., Nordo, G.: A note on dimension and gaps in digital geometry. Filomat **31**(5), 1215–1227 (2017)
23. Nordo, G., Maimone, A.: 0-gaps on 3D digital curves. CoRR abs/2109.13341 (2021)
24. Sossa, H.: On the number of holes of a 2-D binary object. In: 14th IAPR International Conference on Machine Vision Applications, MVA, pp. 299–302 (2015)
25. Yagel, R., Cohen, D., Kaufman, A.E.: Discrete ray tracing. IEEE Comput. Graph. Appl. **12**(5), 19–28 (1992)

On Density Extrema for Digital Discs

Nilanjana G. Basu[1], Partha Bhowmick[2(✉)], and Subhashis Majumder[1(✉)]

[1] Department of Computer Science and Engineering, Heritage Institute of Technology, Kolkata, India
{nilanjanag.basu,subhashis.majumder}@heritageit.edu

[2] Department of Computer Science and Engineering, Indian Institute of Technology, Kharagpur, India
pb@cse.iitkgp.ac.in

https://www.heritageit.edu/CSE.aspx, https://cse.iitkgp.ac.in/~pb/

Abstract. The act of characterizing and measuring different attributes of primitive shapes is of paramount importance in the subject of discrete geometry. A very rich collection of work can be found in this domain, which is predominantly focused on the Euclidean space. The digital space, on the contrary, is relatively unexplored, possibly because of the fact that it has evolved much later and it does not readily migrate to the continuous space. This work studies the unique problem of characterizing density extrema (of integer points) for discs moving on the integer plane. To the best of our knowledge, there has not been a significant study in this direction, which motivates us to look into this problem. As 'density' provides a notion of the relative concentration or rarefaction of a collection of points within a given shape or region, it has applications in image analysis and related areas, apart from different branches of physical science. We present some novel results, which are fundamental to understanding density minima and maxima for circular shapes in digital space. We have also pointed out a few more interesting problems that might advance this study further ahead.

Keywords: Digital disc · Digital geometry · Pixel density · Geometry of numbers · Number theory

1 Introduction

The subject of Euclidean distance geometry is concerned with the geometry based on a distance function and has uses in a wide variety of applications, ranging from identification of molecular conformations in computational chemistry [16] to pattern recognition and image processing [25]. Different types of mathematical transforms are used to generate efficient algorithms that find applications in various problems [10,11,24]. Exploring the nature of an ensemble of points and thereby analyzing their properties have gathered sufficient interest among researchers in different domains. Among many other properties, the

R. P. Barneva et al. (Eds.): IWCIA 2022, LNCS 13348, pp. 56–70, 2023.
https://doi.org/10.1007/978-3-031-23612-9_4

knowledge of a particular fact – how clustered or how much dense the points are in the corresponding space turned out to be of remarkable importance. It has been quite conspicuous in the realms of Social Networking and Complex Networks, which got established as some of the hugely popular areas in recent times.

1.1 Existing Work

In the domain of computational and discrete geometry, some of the early works tried to answer the question of whether a particular subset of points is more concentrated or dispersed with respect to a given set. Some of the geometers used the term 'discrepancy of points' [1] to describe it, whereas some others preferred to introduce a new concept called 'density of points' [18]. For 2D applications, 'discrepancy' has mostly been used in a rectangular axes-parallel setting [6,9], while 'density' though first defined for an axes-parallel rectangle [17], is more open to accommodating different shapes or configurations in consideration for containing the set of points. A short account of why the two measures, though appear to be similar, are different from each other can be seen in [18]. Typically, for the case of unweighted (weighted) points, *density* in a 2D region is expressed as the count (sum) of the points (weights) present in that region per unit area.

As shown in [18], the axis-parallel region of maximum (minimum) density would always contain only two (one) of the points from the given set of points. An efficient algorithm has been proposed in [18] to identify the rectangular region of maximum density in $\mathbb{R}^2$ when the points are of uniform weight. For higher dimensions, algorithms for finding maximum- and minimum-density regions have been proposed later in [4].

One of the topmost motivations for characterizing or computing density function comes from the VLSI domain. Identification of hotspots on a chip is a burning issue, as the dimensions of the chips keep on shrinking every year by leaps and bounds. Hence, locating the areas on the chip where the thermal sources are present with the highest concentration is a crucial problem. The rationale to locate the regions of minimum density is to find alternate places where the circuitry with the trouble-making heat sources can be moved so that the thermal gradient of the chip gets balanced again. As VLSI floorplans are generally rectangular, rectangular regions are usually considered for calculating densities. However, as discussed in [18], awkwardly thin rectangles pose unwanted problems as their densities become unnecessarily high and hence incorrectly reflect the exact physical scenario of thermal transmission.

A representative example is illustrated in Fig. 1. It shows two axis-parallel rectangles, each of 16 square units of area. By the definition of density, both of them would be considered to have the same density. However, in reality, the two heat sources placed at the two corner points of the thin rectangle are far more distant than those of the square, and hence should not cause the same type/level of damage to a VLSI chip. So the method used for the calculation of the density of the rectangle will unnecessarily flag a higher value, thus raising a false alarm. This gave us the motivation to look for an alternative to the

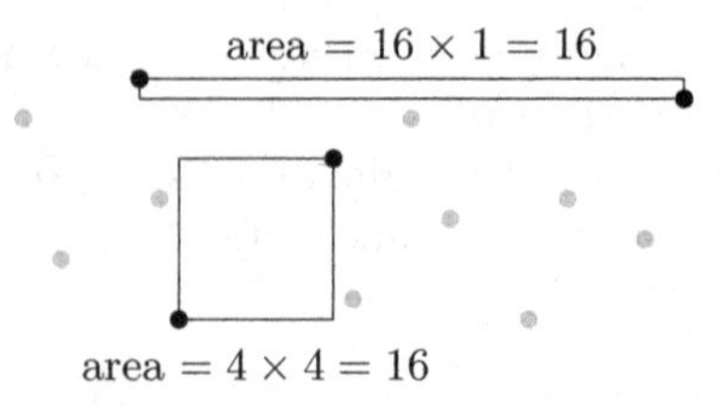

Fig. 1. Anomalies with densities for rectangles.

notion of using axes-parallel regions. Note that, for finding maximum density, if we relaxed the criterion of axis-parallelism and in turn tried to find rectangles of any orientation that will be of minimum area, then a region containing a set of only two points will have zero area, as the corresponding rectangle will degenerate to a straight line segment. Though the idea of using a convex hull instead of rectangles is suggested in [18] to avoid the anomaly posed by thin rectangles, from the perspective of symmetry, we thought considering a circular region would be an appropriate way to tackle it.

In the literature of digital geometry, there is a good collection of work related to digital disks and digital balls defined on a square or non-square grid. Some of these may be seen in [15,19,21] and in the bibliographies therein. Another geometric primitive that is closely related to digital disc is digital circle, and there are even more research work on its analytical description and algorithmic construction. We refer to [2,3,7,20,23] for different algorithmic techniques and their analyses related to the construction of digital circles. In [8], a theoretical analysis and experimental findings for polygonal covers of digital discs are presented. In connection with digital image processing, there are various works on the recognition of digital discs, e.g., [12]. Also, about the counting of digital discs and digital circles, their encoding and recognition, and on estimation and measures of related parameters, there are several research papers, such as [13,14,20,22,23,26,27].

It may be noted that the concept of density estimation or density extrema for digital discs is not found in the existing literature, some of which are mentioned above. This motivates us to look into this problem, and our contribution is highlighted in the following section.

1.2 Our Contribution

Finding optimally dense circular regions is a different ball game altogether from that of identifying axes-parallel regions, especially when the points under consideration can be located anywhere in $\mathbb{R}^2$ and we address this problem elsewhere. In this work, we consider that the input points are all grid points of a square grid and have a uniform weight. The grid is conceived as the integer plane, i.e., $\mathbb{Z}^2$, for simplicity. We prove that for a circular region whose center is aligned with a grid point, and the radius is an integral multiple of the grid unit, the maximum-

and the minimum-density circular regions respectively occur when the radii are one and seven. We have also identified the location of the discs with maximum and minimum densities, when the radius is relaxed to be any real number but the center remains pinned to a grid point. We also found the maximum-density disc on this grid set-up, when the center can be located anywhere on the plane and the radius can be any real number. The only combination that remains unexplored here is the case of minimum-density discs with real values of center and radius, which we plan to do in near future.

2 Maximum Density

An *integer point* or *pixel* is an element of $\mathbb{Z}^2$; that is, it is a two-dimensional point with integer coordinates. We denote by $D_{c,r}$ the real disc centered at c and of radius r. When $c \in \mathbb{Z}^2$, our analysis and result are independent of the coordinates of c, and hence for brevity we will drop 'c' from our notations; that is, we will use D_r instead of $D_{c,r}$, and will use simplified notations in what follows. In particular, we will fix the center at $(0,0)$. When the center is an arbitrary point in $\mathbb{R}^2$, we will modify the notations accordingly with a note in the relevant section.

The set of pixels contained in D_r is denoted by $\mathbb{D}_r := D_r \cap \mathbb{Z}^2$ and is referred to as a *digital disc*. The cardinality of $\mathbb{D}_r$ is denoted by $|\mathbb{D}_r|$, and the density of pixels in $\mathbb{D}_r$ is given by $d_r := \frac{|\mathbb{D}_r|}{\pi r^2}$.

Let S_r be the real axis-parallel square that circumscribes D_r; that is, $S_r = \{(x,y) \in \mathbb{R}^2 : \max(|x|,|y|) \leq r\}$. Let $\mathbb{S}_r := S_r \cap \mathbb{Z}^2$ be the *digital square*, defined as the set of pixels contained in S_r. An illustration is given in Fig. 2.

2.1 Integer Center and Integer Radius

Let r be a positive integer. Then, $\sqrt{2}(r-1) < r$ if and only if $r \leq 3$, which implies $S_{r-1} \subset D_r$ if and only if $r \leq 3$. This leads to the following observation.

Observation 1. *For a positive integer r,*

$$\mathbb{S}_{r-1} \setminus \mathbb{D}_r = \varnothing \quad \textit{if } r \leq 3 \tag{1a}$$

$$\mathbb{S}_{r-1} \setminus \mathbb{D}_r \neq \varnothing \quad \textit{otherwise.} \tag{1b}$$

Further, for any positive integer r, it is easy to see that the following two statements are true.

1. The set $\mathbb{D}_r \setminus \mathbb{S}_{r-1}$ is identical to the set $\mathbb{D}_r \cap (\mathbb{S}_r \setminus \mathbb{S}_{r-1})$.
2. $\left|\mathbb{D}_r \cap (\mathbb{S}_r \setminus \mathbb{S}_{r-1})\right| = 4$.

An example is shown in Fig. 2 where the pixels in $\mathbb{D}_r \setminus \mathbb{S}_{r-1}$ for $r = 4$ are colored red. Combining the above two statements, we have the following observation.

Observation 2. *For any positive integer r, $\left|\mathbb{D}_r \setminus \mathbb{S}_{r-1}\right| = 4$.*

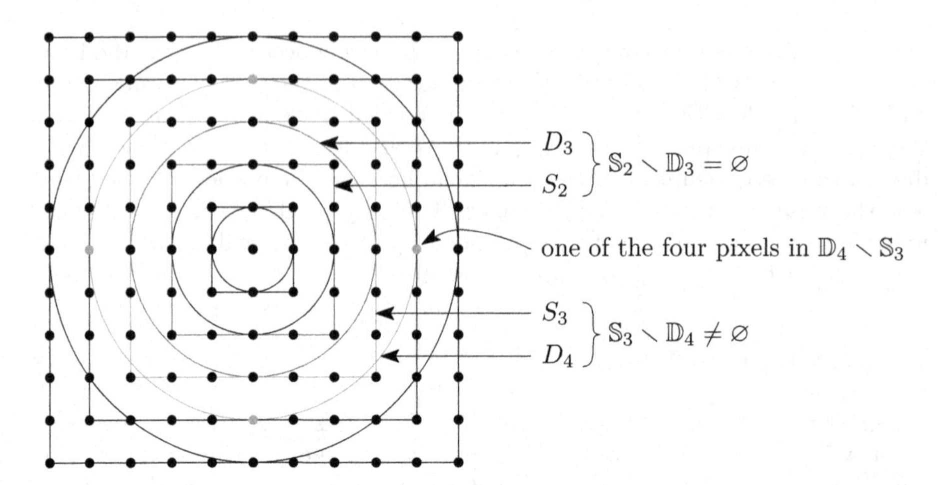

Fig. 2. Concentric discs with integer center and integer radii, and their circumscribing squares.

Based on the above results, we have the following lemma on the density of pixels in a disc centered at a pixel and having an integer radius.

Theorem 1. *In the collection of all discs with integer centers and integer radii,* $d_1 > d_r \ \forall r \geq 2$.

Proof. Let us first consider that $r \geq 4$. Then, from (1b) of Observation 1 and from Observation 2, it follows that $|\mathbb{D}_r| < |\mathbb{S}_{r-1}| + 4$, which implies

$$\begin{aligned} d_r &< \frac{|\mathbb{S}_{r-1}| + 4}{\pi r^2} = \frac{(2r-1)^2 + 4}{\pi r^2} = \frac{1}{\pi}\left(4 - \frac{4}{r} + \frac{5}{r^2}\right) \\ \implies d_r - d_1 &< \frac{1}{\pi}\left(4 - \frac{4}{r} + \frac{5}{r^2}\right) - \frac{5}{\pi} = \frac{1}{\pi}\left(-1 - \frac{4}{r} + \frac{5}{r^2}\right) \\ &= \frac{5 - 4r - r^2}{\pi r^2} = \frac{(1-r)(5+r)}{\pi r^2} < 0 \\ \implies \quad & d_1 > d_r. \end{aligned}$$

Now, from Fig. 2, we can see that $d_1 = \frac{5}{\pi}, d_2 = \frac{13}{4\pi}, d_3 = \frac{29}{9\pi}$, which implies $d_1 > d_2 > d_3$, and hence the proof. $\square$

2.2 Integer Center and Real Radius

For discs with integer centers and (positive) real radii, we modify some of the previous notations. We now denote by $D_{x,y}$ the real disc centered at $(0,0)$ and passing through the point $(x, y) \in \mathbb{R}^2$. Accordingly, we denote by $\mathbb{D}_{x,y}$ the digital disc corresponding to $D_{x,y}$, i.e., $\mathbb{D}_{x,y} := D_{x,y} \cap \mathbb{Z}^2$, and by $d_{x,y}$ the density of pixels in $D_{x,y}$, i.e., $d_{x,y} = \frac{|\mathbb{D}_{x,y}|}{\pi(x^2+y^2)}$. Note that, as in Lemma 1, the density remains invariant with the choice of the center as it has integer coordinates, and hence w.l.o.g. we consider $(0,0)$ as the center. Further, as the density of such a disc can be made arbitrarily large by making its radius infinitesimally small, we consider here discs with at least two pixels.

As we are concerned with the maximum density of pixels over all real-valued radii, we now make an important observation that comes to use in narrowing down our attention to a countable collection of discs and subsequently in proving the next lemma. Let $D_{x,y}$ be a disc that does not contain any pixel on its boundary. Let $(i,j) \in \mathbb{D}_{x,y}$ be a pixel lying farthest from $(0,0)$. Then, $D_{i,j} \subsetneq D_{x,y}$ and $\mathbb{D}_{i,j} = \mathbb{D}_{x,y}$, which implies that the density for $D_{i,j}$ is larger than that for $D_{x,y}$, whence the following observation.

Observation 3. *For any disc without any pixel on its boundary, there always exists a higher-density disc with a pixel on its boundary.*

Observation 3 implies that the maximum density over all discs with real radii will be the density of a disc from the countable collection $\{D_{i,j} : (i,j) \in \mathbb{Z}^2\}$, i.e.,

$$\max\{d_{x,y} : (x,y) \in \mathbb{R}^2\} = \max\{d_{i,j} : (i,j) \in \mathbb{Z}^2\}.$$

Now, consider any disc $D_{i,j}$ with $(i,j) \in \mathbb{Z}^2 \setminus \{(0,0)\}$. If $i = j$, then its boundary will contain four pixels, namely $\{(a,b) : |a| = |b| = |i|\}$. If $i \neq j$, then its boundary will contain four or eight pixels — four if $j = 0$ and eight if $j \neq 0$ — comprising the set $\{(a,b) : \{|a|\} \cup \{|b|\} = \{|i|, |j|\}\}$. In either case, the boundary may contain more pixels—when $i^2 + j^2$ is expressible as the sum of squares of a pair of positive integers other than $(|i|, |j|)$. Out of all such pixels that lie on the boundary of $D_{i,j}$, we fix (i,j) as the *defining pixel* of $D_{i,j}$ if $0 \leq j \leq i$. An illustration is given in Fig. 3. Thus, the aforesaid countable collection of discs reduces to a smaller countable collection of discs, namely $\mathscr{D} := \{D_{i,j} : (i,j) \in \mathbb{Z}^2, 0 \leq j \leq i\}$.[1] In what follows next, we assume that each disc $D_{i,j}$ belongs to $\mathscr{D}$.

As before, we denote by S_r the set $\{(x,y) \in \mathbb{R}^2 : \max(|x|, |y|) \leq r\}$, and by $\mathbb{S}_r$ the set $S_r \cap \mathbb{Z}^2$.

For any integer $i \geq 0$, we define

$$\mathbb{S}_{\Delta_i} := \begin{cases} \mathbb{S}_0 := \{(0,0)\} & \text{if } i = 0, \\ \mathbb{S}_i \setminus \mathbb{S}_{i-1} & \text{otherwise.} \end{cases}$$

Notice that the sets $\mathbb{S}_{\Delta_i}$ and $\mathbb{S}_{\Delta_{i+1}}$ are pairwise disjoint, and the set $\mathbb{Z}^2$ admits the partition $\{\mathbb{S}_{\Delta_i}\}_{i=0}^{\infty}$. As a result, we get

$$\mathscr{D} = \left\{ D_{i,j} : (i,j) \in \biguplus_{i=0}^{\infty} \mathbb{S}_{\Delta_i}, 0 \leq j \leq i \right\}. \tag{2}$$

[1] The countability of $\mathscr{D}$ follows from the fact that $\mathbb{Z}^2$ is countable. However, we call it a "countable collection" instead of "countable set" because a disc may appear more than once; e.g., $D_{5,0}$ and $D_{4,3}$ are identical discs but defined twice—once by $(5,0)$ and once by $(4,3)$.

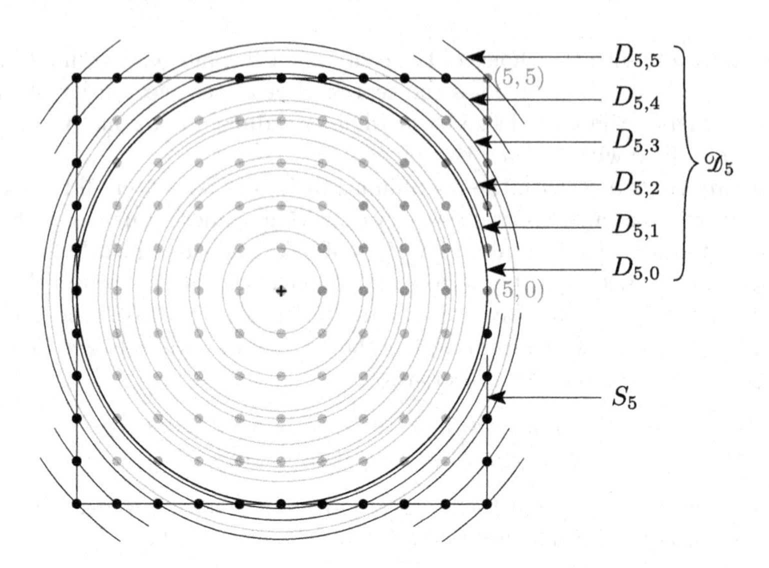

Fig. 3. Digital discs with their boundaries defined by pixels (shown in red) on the boundaries of squares. For example, the discs in $\mathscr{D}_5$ are defined by the red pixels of $\mathbb{S}_{\Delta_5} := \mathbb{S}_5 \setminus \mathbb{S}_4$. For clarity, the other defining pixels of $\mathbb{S}_5$ are shown in light red. (Color figure online)

For a given integer $i > 0$, we denote by $\mathscr{D}_i$ the sub-collection of $\mathscr{D}$ that comprises the discs defined by $\mathbb{S}_{\Delta_i}$, i.e., $\mathscr{D}_i := \{D_{i,j} : (i,j) \in \mathbb{S}_{\Delta_i}, 0 \leq j \leq i\}$.

Theorem 2. *In the collection of all discs with integer centers and real radii,* $d_1 > d_r \; \forall r > 1$.

Proof. Let i, j be two integers such that $i \geq 1$ and $0 \leq j \leq i$. Let $r = \sqrt{i^2 + j^2}$. Then, S_r circumscribes $D_{i,j}$, which implies $\mathbb{D}_{i,j} \subseteq \mathbb{S}_r$. Thus,

$$|\mathbb{D}_{i,j}| \leq |\mathbb{S}_r| \leq (2r+1)^2 = 4r^2 + 4r + 1.$$

An example is shown in Fig. 4.
This gives

$$d_r = \frac{|\mathbb{D}_{i,j}|}{\pi r^2} \leq \frac{4}{\pi} + \frac{4}{\pi r} + \frac{1}{\pi r^2}.$$

A manual counting of pixels shows that $d_1 = \frac{5}{\pi}$ is the unique maximum over all discs $D_{i,j} \in \bigcup_{r=1}^{4} \mathscr{D}_r$ (see Fig. 3 and Table 1).
For $D_{i,j} \in \bigcup_{r=5}^{\infty} \mathscr{D}_r$, we have $r \geq 5$, and so we get

$$\frac{4}{\pi r} + \frac{1}{\pi r^2} \leq \frac{4}{5\pi} + \frac{1}{25\pi} = \frac{21}{25\pi} \implies d(r) < \frac{5}{\pi} = d_1.$$

□

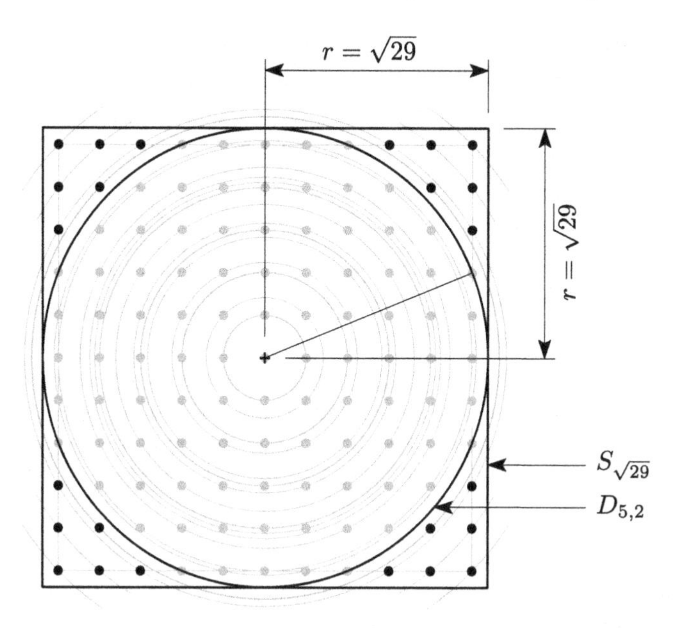

Fig. 4. An example with for which $\mathbb{D}_{i,j} \subsetneq \mathbb{S}_r$. Here $(i,j) = (5,2)$, and so $r = \sqrt{i^2 + j^2} = \sqrt{29}$. The pixels in $\mathbb{D}_{5,2}$ are shown in gray, and those in $\mathbb{S}_{\sqrt{29}} \setminus \mathbb{D}_{5,2}$ in black.

Table 1. Densities in $D_{i,j} \in \bigcup_{r=1}^{4} \mathscr{D}_r$.

(i,j)	r	$\lvert\mathbb{D}_r\rvert$	πd_r	(i,j)	r	$\lvert\mathbb{D}_r\rvert$	πd_r	(i,j)	r	$\lvert\mathbb{D}_r\rvert$	πd_r	(i,j)	r	$\lvert\mathbb{D}_r\rvert$	πd_r
$(1,0)$	1	5	5	$(2,0)$	2	13	$\frac{13}{4}$	$(3,0)$	3	29	$\frac{29}{9}$	$(4,0)$	4	49	$\frac{49}{16}$
$(1,1)$	$\sqrt{2}$	9	$\frac{9}{2}$	$(2,1)$	$\sqrt{5}$	21	$\frac{21}{5}$	$(3,1)$	$\sqrt{10}$	37	$\frac{37}{10}$	$(4,1)$	$\sqrt{17}$	57	$\frac{57}{17}$
				$(2,2)$	$2\sqrt{2}$	25	$\frac{25}{8}$	$(3,2)$	$\sqrt{13}$	45	$\frac{45}{13}$	$(4,2)$	$2\sqrt{5}$	69	$\frac{69}{20}$
								$(3,3)$	$3\sqrt{2}$	61	$\frac{61}{18}$	$(4,3)$	5	81	$\frac{81}{25}$
												$(4,4)$	$4\sqrt{2}$	89	$\frac{89}{32}$

2.3 Unrestricted Center and Radius

We have a few observations when centers and radii are all in the real space. As in the previous sections, we consider discs containing at least two pixels, because one-pixel containment is trivial and degenerates to the limiting case of infinite density. Here, "k-pixel containment" refers to those cases where any disc contains exactly k pixels.

We start with the following observation, which is needed to conceive later observations. Its rationale follows from elementary geometry and, for quick comprehension, is illustrated in Fig. 5.

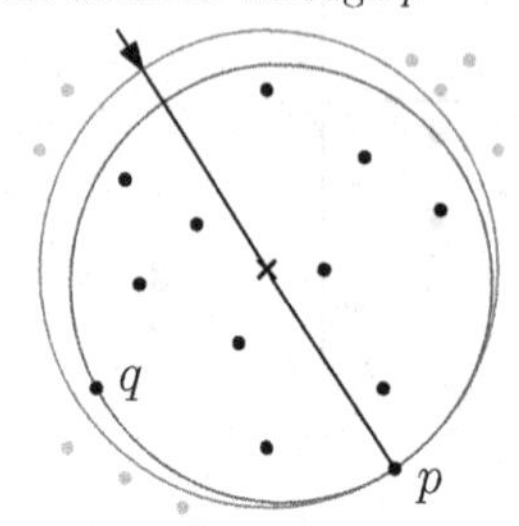

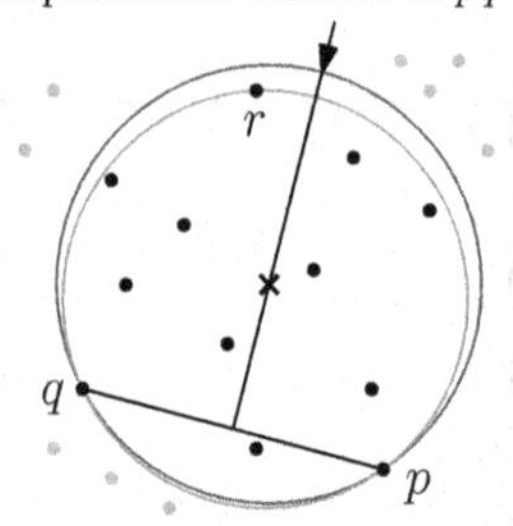

Fig. 5. Left: We can reduce the enclosing circle passing through a single point p by shifting its center along the diameter through p, towards p, until it touches a second point, q. **Right:** When the circle passes through exactly two points, p and q, such that pq is not a diameter, we can reduce it further by shifting its center towards pq until it touches a third point, r.

Observation 4. *Out of all discs containing a given set S of two or more pixels, the smallest disc is unique and contains at least two pixels from S on its boundary. Further, if exactly two pixels lie on the boundary, then they are diametrically opposite.*

An immediate outcome of Observation 4 is that for 2-pixel containment, both the pixels are boundary pixels, and so we have the following observation.

Observation 5. *For 2-pixel containment, the densest disc has unit diameter with density $\frac{8}{\pi}$.*

As shown later, $\frac{8}{\pi}$ is, in fact the maximum density.

Three-pixel containment is not possible, because either two or three pixels will lie on the boundary of the disc, and hence more pixels get covered by the disc (Fig. 6). So we make the following observation.

Observation 6. *The densest disc contains either two or at least four pixels.*

For 4- and 5-pixel containments, simple geometric arguments yield the following.

Observation 7. *The smallest disc containing 4 pixels has radius $\frac{1}{\sqrt{2}}$, and that with 5 pixels has radius 1.*

From the above results, we get the following theorem.

Theorem 3. *Maximum density over all discs containing at least two pixels is $\frac{8}{\pi}$, which occurs for 2-pixel and 4-pixel containments only, corresponding to the radii $\frac{1}{2}$ and $\frac{1}{\sqrt{2}}$ respectively.*

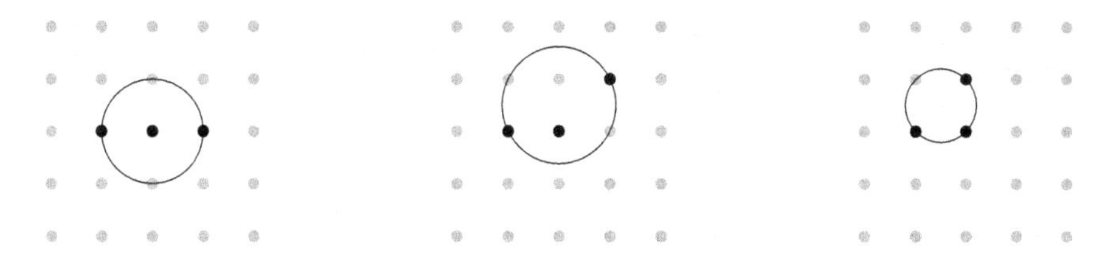

Fig. 6. Three possible (rotationally asymmetric) configurations of three black pixels. In every configuration, the smallest disc containing the black pixels must contain some other (i.e., gray) pixel.

Fig. 7. Densest discs for 6- and 7-pixel containments.

Proof. From Observation 5 and Observation 7, we see that up to $k = 5$, maximum density is $\frac{8}{\pi}$, which occurs for 2-pixel and 4-pixel containments only. For 6-pixel containment, it can be verified that the smallest disc has radius $\frac{\sqrt{5}}{2}$ (Fig. 7), and so its density is $\frac{24}{5\pi}$. We also verify that for $k = 7$, the smallest disc has radius $\frac{5}{4}$ (Fig. 7) and density less than $\frac{8}{\pi}$.

For $k \geq 8$, radius r of the smallest disc D_r will be no less than that corresponding to $k = 7$, i.e., $r \geq \frac{5}{4}$. As D_r contains no more than $(2r+1)^2$ pixels, we have

$$d_r \leq \frac{(2r+1)^2}{\pi r^2} \leq \frac{4}{\pi} + \frac{16}{5\pi} + \frac{16}{25\pi} = \frac{196}{25\pi} < \frac{8}{\pi}.$$

Now, using the maximum density values for other types of discs (Sect. 2.1 and Sect. 2.2), the proof concludes. □

3 Minimum Density

Finding the minimum density seems to be more difficult. We present here some results for discs with integer center—first for integer radius and then for real radius. For maximum density over discs having real centers, we have shown results in Sect. 2.3. However, similar results for minimum density is not yet known to us.

Table 2. Densities in D_r for $r \leq 45$.

r	$\lvert\mathbb{D}_r\rvert$	πd_r	r	$\lvert\mathbb{D}_r\rvert$	πd_r	r	$\lvert\mathbb{D}_r\rvert$	πd_r	r	$\lvert\mathbb{D}_r\rvert$	πd_r	r	$\lvert\mathbb{D}_r\rvert$	πd_r
1	5	5	10	317	$\frac{317}{100}$	19	1129	$\frac{1129}{361}$	28	2453	$\frac{2453}{784}$	37	4293	$\frac{4293}{1369}$
2	13	$\frac{13}{4}$	11	377	$\frac{377}{121}$	20	1257	$\frac{1257}{400}$	29	2629	$\frac{2629}{841}$	38	4513	$\frac{4513}{1444}$
3	29	$\frac{29}{9}$	12	441	$\frac{441}{144}$	21	1373	$\frac{1373}{441}$	30	2821	$\frac{2821}{900}$	39	4777	$\frac{4777}{1521}$
4	49	$\frac{49}{16}$	13	529	$\frac{529}{169}$	22	1517	$\frac{1517}{484}$	31	3001	$\frac{3001}{961}$	40	5025	$\frac{5025}{1600}$
5	81	$\frac{81}{25}$	14	613	$\frac{613}{196}$	23	1653	$\frac{1653}{529}$	32	3209	$\frac{3209}{1024}$	41	5261	$\frac{5261}{1681}$
6	113	$\frac{113}{36}$	15	709	$\frac{709}{225}$	24	1793	$\frac{1793}{576}$	33	3409	$\frac{3409}{1089}$	42	5525	$\frac{5525}{1764}$
7	149	$\frac{149}{49}$	16	797	$\frac{797}{256}$	25	1961	$\frac{1961}{625}$	34	3625	$\frac{3625}{1156}$	43	5789	$\frac{5789}{1849}$
8	197	$\frac{197}{64}$	17	901	$\frac{901}{289}$	26	2121	$\frac{2121}{676}$	35	3853	$\frac{3853}{1225}$	44	6077	$\frac{6077}{1936}$
9	253	$\frac{253}{81}$	18	1009	$\frac{1009}{324}$	27	2289	$\frac{2289}{729}$	36	4053	$\frac{4053}{1296}$	45	6361	$\frac{6361}{2025}$

3.1 Integer Center and Integer Radius

Let the radius r be a positive integer. As before, w.l.o.g., we consider the center at $(0, 0)$. We denote by D_r the real disc centered at $(0, 0)$ and having radius r.

Let p be any pixel, and let u_p denote the cell (i.e., the unit-length axis-parallel square) centered at p. Clearly, the (Euclidean) distance of p from any point in u_p is at most $\frac{1}{\sqrt{2}}$. So if $p \in \mathbb{Z}^2 \setminus \mathbb{D}_r$, then the distance of p from $D_{r-\frac{1}{\sqrt{2}}}$ is greater than $\frac{1}{\sqrt{2}}$, which implies $u_p \cap D_{r-\frac{1}{\sqrt{2}}} = \varnothing$. An illustration is given in Fig. 8. This results to the following observation.

Observation 8. *$D_{r-\frac{1}{\sqrt{2}}}$ lies in the interior of $\bigcup_{p \in \mathbb{D}_r} u_p$.*

The above observation is needed to prove the following theorem.

Theorem 4. *In the collection of all discs with integer centers and integer radii, $d_7 < d_r \; \forall r \geq 1$.*

Proof. By brute-force counting (Table 2), we get $d_7 = \frac{149}{49\pi}$, which, in fact, is the unique smallest in the set $\{d_r : 1 \leq r \leq 45\}$. So, what remains to show now is $d_7 < d_r \; \forall r \geq 46$, and its proof goes as follows.

The area of the (closed) region $\bigcup_{p \in \mathbb{D}_r} u_p$ is numerically equal to $|\mathbb{D}_r|$, and so by Observation 8 we have

$$|\mathbb{D}_r| > \pi \left(r - \frac{1}{\sqrt{2}} \right)^2 \implies d_r > \left(1 - \frac{1}{\sqrt{2}r} \right)^2 \implies d_r > \left(\frac{63}{64} \right)^2 \; \forall r \geq 46.$$

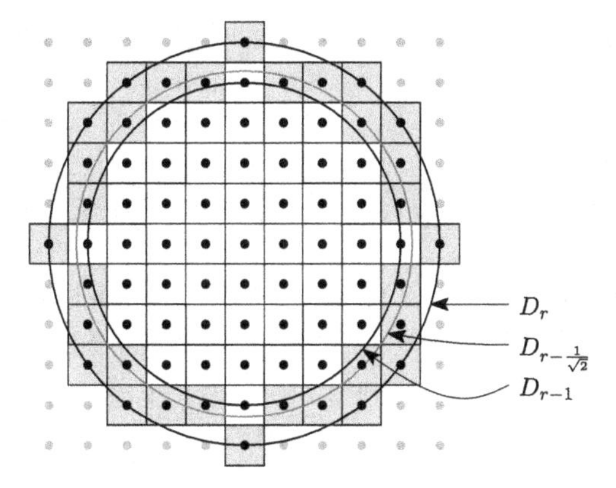

Fig. 8. Disc $D_{r-\frac{1}{\sqrt{2}}}$ lies in the interior of $\bigcup_{p\in\mathbb{D}_r} u_p$. Black pixels comprise $\mathbb{D}_r$, gray pixels comprise $\mathbb{Z}^2 \setminus \mathbb{D}_r$, white cells comprise $\{u_p : p \in \mathbb{D}_{r-1}\}$, and gray cells comprise $\{u_p : p \in \mathbb{D}_r \setminus \mathbb{D}_{r-1}\}$.

From the continued fraction representation of π [5], we have

$$\pi = \cfrac{4}{1+\cfrac{1^2}{3+\cfrac{2^2}{5+\cfrac{3^2}{7+\cfrac{5^2}{\ddots}}}}} = (4-1)+\left(\frac{1}{6}-\frac{1}{34}\right)+\left(\frac{16}{3145}-\frac{4}{4551}\right)+\left(\frac{1}{6601}-\frac{1}{38341}\right)+\cdots,$$

which implies

$$\pi > 3+\left(\frac{1}{6}-\frac{1}{34}\right)+\left(\frac{16}{3145}-\frac{4}{4551}\right)=\frac{644}{205}.$$

So, we get

$$d_7 < \frac{149}{49\times\frac{644}{205}} = \frac{30545}{31556} < \frac{63^2}{64^2} < d_r \ \forall r \geq 46,$$

and hence the proof. □

3.2 Integer Center and Real Radius

As before, w.l.o.g., we consider the center at $(0, 0)$. Let ε be a real number. We have the following theorem.

Theorem 5. *In the collection of all discs with integer centers and real radii,* $\lim\limits_{\varepsilon\to 0} d_{1-|\varepsilon|} < d_r \ \forall\, r > 0$.

Proof. Note that the minimum distance between two points, one on the boundary of D_r and the other on the boundary of $D_{r-\frac{1}{\sqrt{2}}}$ is $\frac{1}{\sqrt{2}}$, and so Observation 8 holds just as well if r is real. That means, here, in particular, for any real $r > \frac{1}{\sqrt{2}}$, $D_{r-\frac{1}{\sqrt{2}}}$ lies in the interior of $\bigcup_{p\in\mathbb{D}_r} u_p$. So, we have

$$|\mathbb{D}_r| > \pi\left(r - \frac{1}{\sqrt{2}}\right)^2 \implies d_r > \left(1 - \frac{\frac{1}{\sqrt{2}}}{r}\right)^2 > \frac{1}{\pi} \ \forall r \geq 2.$$

It is easy to see that for $r < 2$ there exists only the following maximal discs with no points on their boundary, and their densities are as follows.

$$\begin{aligned}
\lim_{\varepsilon\to 0} d_{1-|\varepsilon|} &= \frac{1}{\pi\cdot 1^2} &&= \frac{1}{\pi},\\
\lim_{\varepsilon\to 0} d_{\sqrt{2}-|\varepsilon|} &= \frac{5}{\pi\cdot(\sqrt{2})^2} &&= \frac{5}{2\pi},\\
\lim_{\varepsilon\to 0} d_{2-|\varepsilon|} &= \frac{9}{\pi\cdot 2^2} &&= \frac{9}{4\pi}.
\end{aligned}$$

Note that the least among them is $\frac{1}{\pi}$ and so for all real r, $d_r \geq \frac{1}{\pi}$. Hence, the minimum-density disc with integer center and real radius is the one that contains only one pixel and has a radius just less than unity. □

4 Conclusion and Future Work

As per our investigations so far, finding minimum-density discs for real-valued specification of center and radius remains an open problem. In addition, and more importantly, the following questions are quite natural as a followup to what is presented in this paper.

1. Given a range of radius, what would be a/the radius for which density is maximum (or for minimum) in that range? Further, how does it differ between real and integer specification, and how with the position of the center?
2. How does the result presented in this paper come around when the distance metric is not Euclidean but some other norm (e.g., l_∞ or l_1)? This might play a crucial role to generalize the result of density extrema from 'discs' to 'balls'.
3. How would the techniques used to derive the density extrema for discs in 2D be commensurate with mathematical deductions in higher dimensions?

References

1. Alexander, J.R., Beck, J., Chen, W.W.L.: Handbook of Discrete and Computational Geom, 2nd edn. CRC Press, Boca Raton (1997)
2. Andres, E., Roussillon, T.: Analytical description of digital circles. In: Debled-Rennesson, I., Domenjoud, E., Kerautret, B., Even, P. (eds.) DGCI 2011. LNCS, vol. 6607, pp. 235–246. Springer, Heidelberg (2011). https://doi.org/10.1007/978-3-642-19867-0_20

3. Barrera, T., Hast, A., Bengtsson, E.: A chronological and mathematical overview of digital circle generation algorithms - introducing efficient 4- and 8-connected circles. Int. J. Comput. Math. **93**(8), 1241–1253 (2016). https://doi.org/10.1080/00207160.2015.1056170
4. Basu, N.G., Majumder, S., Hon, W.K.: On finding the maximum and minimum density axis-parallel regions in $\mathbb{R}^d$. Fundam. Informaticae **152**(1), 1–12 (2017)
5. Beckmann, P.: A History of Pi. St. Martin's Press Inc., Manhattan (1971)
6. Berg, M.D., Kreveld, M.V., Overmars, M., Schwarzkopf, O.: Computational Geometry: Algorithms and Applications. Springer, Heidelberg (1997)
7. Bhowmick, P., Bhattacharya, B.B.: Number-theoretic interpretation and construction of a digital circle. Discret. Appl. Math. **156**(12), 2381–2399 (2008). https://doi.org/10.1016/j.dam.2007.10.022
8. Bhowmick, P., Bhattacharya, B.B.: Real polygonal covers of digital discs - some theories and experiments. Fundam. Informaticae **91**(3–4), 487–505 (2009). https://doi.org/10.3233/FI-2009-0053
9. Doerr, C., Gnewuch, M., Wahlström, M.: Calculation of discrepancy measures and applications. In: Chen, W., Srivastav, A., Travaglini, G. (eds.) A Panorama of Discrepancy Theory. LNM, vol. 2107, pp. 621–678. Springer, Cham (2014). https://doi.org/10.1007/978-3-319-04696-9_10
10. Dokmanic, I., Parhizkar, R., Ranieri, J., Vetterli, M.: Euclidean distance matrices: essential theory, algorithms, and applications. IEEE Signal Process. Mag. **32**(6), 12–30 (2015). https://doi.org/10.1109/MSP.2015.2398954
11. Fabbri, R., Costa, L., Torelli, J., Bruno, O.: 2D euclidean distance transform algorithms: a comparative survey. ACM Comput. Surv. **40**, 1–44 (2008). https://doi.org/10.1145/1322432.1322434
12. Fisk, S.: Separating point sets by circles, and the recognition of digital disks. IEEE Trans. Pattern Anal. Mach. Intell. **8**(4), 554–556 (1986). https://doi.org/10.1109/TPAMI.1986.4767821
13. Huxley, M.N., Zunic, J.D.: The number of n-point digital discs. IEEE Trans. Pattern Anal. Mach. Intell. **29**(1), 159–161 (2007). https://doi.org/10.1109/TPAMI.2007.250606
14. Kim, C.E., Anderson, T.A.: Digital disks and a digital compactness measure. In: DeMillo, R.A. (ed.) Proceedings of the 16th Annual ACM Symposium on Theory of Computing, 30 April - 2 May 1984, Washington, DC, USA, pp. 117–124. ACM (1984). https://doi.org/10.1145/800057.808673
15. Kovács, G., Nagy, B., Vizvári, B.: Weighted distances and digital disks on the khalimsky grid. J. Math. Imaging Vision **59**(1), 2–22 (2017). https://doi.org/10.1007/s10851-016-0701-5
16. Liberti, L., Lavor, C., Maculan, N., Mucherino, A.: Euclidean distance geometry and applications. SIAM Rev. **56**(1), 3–69 (2014). https://doi.org/10.1137/120875909
17. Majumder, S., Bhattacharya, B.B.: Density or discrepancy: a VLSI designer's dilemma in hot spot analysis. In: Proceedings of the 17th Canadian Conference on Computational Geometry, CCCG'05, University of Windsor, Ontario, Canada, 10–12 August 2005, pp. 167–170 (2005). http://www.cccg.ca/proceedings/2005/22.pdf
18. Majumder, S., Bhattacharya, B.B.: On the density and discrepancy of a 2D point set with applications to thermal analysis of VLSI chips. Inf. Process. Lett. **107**(5), 177–182 (2008)

19. Matic-Kekic, S., Acketa, D.M., Zunic, J.D.: An exact construction of digital convex polygons with minimal diameter. Discret. Math. **150**(1–3), 303–313 (1996). https://doi.org/10.1016/0012-365X(95)00195-3
20. Nagy, B.: An algorithm to find the number of the digitizations of discs with a fixed radius. Electron. Notes Discret. Math. **20**, 607–622 (2005). https://doi.org/10.1016/j.endm.2005.04.006
21. Nagy, B.: Number of words characterizing digital balls on the triangular tiling. In: Normand, N., Guédon, J., Autrusseau, F. (eds.) DGCI 2016. LNCS, vol. 9647, pp. 31–44. Springer, Cham (2016). https://doi.org/10.1007/978-3-319-32360-2_3
22. Nakamura, A., Aizawa, K.: Digital circles. Comput. Vis. Graph. Image Process. **26**(2), 242–255 (1984). https://doi.org/10.1016/0734-189X(84)90187-7
23. Pham, S.: Digital circles with non-lattice point centers. Vis. Comput. **9**(1), 1–24 (1992). https://doi.org/10.1007/BF01901025
24. Tasissa, A., Lai, R.: Exact reconstruction of euclidean distance geometry problem using low-rank matrix completion. IEEE Trans. Inf. Theory **65**(5), 3124–3144 (2019). https://doi.org/10.1109/TIT.2018.2881749
25. Wang, J., Tan, Y.: Efficient euclidean distance transform using perpendicular bisector segmentation. In: CVPR 2011, pp. 1625–1632 (2011). https://doi.org/10.1109/CVPR.2011.5995644
26. Zunic, J.D.: On the number of digital discs. J. Math. Imaging Vis. **21**(3), 199–204 (2004). https://doi.org/10.1023/B:JMIV.0000043736.15525.ed
27. Žunić, J., Sladoje, N.: A characterization of digital disks by discrete moments. In: Sommer, G., Daniilidis, K., Pauli, J. (eds.) CAIP 1997. LNCS, vol. 1296, pp. 582–589. Springer, Heidelberg (1997). https://doi.org/10.1007/3-540-63460-6_166

Sufficient Conditions for Topology-Preserving Parallel Reductions on the BCC Grid

Kálmán Palágyi(✉), Gábor Karai, and Péter Kardos

Department of Image Processing and Computer Graphics, University of Szeged, Szeged, Hungary
{palagyi,karai,pkardos}@inf.u-szeged.hu

Abstract. Parallel reductions transform binary pictures only by changing a set of black points to white ones simultaneously. Topology preservation is a major concern of some topological algorithms composed of parallel reductions. For 3D binary pictures sampled on the body-centered cubic (BCC) grid, we propose a new sufficient condition for topology-preserving parallel reductions. This condition takes some configurations of deleted points into consideration, and it provides a method of verifying that formerly constructed parallel reductions preserve the topology. We present two further sufficient conditions that investigate individual points, directly provide deletion rules of topology-preserving parallel reductions, and allow us to construct parallel thinning algorithms.

Keywords: Digital topology · Topology preservation · BCC grid · Thinning

1 Introduction

A binary digital *picture* is composed of *black* and *white* points that form black and white *components* [8]. A *parallel reduction* is an operation that transforms a picture only by changing a set of black pixels to white ones at a time, which is referred to as deletion [4].

Topology preservation is a major concern of thinning algorithms composed of parallel reductions. A parallel reduction in a 2D picture does *not* preserve topology if any black component in the input picture is split or is completely deleted, any white component in the input picture is merged with another white component, or a white component is created where there was none in the input picture [8]. There is an additional concept called *hole* in 3D pictures. A hole (which donuts have) is formed of white points, but it is not a white component. Topology preservation in 3D implies that eliminating or creating any hole is not allowed [8].

Methods of verifying that a given topological algorithm is topology-preserving (i.e., it performs topology-preserving reductions for all possible pictures) are well-established for 2D pictures on the three possible regular grids [6,7,9,12,14] and the 3D pictures sampled on the conventional cubic grid

R. P. Barneva et al. (Eds.): IWCIA 2022, LNCS 13348, pp. 71–83, 2023.
https://doi.org/10.1007/978-3-031-23612-9_5

[1,9,10,13]. It cannot be stated for 3D pictures on the *body-centered cubic (BCC) grid* which tessallates the space into truncated octahedra [8], however, the importance of the BCC grid shows an upward tendency due to its advantages of geometrical and topological properties [2,3,11,15,17,18].

In this work, we propose a new sufficient condition for topology-preserving parallel reductions on the BCC grid. Our condition takes some configurations of deleted points into consideration, and it provides a method of verifying that formerly constructed parallel reductions preserve the topology. Two further sufficient conditions are stated that investigate individual points, directly provide deletion rules of topology-preserving parallel reductions, and allow us to construct parallel thinning algorithms.

The rest of this paper is organized as follows: Sect. 2 reviews the basic notions and results. Then, in Sect. 3 we recall the only existing configuration-based sufficient condition for topology-preserving parallel reductions on the BCC grid, and its simplified version is proposed. Section 4 presents our symmetric and asymmetric point-based sufficient conditions, and we generate directly two topology-preserving parallel reductions in Sect. 5. Finally, we round off this work with some concluding remarks.

2 Basic Notions and Results

Next, we recall the basic notions and results concerning the BCC grid [8].

The BCC grid is the following subset of $\mathbb{Z}^3$:

$$\mathbb{B} = \{(x, y, z) \in \mathbb{Z}^3 \mid x \equiv y \equiv z \pmod 2\}.$$

Let $B \subset \mathbb{B}$ be the the set of *black points* in a binary digital picture sampled on the BCC grid, and each point in $\mathbb{B} \setminus B$ is said to be a *white point* in this picture. For practical purposes we assume that B contains finitely many points.

The same adjacency relation called 14-neighborhood is assigned to the sets of black and white points, and let $N_{14}(p)$ denote the set of points that are 14-adjacent to p, see Fig. 1.

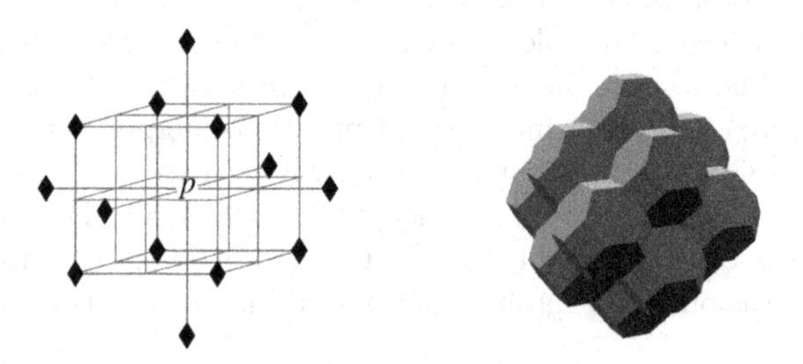

Fig. 1. The studied adjacency relation on $\mathbb{B}$ (left). The 14 points marked '♦' form the set $N_{14}(p)$. (Note that unmarked elements in $\mathbb{Z}^3$ are not points in $\mathbb{B}$.) The voxel-representation of $N_{14}(p)$, where each voxel is a truncated octahedron (right).

Since the considered adjacency relation is symmetric, its reflexive-transitive closure generates an equivalence relation, and its equivalence classes are called *components*. A *black component* or an *object* is a component of B, while a *white component* is a component of $\mathbb{B} \setminus B$.

A point $p \in B$ is an *interior point* for B if $N_{14}(p) \subset B$, p is called a *border point* if it is not an interior point, and p is said to be an *isolated point* if it forms a singleton object (i.e., $N_{14}(p) \cap B = \emptyset$).

A single black point is said to be *simple* if its deletion is a topology-preserving reduction. Now we will make use of the following characterization of simple points:

Theorem 1. [16] *A point $p \in \mathbb{B}$ is* simple *for the set of black points B if and only if the following conditions hold:*

1. *Set of black points $N_{14}(p) \cap B$ contains exactly one component.*
2. *Set of white points $N_{14}(p) \setminus B$ contains exactly one component.*

It is an easy consequence of Theorem 1 that only non-isolated border points may be simple, and the simpleness of a point p can be decided by examining $N_{14}(p)$ (i.e., its small local neighborhood). Figure 2 gives some illustrative examples of simple and non-simple points.

Deleting a single black point p preserves the topology if and only if p is simple. Since parallel reductions can delete a set of points simultaneously, we need a precise definition of what is meant by topology preservation when a number of points are deleted at a time.

Here, we need to define the concepts of a simple set and a simple sequence.

Definition 1. [9,10] *Let* **P** *be an arbitrary picture. A set of n black points Q is a* simple set *in* **P** *if it is possible to arrange the elements of Q in a sequence $\langle q_1, \ldots, q_n \rangle$ such that q_1 is simple in* **P** *and each q_i is simple after the set of points $\{q_1, \ldots, q_{i-1}\}$ is deleted ($i = 2, \ldots, n$). Such a sequence is called a* simple sequence. *(And let the empty set be simple.)*

Figure 3 gives examples of simple and non-simple sets in a picture on the BCC grid.

3 Configuration-Based Conditions

One of the authors established the very first sufficient condition for topology-preserving parallel reductions on the BCC grid:

Theorem 2. [5] *A parallel reduction $\mathcal{R}$ is topology-preserving if it fulfills the following conditions:*

1. *Any set of at most three mutually* 14*-adjacent black points deleted by $\mathcal{R}$ is simple.*

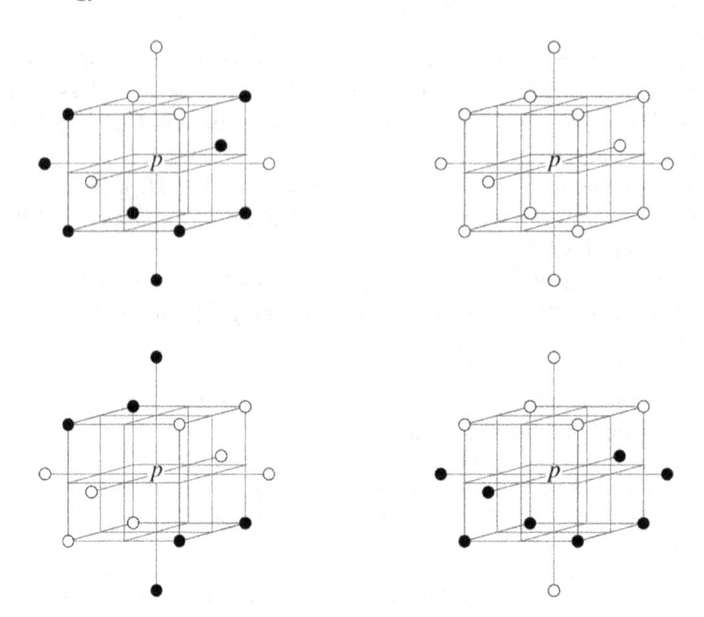

Fig. 2. Examples for simple and non-simple points. The positions marked '•' and '∘' represent black and white points, respectively. Black point p is simple only in the top left configuration. In the top right example, p is an isolated black point, while in the bottom left figure we can find two 14-components in $N_{14}(p) \cap B$, hence in both cases Condition 1 of Theorem 1 is violated. The bottom right configuration depicts a case where there exist two 14-components in $N_{14}(p) \setminus B$, thus Condition 2 of Theorem 1 does not hold.

2. $\mathcal{R}$ does not delete completely any object composed of at most four mutually 14-adjacent points.

We can state that Theorem 2 takes configurations of at most four mutually 14-adjacent points into consideration. Thus that theorem states a *configuration-based* sufficient condition for topology-preserving parallel reductions.

Notice that, by Theorem 2, it is difficult to verify that a previously designed parallel reduction preserves the topology. That is why we simplify this very first condition.

Firstly we need to recall an absolutely general lemma stated by Kardos and Palágyi:

Lemma 1. [7] *Let p and q be two black simple points in an arbitrary picture. If p remains simple after the deletion of q, q remains simple after the deletion of p.*

Lemma 1 can be rephrased as follows:

– If a simple set is formed by two simple points, then both possible sequences of its elements are simple sequences.

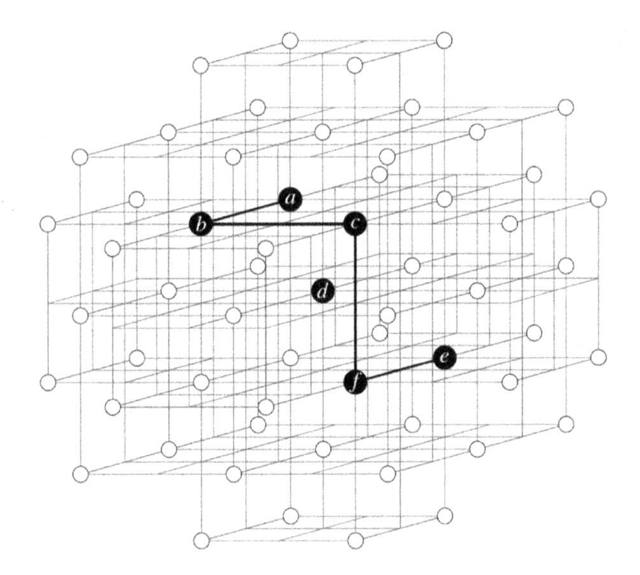

Fig. 3. Examples of simple and non-simple sets. The set of four black points $\{a, b, c, d\}$ is simple since the 10 sequences (of the possible 24 ones) $\langle a, b, c, d\rangle$, $\langle a, b, d, c\rangle$, $\langle a, c, b, d\rangle$, $\langle a, d, b, c\rangle$, $\langle b, a, c, d\rangle$, $\langle b, a, d, c\rangle$, $\langle b, c, a, d\rangle$, $\langle c, a, b, d\rangle$, $\langle c, b, a, d\rangle$, $\langle d, a, b, c\rangle$ are all simple. The set of black points $\{c, d\}$ is non-simple, since both sequences $\langle c, d\rangle$ and $\langle d, c\rangle$ are non-simple. Note that $\{a, b, c, e, f\} \subset N_{14}(d)$ and thick lines connect 14-adjacent points in $\{a, b, c, e, f\}$.

- If two black points p and q are simple in a picture, then the following two statements are equivalent:
  - p is simple after the deletion of q (i.e., $\langle q, p\rangle$ is a simple sequence).
  - q is simple after the deletion of p (i.e., $\langle p, q\rangle$ is a simple sequence).

In other words, the simpleness of a set of two simple points can be decided by testing just one sequence of its elements.

The following proposition is a straightforward consequence of Definition 1:

Proposition 1. *Let $\langle q_1, \ldots, q_n\rangle$ be a simple sequence of a set of $n \geq 0$ black points Q. If a black point $p \notin Q$ is simple after the deletion of Q, the set of $n+1$ points $\{q_1, \ldots, q_n, p\}$ is simple.*

We will also make use of the following proposition:

Proposition 2. *If a set of black points Q forms an object, Q is not a simple set.*

Proof. Let us investigate an arbitrary sequence $\langle q_1, \ldots, q_{n-1}, q_n\rangle$ of the elements of the set of n points Q $(n \geq 1)$. It is obvious that q_n is an isolated point after the deletion of the set $\{q_1, \ldots, q_{n-1}\}$. Since only non-isolated border points may be simple, $\langle q_1, \ldots, q_{n-1}, q_n\rangle$ is not a simple sequence. Thus Q is not a simple set. □

With the help of Lemma 1, Proposition 1, and Proposition 2, Theorem 2 can be rephrased as follows:

Theorem 3. *A parallel reduction is topology-preserving for* $B \subset \mathbb{B}$ *if the following conditions hold:*

1. *Only simple points for* B *are deleted.*
2. *If two* 14*-adjacent points* p *and* q *are deleted,* p *is simple for* $B \backslash \{q\}$.
3. *If three mutually* 14*-adjacent points* p, q, *and* r *are deleted,* p *is simple for* $B \backslash \{q, r\}$, *or* q *is simple for* $B \backslash \{p, r\}$, *or* r *is simple for* $B \backslash \{p, q\}$.
4. *No object consisting of exactly four mutually* 14*-adjacent points is deleted completely.*

Proof. Let us suppose that a parallel reduction satisfies all conditions of this theorem, and it deletes the set of points $D \subset B$. To prove this theorem, we must show that both conditions of Theorem 2 hold.

- Let $Q \subseteq D$ be a set of at most three mutually 14-adjacent black points. Then the following three points are to be investigated:
  - $Q = \{p\}$:
    By Condition 1 of this theorem, point p is simple for B. Thus the singleton set Q is a simple set.
  - $Q = \{p, q\}$:
    By Condition 1 of this theorem, both points p and q are simple for B.
    By Condition 2 of this theorem, p is simple for $B \backslash \{q\}$.
    Consequently the set of two points Q is a simple set. (Note that, by Lemma 1, q is also simple after the deletion of p. That is why we do not need to distinguish p and q.)
  - $Q = \{p, q, r\}$:
    By Condition 1 of this theorem, all the three points p, q, and r are simple for B.
    By Condition 2 of this theorem and Lemma 1, all the three sets of points $\{p, q\}$, $\{p, q\}$, and $\{p, q\}$ are simple sets, and all the six sequences $\langle p, q \rangle$, $\langle q, p \rangle$, $\langle p, r \rangle$, $\langle r, p \rangle$, $\langle q, r \rangle$, and $\langle r, q \rangle$ are simple sequences. By Condition 3 of this theorem and Proposition 1, at least two of the six sequences $\langle p, q, r \rangle$, $\langle q, p, r \rangle$, $\langle p, r, q \rangle$, $\langle r, p, q \rangle$, $\langle q, r, p \rangle$, and $\langle r, q, p \rangle$ are simple. Thus the set of three points Q is a simple set.

  Since any set of at most three mutually 14-adjacent deleted points is a simple set, Condition 1 of Theorem 2 holds.
- Let us assume that the set of points $Q \subseteq D$ forms an object of at most four mutually 14-adjacent points. Then the following two cases are distinguished:

- If Q contains at most three elements, by Condition 1 of Theorem 2, Q is a simple set. By Proposition 2, Q cannot form an object. Hence we arrive at a contradiction.
- If Q contains exactly four points, by Condition 4 of this theorem, Q cannot be deleted completely.

Thus Condition 2 of Theorem 2 holds.

Since both conditions of Theorem 2 are satisfied, the proof is completed. □

We can state that (similarly to Theorem 2) Theorem 3 provides a *configuration-based* sufficient condition for topology-preserving parallel reductions on the BCC grid.

4 Point-Based Conditions

Theorems 2 and 3 (i.e., the two configuration-based sufficient conditions for topology-preserving parallel reductions) just provide methods of verifying that a previously designed parallel reduction preserves the topology, rather than a methodology, for constructing topology-preserving parallel reductions. That is why we propose *point-based* sufficient conditions that directly provide deletion rules of topology-preserving parallel reductions, and allow us to generate various topology-preserving parallel thinning algorithms.

The following theorem (i.e., the very first point-based condition on the BCC grid) states the deletability of individual points:

Theorem 4. *A parallel reduction is topology-preserving for $B \subset \mathbb{B}$ if each point p deleted by this reduction satisfies the following conditions:*

1. *Point p is simple for B.*
2. *For any point $q \in N_{14}(p) \cap B$ that is simple for B, point p is simple for $B \setminus \{q\}$.*
3. *For any two points $q \in N_{14}(p) \cap B$ and $r \in N_{14}(p) \cap N_{14}(q) \cap B$ that are simple for B, and q is simple for $B \setminus \{r\}$, p is simple for $B \setminus \{q, r\}$.*
4. *Point p is not an element of an object consisting of four mutually 14-adjacent points.*

Proof. Let us suppose that a parallel reduction satisfies all conditions of this theorem, it deletes the set of points $D \subset B$, and a black point p is in D. To prove this theorem, we must show that all conditions of Theorem 3 hold.

- By Condition 1 of this theorem, point p is simple for B. Thus Condition 1 of Theorem 3 holds.
- By Condition 2 of this theorem, for any $q \in B$, the set of two mutually 14-adjacent points $\{p, q\}$ is simple for B. It also holds if $q \in D$. Consequently Condition 2 of Theorem 3 is satisfied.
- By Condition 3 of this theorem, $\langle r, q, p \rangle$ is a simple sequence for any two points q and r. Thus the set of three mutually 14-adjacent points $\{r, q, p\}$ is simple if $q \in D$ and $r \in D$. Therefore Condition 3 of Theorem 3 holds.

– By Condition 4 of this theorem, none of the elements of an object consisting of four mutually 14-adjacent points may be deleted. Since such an object is not deleted completely, Condition 4 of Theorem 3 is satisfied.

Since all conditions of Theorem 3 hold, this theorem is true. □

Conditions of Theorem 4 may be viewed as *symmetric* since elements in the examined sets of at most four mutually 14-adjacent points are not distinguished.

Let us focus on the addressing scheme shown in Fig. 4, which maps every point in $\mathbb{B}$ to a triplet of integer coordinates. The *lexicographical order* relation '$\prec$' between two distinct points $p = (p_x, p_y, p_z)$ and $q = (q_x, q_y, q_z)$ is defined as follows:

$$p \prec q \quad \Leftrightarrow \quad (p_z < q_z) \vee (p_z = q_z \wedge p_y < q_y) \vee (p_z = q_z \wedge p_y = q_y \wedge p_x < q_x).$$

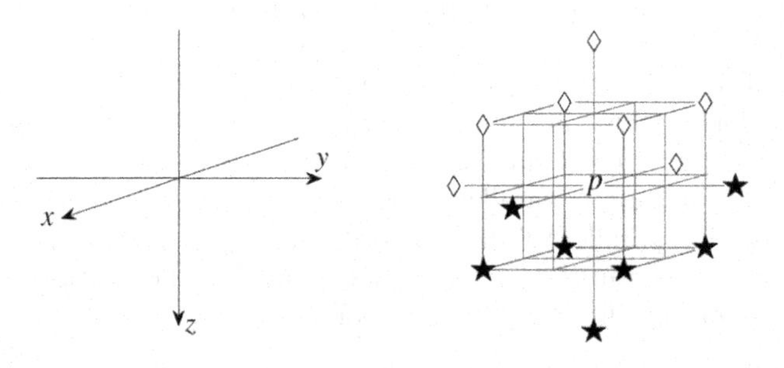

Fig. 4. The considered coordinate system (left) and the proposed ordering scheme for the BCC grid (right). The elements of the set of seven points $\{ q \mid q \in N_{14}(p) \text{ and } p \prec q \}$ are marked '★', and the remaining seven points in $\{ r \mid r \in N_{14}(p) \text{ and } r \prec p \}$ are marked '◊'.

Let $Q \subset \mathbb{B}$ be a finite set of points. Point $p \in Q$ is the *smallest element* of Q if for any $q \in Q \setminus \{p\}$, $p \prec q$.

With the help of the proposed ordering, we state the following *asymmetric point-based condition* for topology-preserving parallel reductions:

Theorem 5. *A parallel reduction is topology-preserving for $B \subset \mathbb{B}$ if each point p deleted by that reduction satisfies the following conditions:*

1. *Point p is simple for B.*
2. *For any point $q \in N_{14}(p) \cap B$ that is simple for B, point p is simple for $B \setminus \{q\}$, or $q \prec p$.*
3. *For any two points $q \in N_{14}(p) \cap B$ and $r \in N_{14}(p) \cap N_{14}(q) \cap B$ that are simple for B, and q is simple for $B \setminus \{r\}$, point p is simple for $B \setminus \{q, r\}$, or p is not the smallest element of set $\{p, q, r\}$.*
4. *Point p is not the smallest element of an object formed by four mutually 14-adjacent points.*

Proof. Let us suppose that a parallel reduction satisfies all conditions of this theorem, and it deletes the set of points $D \subset B$. To prove this theorem, we must show that all the four conditions of Theorem 3 hold.

- Let $Q \subseteq D$ be a set of at most three mutually 14-adjacent black points. Then the following three points are to be investigated:
  - $Q = \{p\}$:
    By Condition 1 of this theorem, point p is simple for B. Thus Condition 1 of Theorem 3 holds.
  - $Q = \{p, q\}$:
    By Condition 1 of this theorem, both points p and q are simple for B. By Condition 2 of this theorem, p is simple for $B \backslash \{q\}$, or $q \prec p$.
    * If p is simple for $B \backslash \{q\}$, Condition 2 of Theorem 3 is satisfied.
    * If p is not simple for $B \backslash \{q\}$ and $q \prec p$, by Lemma 1, q is not simple for $B \backslash \{p\}$. Since $q \in Q \subseteq D$, Condition 2 of this theorem is violated, and we arrive at a contradiction.
  - $Q = \{p, q, r\}$:
    Let us assume that p is the smallest element of set $\{p, q, r\}$. Since $p \in Q \subseteq D$, by Condition 3 of this theorem, p is simple for $B \backslash \{q, r\}$. Thus Condition 3 of Theorem 3 holds.
- By Condition 4 of this theorem, the smallest element of an object formed by four mutually 14-adjacent points cannot be deleted. Thus that object cannot be deleted completely, and Condition 4 of Theorem 3 is satisfied.

Since all the four conditions of Theorem 3 hold, the proof is completed. □

5 Generating Topology-Preserving Parallel Reductions

In this section we show that the point-based sufficient conditions (see Theorems 4 and 5) allow us to generate directly topology-preserving parallel reductions.

Definition 2. *A black point is deleted by the parallel reduction* $\mathcal{R}_{\mathrm{symm}}$ *if it satisfies all conditions of Theorem 4.*

Definition 3. *A black point is deleted by the parallel reduction* $\mathcal{R}_{\mathrm{asymm}}$ *if it satisfies all conditions of Theorem 5.*

The *support* of a parallel reduction is the minimal set of points whose values determine whether a point is deleted [4]. The support of $\mathcal{R}_{\mathrm{symm}}$ contains 64 points (see Fig. 5), and the count of points is 47 in the support of $\mathcal{R}_{\mathrm{asymm}}$ (see Fig. 6). Note that the symmetric reduction has a symmetric support, and the support of the asymmetric reduction is asymmetric.

By Theorems 4 and 5, it is obvious that both derived parallel reductions $\mathcal{R}_{\mathrm{symm}}$ and $\mathcal{R}_{\mathrm{asymm}}$ are topology-preserving.

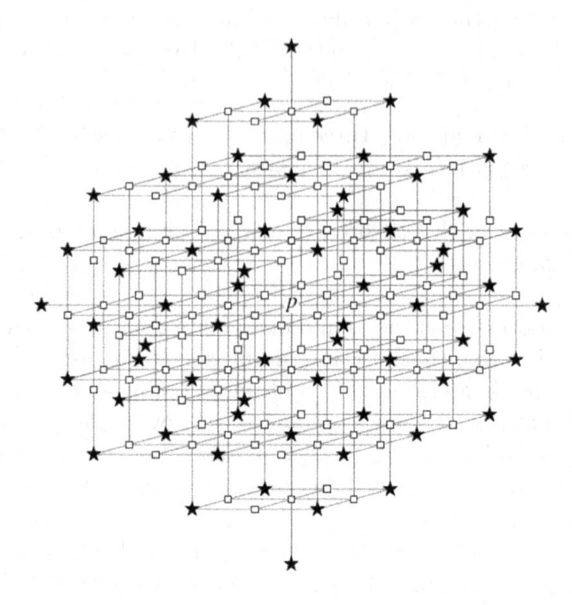

Fig. 5. The 64 points marked '★' are in the symmetric support of $\mathcal{R}_{\mathrm{symm}}$. Note that points marked '□' are in $\mathbb{Z}^3 \setminus \mathbb{B}$.

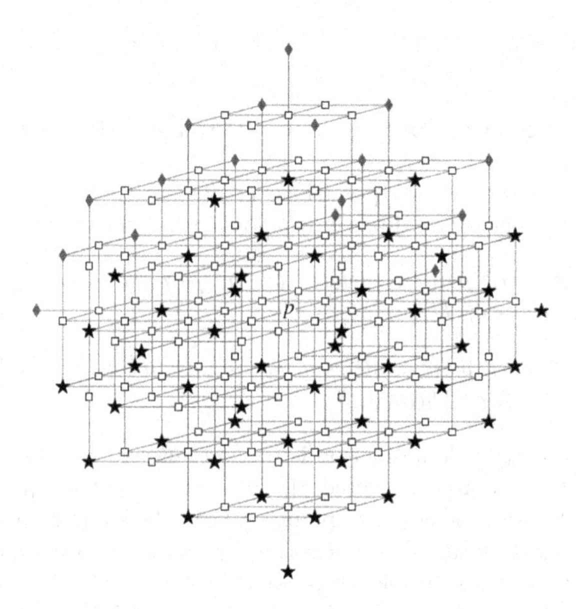

Fig. 6. The 47 points marked '★' are in the asymmetric support of $\mathcal{R}_{\mathrm{asymm}}$. Note that the points depicted '♦' are only contained in the symmetric support (see Fig. 5), and the points marked '□' are in $\mathbb{Z}^3 \setminus \mathbb{B}$.

By Lemma 1, if a set of two simple points $\{p, q\}$ is simple, both sequences $\langle p, q \rangle$ and $\langle q, p \rangle$ are simple. Similarly, if a set of two simple points $\{p, q\}$ is *not* simple, both sequences $\langle p, q \rangle$ and $\langle q, p \rangle$ are *not* simple. If the set of two mutually 14-adjacent simple points $\{p, q\}$ is *not* simple, none of them can be deleted by parallel reduction $\mathcal{R}_{\text{symm}}$ (since Condition 2 of Theorem 4 is violated). On the contrary, in the same case it is possible to delete point $p \prec q$ by $\mathcal{R}_{\text{asymm}}$ (see Condition 2 of Theorem 5). Conditions 3 and 4 of Theorem 5 also involve that reduction $\mathcal{R}_{\text{asymm}}$ can delete more points from a picture than reduction $\mathcal{R}_{\text{symm}}$ does. Figure 7 illustrates the different behaviors of these two generated parallel reductions.

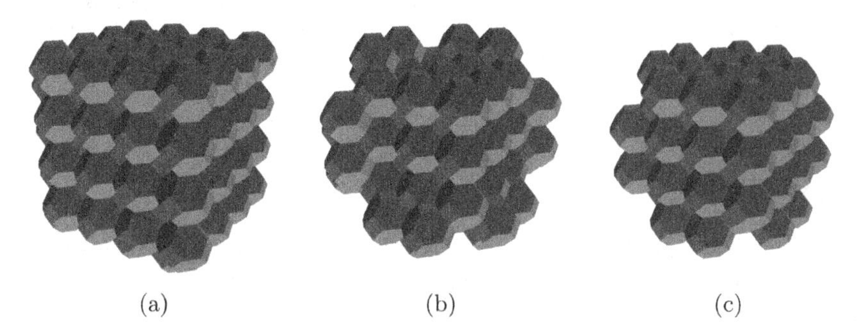

(a) (b) (c)

Fig. 7. An image of a small cube containing 128 black points/voxels (a) and the two different objects produced by parallel reductions $\mathcal{R}_{\text{symm}}$ (b) and $\mathcal{R}_{\text{asymm}}$ (c). Reductions $\mathcal{R}_{\text{symm}}$ and $\mathcal{R}_{\text{asymm}}$ can delete 50 and 62 points/voxels, respectively.

The otherness of the two generated reductions $\mathcal{R}_{\text{symm}}$ and $\mathcal{R}_{\text{asymm}}$ is markedly illustrated by Figs. 8 and 9. In these figures these two reductions are iterated until stability is reached.

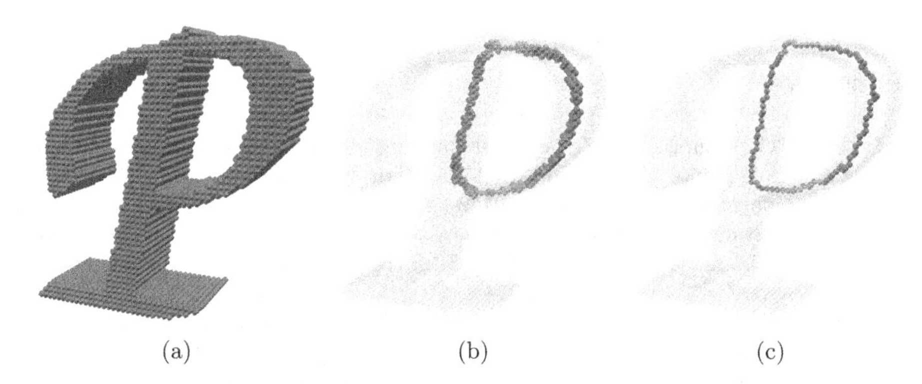

(a) (b) (c)

Fig. 8. An image of a 'P' containing 19 362 black points/voxels (a) and the objects produced by the iteratively repeated parallel reductions $\mathcal{R}_{\text{symm}}$ (b) and $\mathcal{R}_{\text{asymm}}$ (c). Iterated reductions $\mathcal{R}_{\text{symm}}$ and $\mathcal{R}_{\text{asymm}}$ lead to objects containing 264 and 97 points/voxels, respectively.

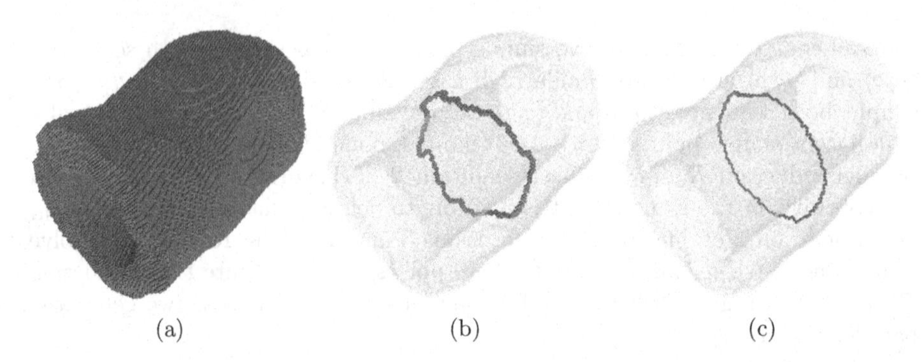

Fig. 9. An image of a tube containing 269 336 black points/voxels (a) and the objects produced by the iteratively repeated parallel reductions $\mathcal{R}_{\text{symm}}$ (b) and $\mathcal{R}_{\text{asymm}}$ (c). Iterated reductions $\mathcal{R}_{\text{symm}}$ and $\mathcal{R}_{\text{asymm}}$ lead to objects containing 597 and 198 points/voxels, respectively.

6 Conclusions

In this paper, a simplified version of the only existing configuration-based sufficient condition for topology-preserving parallel reductions on the BCC grid is reported. We established two further sufficient conditions that investigate the deletability of individual points. These point-based conditions directly provide deletion rules of two topology-preserving parallel reductions.

In a future work, we are to combine our point-based conditions with parallel thinning strategies and geometric constraints to generate a family of topology-preserving parallel thinning algorithms on the BCC grid.

Acknowledgments. Project no. TKP2021-NVA-09 has been implemented with the support provided by the Ministry of Innovation and Technology of Hungary from the National Research, Development and Innovation Fund, financed under the TKP2021-NVA funding scheme.

References

1. Bertrand, G., Couprie, M.: On parallel thinning algorithms: minimal non-simple sets, P-simple points and critical kernels. J. Math. Imaging Vis. **35**, 23–35 (2009). https://doi.org/10.1007/s10851-009-0152-3
2. Čomić, L., Nagy, B.: A combinatorial coordinate system for the body-centered cubic grid. Graph. Models **87**, 11–22 (2016). https://doi.org/10.1016/j.gmod.2016.08.001
3. Csébfalvi, B.: Cosine-weighted B-spline interpolation: a fast and high-quality reconstruction scheme for the body-centered cubic lattice. IEEE Trans. Visual. Comput. Graph. **19**, 1455–1466 (2013). https://doi.org/10.1109/TVCG.2013.7
4. Hall, R.W.: Parallel connectivity-preserving thinning algorithms. In: Kong, T.Y., Rosenfeld, A. (eds.) Topological Algorithms for Digital Image Processing, pp. 145–179, Elsevier Science (1996). https://doi.org/10.1016/S0923-0459(96)80014-0

5. Kardos, P.: Topology preservation on the BCC grid. J. Comb. Optim. **44**, 1981–2995 (2021). https://doi.org/10.1007/s10878-021-00828-9
6. Kardos, P., Palágyi, K.: Topology-preserving hexagonal thinning. Int. J. Comput. Math. **90**, 1607–1617 (2013). https://doi.org/10.1080/00207160.2012.724198
7. Kardos, P., Palágyi, K.: Topology preservation on the triangular grid. Ann. Math. Artif. Intell. **75**, 53–68 (2015). https://doi.org/10.1007/s10472-014-9426-6
8. Kong, T.Y., Rosenfeld, A.: Digital topology: introduction and survey. Comput. Visi. Graph. Image Process. **48**, 357–393 (1989). https://doi.org/10.1016/0734-189X(89)90147-3
9. Kong, T.Y.: On topology preservation in 2-D and 3-D thinning. Int. J. Pattern Recognit Artif Intell. **9**, 813–844 (1995). https://doi.org/10.1142/S0218001495000341
10. Ma, C.M.: On topology preservation in 3D thinning. CVGIP: Image Underst. **59**, 328–339 (1994). https://doi.org/10.1006/ciun.1994.1023
11. Matej, S., Lewitt, R.M.: Efficient 3D grids for image reconstruction using spherically-symmetric volume elements. IEEE Trans. Nucl. Sci. **42**, 1361–1370 (1995). https://doi.org/10.1109/23.467854
12. Németh, G., Palágyi, K.: Topology-preserving hexagonal thinning. Int. J. Comput. Math. **90**, 1607–1617 (2013). https://doi.org/10.1007/s10472-014-9426-6
13. Palágyi, K., Németh, G., Kardos, P.: Topology preserving parallel 3D thinning algorithms. In: Brimkov, V.E., Barneva, R.P. (eds.) Digital Geometry Algorithms: Theoretical Foundations and Applications to Computational Imaging, pp. 165–188. Springer, Heidelberg (2012). https://doi.org/10.1007/978-94-007-4174-4_6
14. Ronse, C.: Minimal test patterns for connectivity preservation in parallel thinning algorithms for binary digital images. Discrete Appl. Math. **21**, 67–79 (1988). https://doi.org/10.1016/0166-218X(88)90034-0
15. Strand, R.: Surface skeletons in grids with non-cubic voxels. In Proceedings of 17th International Conference on Pattern Recognition, ICPR 2004, pp. 548–551 (2004). https://doi.org/10.1109/ICPR.2004.1334195
16. Strand, R., Brunner, D.: Simple points on the body-centered cubic grid. Technical report 42, Centre for Image Analysis, Uppsala University, Uppsala, Sweden (2006)
17. Strand, R., Nagy, B.: Weighted neighbourhood sequences in non-standard three-dimensional grids - Metricity and algorithms. In Proceedings of 14th IAPR International Conference on Discrete Geometry for Computer Imagery, DGCI 2008, pp. 201–212 (2008). https://doi.org/10.1007/978-3-540-79126-3_19
18. Theussl, T., Möller, T., Grölle, M.E.: Optimal regular volume sampling. In Proceedings of IEEE Visualization, VIS 2001, pp. 91–98 (2001). https://doi.org/10.1109/VISUAL.2001.964498

On the Construction of Planar Embedding for a Class of Orthogonal Polyhedra

Nilanjana Karmakar[1], Arindam Biswas[2(✉)], Subhas C. Nandy[3], and Bhargab B. Bhattacharya[4]

[1] Department of Information Technology, St. Thomas' College of Engineering and Technology, Kolkata, India
nilanjana.nk2@gmail.com

[2] Department of Information Technology, Indian Institute of Engineering Science and Technology, Shibpur, India
barindam@gmail.com

[3] Advanced Computing and Microelectronics Unit, Indian Statistical Institute, Kolkata, India
nandysc@isical.ac.in

[4] Department of Computer Science and Engineering, Indian Institute of Technology Kharagpur, Kharagpur, India
bbbiitkgp@gmail.com

Abstract. 2D-representations of 3D digital objects find versatile applications to computer vision, robotics, medical imaging, and in discrete geometry. This work presents an algorithm for constructing a planar embedding with only straight-line edges for a general non-intersecting orthogonal polyhedron that has genus 0. We discover certain characterizations of vertices and edges of a polyhedron that lead to efficient graph-drawing on the 2D plane. The original orthogonal polyhedron can be fully reconstructed from this graph provided the information regarding the coordinates of vertices, are preserved. The time complexity of the proposed embedding is linear in the number of edges of the orthogonal polyhedron.

Keywords: Orthogonal polyhedron · Planar graph · Graph drawing

1 Introduction

Analysis of polyhedrons using graph-theoretic tools has always drawn considerable attention in the field of combinatorics. In the words of Grunbaum, Steinitz's theorem "is the most important and deepest known result on 3-polytopes" [15]. 1-skeleton of a polyhedron is the graph consisting only of its vertices and edges. Steinitz's theorem may be stated in terms of graph theory as: A graph G is isomorphic to the 1-skeleton of a 3-dimensional convex polyhedron $\mathcal{P}$ if and only if G is planar and 3-connected [17]. In other words, a given convex polyhedron

R. P. Barneva et al. (Eds.): IWCIA 2022, LNCS 13348, pp. 84–104, 2023.
https://doi.org/10.1007/978-3-031-23612-9_6

can be represented as a polyhedral graph, i.e., a 3-connected planar graph. A cubic bipartite polyhedral graph always corresponds to some specific type of orthogonal polyhedron termed as simple orthogonal polyhedron [14,15]. This type of orthogonal polyhedron excludes those not having the topology of sphere, or containing more than three edges incident at each vertex. Therefore, the graph representing an orthogonal polyhedron need not necessarily be a cubic bipartite polyhedral graph. The current work generalizes previous approaches and shows how every closed non self-intersecting orthogonal polyhedron that has genus 0, can be embedded on the 2D plane using only non-crossing straight-line edges. Note that given such a polyhedron, the use of direct (i.e., without re-embedding it on a sphere) stereographic projections may lead to edge-crossovers on the plane.

Certain graphic standards are generally maintained to produce aesthetically desirable graphs for any application. For instance, edge crossings and bends in edges are avoided, uniformity of edge lengths and distribution of vertices are maintained, etc. Optimization of such standards have been found to be NP-hard [5]. Even then aesthetic criteria play an important role in graph drawing to maintain readability of the graph, thereby optimizing graphic standards with certain trade-offs [9]. One such aesthetically desirable graph of importance is planar graph [32]. Several works have contributed to the testing of planarity of a graph by vertex addition method, path addition method, etc. [20,24]. Planar graph drawing has been categorized as straight line drawing, polyline drawing, orthogonal drawing etc. as mentioned in [4,12,25,30,31]. In the current work, straight line drawing of planar graphs has been adopted to represent a non-self-intersecting orthogonal polyhedron of genus 0.

In the orthogonal domain, it has been proved that it is NP-hard to decide whether a graph with fixed combinatorial embedding, edge lengths, and facial angles, is the graph of an orthogonal polyhedron [7]. However, it is possible to decide the same, and hence reconstruct, in case of orthogonal convex polyhedron [7]. Reconstruction of an orthogonal polyhedron based on its dual graph, dihedral angles, facial angles, edge lengths, or vertex coordinates has been studied in detail [16]. A famous theorem by Cauchy states that a convex polyhedron can be uniquely determined given its incidence structure and face polygons [8]. The limitation of this theorem to convex polyhedra has been overcome by proving that Cauchy's theorem holds for orthogonal polyhedra of genus 0, considering faces without holes [8]. A graph is said to be 3-connected if it has more than 3 vertices and the result of deleting any set of fewer than 3 vertices is a connected graph. Algorithms to produce polyhedral representations from 3-connected planar graphs have also been dealt with in [26]. Other works regarding polyhedral graphs, planar graphs, and construction and unfolding of 3D polytopes are available in [2,3,11,13,18] and [27–29,33].

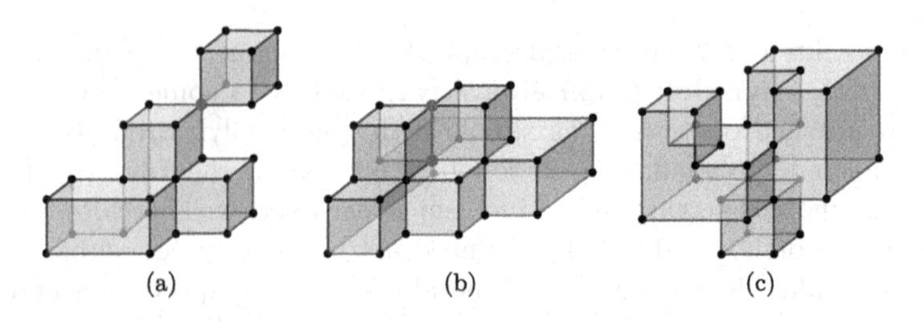

Fig. 1. (a,b) Self-intersecting orthogonal polyhedrons. (c) A non self-intersecting orthogonal polyhedron of genus 0 (Color figure online).

Contribution of Our Work: We propose a straight line planar graph drawing of a given orthogonal polyhedron with genus 0. The polyhedron may have any number of mutually perpendicular axis parallel edges incident at a vertex and may have faces with cavities (holes). A genus 0 polyhedron represents a connected set that is disconnected by a single cut. By non self-intersecting orthogonal polyhedron we mean an orthogonal polyhedron that does not intersect itself along a vertex or an edge. Note that all genus-0 polyhedrons are non self-intersecting. In Fig. 1(a,b) some self-intersecting orthogonal polyhedrons are shown and an example of a non self-intersecting orthogonal polyhedron of genus 0 is shown in Fig. 1(c).

Figure 2(a)(left) shows a non self-intersecting orthogonal polyhedron of genus 0 with a vertex (the red point) incident with more than three axis-parallel edges. The corresponding graph is planar as shown in Fig. 2(a)(right). Also, if an orthogonal polyhedron contains convex or concave regions that touch the polyhedron at a single face and not at its boundary edges (Fig. 2(b)(left)), then the corresponding graph is a disconnected one as shown in Fig. 2(b)(right). Each of the vertices and edges in the graph corresponds to a vertex or edge in the orthogonal polyhedron thereby preserving the incidence structure. Although edge lengths in the graph may not match with those of the polyhedron, they can easily be computed from vertex information stored in the graph. The number of faces in the graph is also equal to that in the orthogonal polyhedron where the face that appears furthest while viewing the polyhedron corresponds to the unbounded outer face of the graph. The incidence structure of the polyhedron faces is maintained in the graph whereas the shape and area of the faces, though visibly different, may be calculated from the information stored in the graph.

The rest of the paper is continued as follows. Section 2 may be referred to recall some relevant definitions and notation. The characterization of an orthogonal polyhedron in terms of its vertex categories is provided in Sect. 3. The graphic standards of planar embedding used in this work are discussed in Sect. 4. The main algorithm is presented in Sect. 5 followed by a detailed discussion in Sect. 6 where several possible cases are analyzed. Finally in Sect. 7, the paper is concluded with future directives.

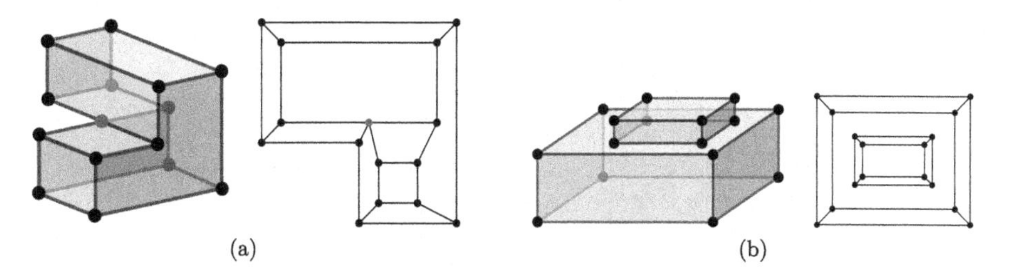

Fig. 2. Two non self-intersecting orthogonal polyhedrons of genus 0 and the corresponding planar graphs. The red point shown in the graph of Fig. 2(a) corresponds to the red-vertex of the polyhedron shown on the left where more than three axis-parallel edges meet. (Color figure online)

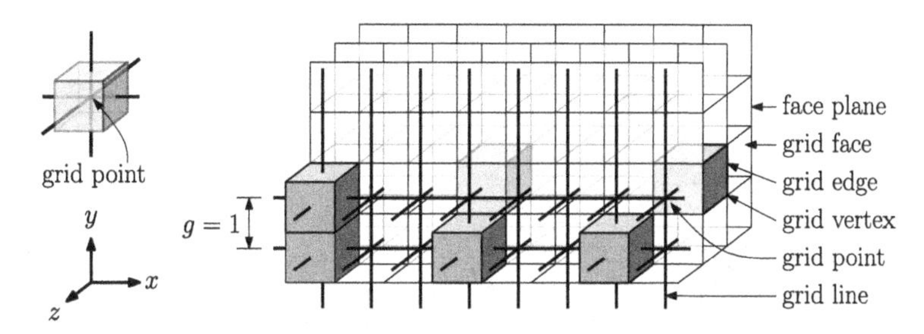

Fig. 3. 3D digital space and 26N [23]. Left: A 3-cell and its corresponding grid point. Right: Three pairs of α-adjacent 3-cells for $\alpha \in \{0, 1, 2\}$, $\alpha \in \{0, 1\}$, and $\alpha = 0$ (from left to right). The 3-cells in each of these three pairs are connected in 26N.

2 Definitions and Preliminaries

Definition 1 (*Unit grid cube*). A unit grid cube (UGC) is a (closed) cube of length g whose vertices are grid vertices, edges constituted by grid edges, and faces constituted by grid faces. Each face of a UGC lies on a face plane (henceforth referred as a UGC-face), which is parallel to one of the three coordinate planes (Fig. 3).

Definition 2 (*Orthogonal polyhedron*). An orthogonal polyhedron imposed on a 3D digital grid $\mathbb{G}$ is a 3D polytope with all its vertices as grid vertices, all its edges made of grid edges, and all its faces lying on face planes (see Fig. 3) [21,22]. Each face of an orthogonal polyhedron is an orthogonal polygon whose alternate edges are axis-parallel and constituted by grid edges of $\mathbb{G}$.

Definition 3 (*Genus*). Genus of a connected set is the minimum number of cuts required to transform the set into a simply connected set (Fig. 4). By a simply connected set in $\mathbb{R}^3$ we mean a connected set that has no tunnels [23] .

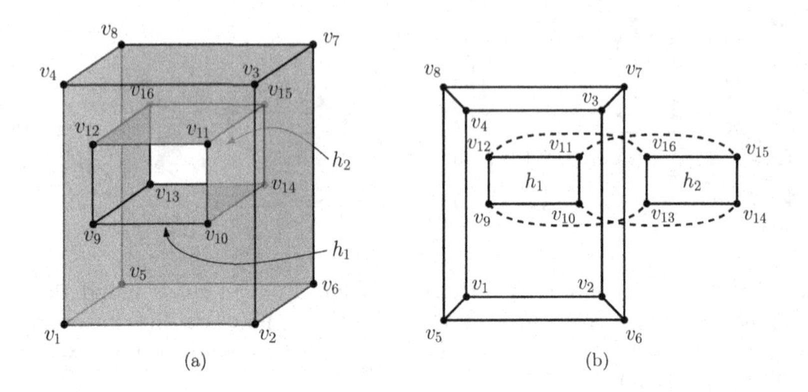

Fig. 4. A sample orthogonal polyhedron of genus 1 and the corresponding graph.

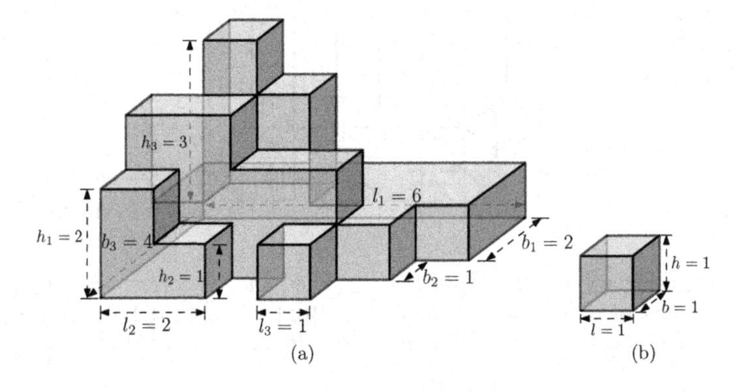

Fig. 5. Determination of unit cube from a given orthogonal polyhedron. (a) $l = GCD(6, 2, 1) = 1$, $b = GCD(2, 1, 4) = 1$, $h = GCD(2, 1, 3) = 1$. (b) unit cube with $l = 1$, $b = 1$, and $h = 1$

Polyhedron as a collection of Unit Cubes

A unit cube is the lowest unit of measurement in three dimensions, which is defined as follows. Given an orthogonal polyhedron, the GCD (Greatest Common Divisor) of the lengths of all of its edges along the three coordinate planes gives the length of the unit cube (Fig. 5(a)). Without loss of generality we equate it with the concept of UGC defined in Sect. 2 (Fig. 5(b)). It is to be noted that the unit cube is selected w.r.t. the polyhedron. It may have length, breadth, and height greater than unity. However, it can be decomposed into a cube of unit length, unit breadth, and unit height and hence it is justified to equate it with a UGC.

Singular Point and Singular Line Segment

Let us consider a point p of the orthogonal polyhedron $\mathcal{P}$. Let $Ball(p, \delta)$ be a ball of radius $\delta(> 0)$ in $\mathbb{R}^3$ centered at p. Then, we use the following definitions.

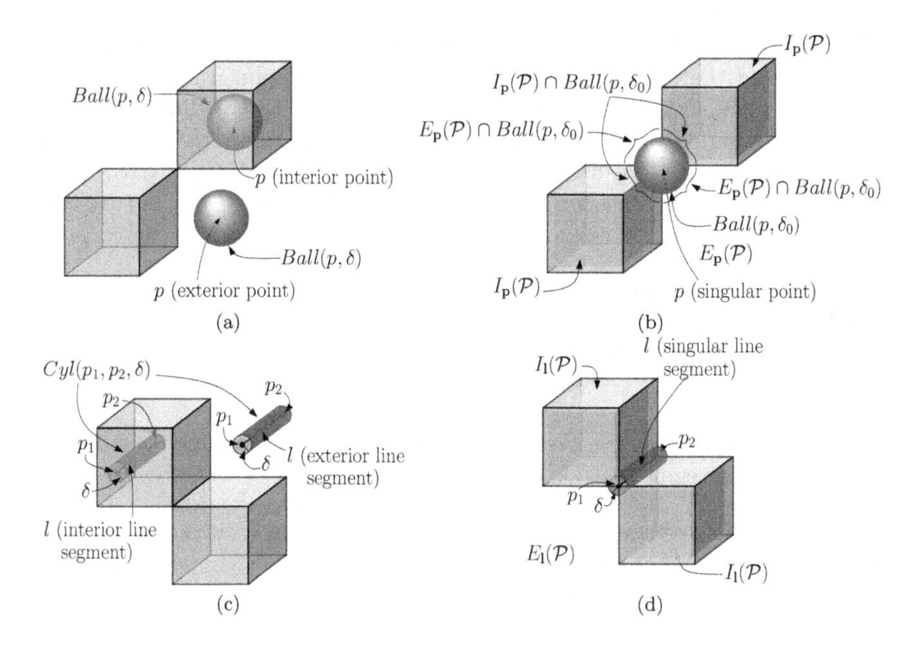

Fig. 6. (a,b) Singular point and (c,d) Singular line segment (Color figure online).

Definition 4 (*Interior point*). A point p is an *interior point* of $\mathcal{P}$ if $Ball(p, \delta) \subseteq \mathcal{P}$ for some $\delta > 0$.

Definition 5 (*Exterior point*). A point p is an *exterior point* of $\mathcal{P}$ if $Ball(p, \delta) \cap \mathcal{P} = \emptyset$ for some $\delta > 0$.

Examples of interior and exterior points are shown in Fig. 6(a). We denote the set of all interior points and the set of exterior points by $I(\mathcal{P})$ and $E(\mathcal{P})$ respectively.

Definition 6 (*Non-singular point*). If $\exists\ \delta_0$ such that $\forall \delta \in (0, \delta_0)$ we have $I(\mathcal{P}) \cap Ball(p, \delta) \neq \emptyset$ and $E(\mathcal{P}) \cap Ball(p, \delta) \neq \emptyset$ and each of $I(\mathcal{P}) \cap Ball(p, \delta)$ and $E(\mathcal{P}) \cap Ball(p, \delta)$ is connected, then p is a *non-singular point* [19].

Definition 7 (*Singular point*). If $\exists\ \delta_0$ such that $\forall \delta \in (0, \delta_0)$ we have $I(\mathcal{P}) \cap Ball(p, \delta) \neq \emptyset$ and $E(\mathcal{P}) \cap Ball(p, \delta) \neq \emptyset$ and either or both of $I(\mathcal{P}) \cap Ball(p, \delta)$ and $E(\mathcal{P}) \cap Ball(p, \delta)$ are disconnected, then p is a *singular point* [19].

A singular point is illustrated in the Fig. 6(b).

Now, let us consider two points p_1 and p_2 belonging to the orthogonal polyhedron $\mathcal{P}$. From the existing definition of non-singular point a concept of non-singular line segment is derived as follows. $Cyl(p_1, p_2, \delta)$ is a cylinder of radius $\delta(>0)$ in $\mathbb{R}^3$ bounded at the two ends by discs of radius δ centered at p_1 and p_2. The line segment $l = [p_1, p_2]$ is an *interior line segment* of $\mathcal{P}$ if $Cyl(p_1, p_2, \delta) \subseteq \mathcal{P}$ for some $\delta > 0$; l is an *exterior line segment* of $\mathcal{P}$ if $Cyl(p_1, p_2, \delta) \cap \mathcal{P} = \emptyset$ for some $\delta > 0$ (Fig. 6(c)).

Definition 8 (*Non-singular Line Segment*). *If* $\exists\ \delta_0$ *such that* $\forall \delta \in (0, \delta_0)$ *we have* $I(\mathcal{P}) \cap Cyl(p_1, p_2, \delta) \neq \emptyset$ *and* $E(\mathcal{P}) \cap Cyl(p_1, p_2, \delta) \neq \emptyset$ *and each of* $I_1(\mathcal{P}) \cap Cyl(p_1, p_2, \delta)$ *and* $E_1(\mathcal{P}) \cap Cyl(p_1, p_2, \delta)$ *is connected, then* l *is denoted as a* non-singular line segment.

Definition 9 (*Singular Line Segment*). *If there is no such* δ_0, *then* l *is denoted as a* singular line segment.

A singular line segment is shown in the Fig. 6(d). It may be noted here that any point on the singular line segment is a singular point.

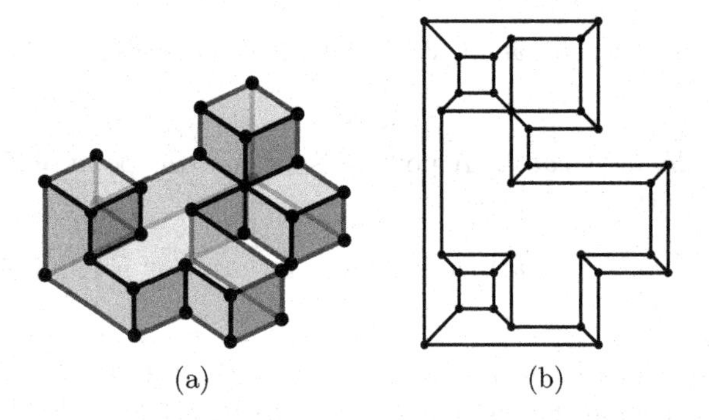

Fig. 7. (a) An orthogonal polyhedron and (b) its corresponding polyhedron graph.

Table 1. Vertex classification for orthogonal polyhedra and orthogonal pseudo-polyhedra.

# incident UGCs	# combinations	Sub-categories	Configuration
0	$\binom{8}{0} = 1$	—	—
1	$\binom{8}{1} = 8$	1 for each UGC	Fig. 9(a)
2	$\binom{8}{2} = 28$	12 for face adjacency	Fig. 9(b)
		12 for edge adjacency	Fig. 8(c)
		4 for vertex adjacency	Fig. 8(a)
3	$\binom{8}{3} = 56$	24 for face-face-edge adjacency	Fig. 9(c)
		24 for face-edge-vertex adjacency	Fig. 8(e)
		8 for edge-edge-edge adjacency	Fig. 8(i)
4	$\binom{8}{4} = 70$	6 combinations	Fig. 9(d)
		24 combinations	Fig. 9(e)
		24 combinations	Fig. 8(h)
		8 combinations	Fig. 9(f)
		6 combinations	Fig. 8(g)
		2 combinations	Fig. 8(k)
5	$\binom{8}{5} = 56$	24 for face-face-edge adjacency	Fig. 9(g)
		24 for face-edge-vertex adjacency	Fig. 8(f)
		8 for edge-edge-edge adjacency	Fig. 8(j)
6	$\binom{8}{6} = 28$	12 for face adjacency	Fig. 9(h)
		12 for edge adjacency	Fig. 8(d)
		4 for vertex adjacency	Fig. 8(b)
7	$\binom{8}{7} = 8$	1 for each set of 7 UGCs	Fig. 9(i)
8	$\binom{8}{8} = 1$	—	Fig. 9(j)

3 Characterization of an Orthogonal Polyhedron

Given a closed orthogonal polyhedron $\mathcal{P}$, its abstract graph $\mathcal{G} = (V, E, F)$ is defined by the set of vertices V, edges E, and faces F of the polyhedron $\mathcal{P}$ such that

- there is only one edge e between two vertices v_1 and v_2,
- there is only one edge e incident with two faces f_1 and f_2,
- there are exactly two faces f_1 and f_2 incident on one edge e.
- every vertex v is incident with at least three faces f_1, f_2, and f_3, and
- every face f is incident with at least four vertices v_1, v_2, v_3, and v_4.

A graph G is planar if it can be drawn on the Euclidean plane with no two edges intersecting each other [10]. As a convention $\mathcal{P}$ is viewed either from the front or from the top whereby the most distant face invisible from the front or top is embedded on the plane as the infinite face.

An orthogonal polyhedron, $\mathcal{P}$, of genus 0 is shown in Fig. 7(a) and its corresponding polyhedron graph is planar (Fig. 7(b)). On the other hand, Fig. 4(a) shows an orthogonal polyhedron, $\mathcal{P}$, of genus one and Fig. 4(b) shows the corresponding graph which is not planar. We devote our study on orthogonal polyhedra of genus 0.

According to the marching cube algorithm, a cube is intersected by a surface such that some of the vertices of the cube lie on or inside the surface while others do not. Therefore, the eight vertices of the cube are classified into two types. An edge between two opposite types of vertices is intersected by the surface. As a cube consists of eight vertices, each in one of the two states (inside or outside the surface), the cube can be intersected by the surface in at most $2^8 = 256$ possible ways. Since at most eight UGCs may be incident at a vertex v, the classification procedure of v is in accordance with the above process of classifying vertices in the marching cube algorithm. The possible number of combinations of UGCs neighboring v that may appear in an orthogonal polyhedron is $2^8 = 256$ [1]. The 256 combinations may be grouped into 22 equivalence classes by using rotational symmetries as shown by the configurations in Fig. 8 and Fig. 9. The 22 configurations are computed as shown in Table 1. For instance, if three UGCs are incident at v with face-face-edge adjacency (Table 1), i.e., two pairs of UGCs are face-adjacent and one pair of UGCs is edge adjacent (Fig. 9(c)), then the configuration can be symmetrically rotated to four sub-categories along each of yz-, zx-, and xy-planes. The set of these 24 configurations comprise an equivalence class.

Of all the possible arrangements for one or more (maximum eight) incident UGCs at a given vertex of $\mathcal{P}$, Fig. 8 shows the exhaustive cases of the singular vertices (Fig. 8 (a) and (b)) and singular line segments (Fig. 8 (c)–(k)). In Fig. 8(a), two UGCs ($\in \mathcal{P}$) are incident at the vertex, and $I(\mathcal{P}) \cap Ball(p, \delta) \neq \emptyset$ and disconnected, hence it is a singular vertex. Similarly, the vertex in Fig. 8(b) is a singular vertex because $E(\mathcal{P}) \cap Ball(p, \delta) \neq \emptyset$ and disconnected. It may be noted here that the arrangement shown in Fig. 8(b) is the complement of the arrangment in Fig. 8(a). The arrangement shown in Fig. 8(c) has a singular line segment (also referred as *singular edge*) as $I(\mathcal{P}) \cap Cyl(p_1, p_2, \delta) \neq \emptyset$ and disconnected. Figure 8(d) shows an example of a singular edge where $E(\mathcal{P}) \cap Cyl(p_1, p_2, \delta) \neq \emptyset$ and disconnected. In each of Figs. 8(c)–(f), there is one singular edge, whereas two singular edges in Figs. 8(g)–(h), three singular edges in Figs. 8(i)–(j), and four singular edges in Fig. 8(k). It is evident that as a singular edge consists of a set of singular points, the end point of a singular edge ($\in \mathcal{P}$), which is a vertex, is a singular vertex. Hence, the following observations hold.

Observation 1. *If one or more edges incident at a vertex, $v \in \mathcal{P}$, is a singular edge then v is a singular vertex.*

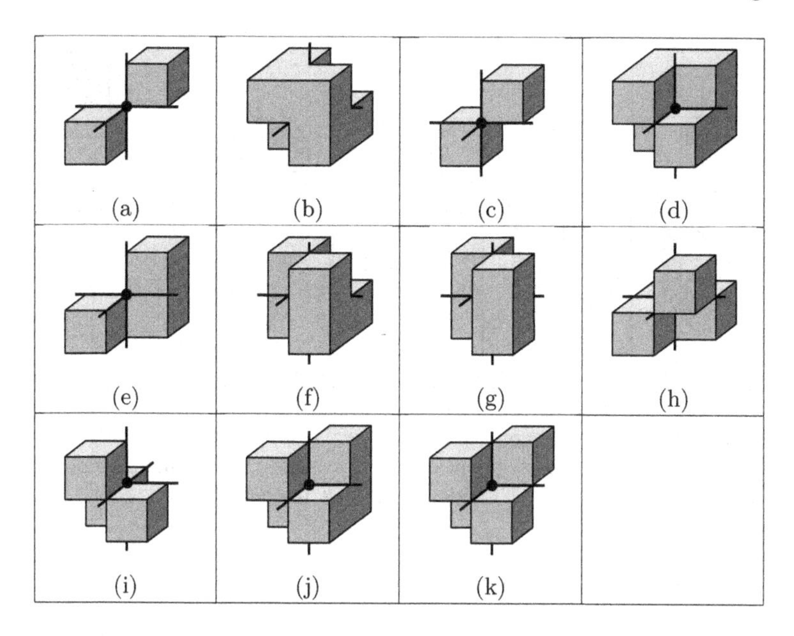

Fig. 8. Configurations of orthogonal polyhedra w.r.t. a vertex v showing all possible cases of singular points and singular line segments.

Observation 2. *If, for a vertex, $v \in \mathcal{P}$, one of the following conditions holds,*

i. only two diagonal UGCs incident at v belong to $\mathcal{P}$ ($\mathcal{P}'$) and the rest six belong to $\mathcal{P}'$ ($\mathcal{P}$).
ii. at least one edge incident at v has exactly four incident faces.

then v is a singular point.

Observation 3. *A singular vertex, $v \in \mathcal{P}$, cannot be embedded in a plane.*

It is evident that if v is singular then the intersection with $Ball(v, \delta)$, disconnects either the interior or the exterior of $\mathcal{P}$. In order to embed v, it is required that both interior and exterior should be connected.

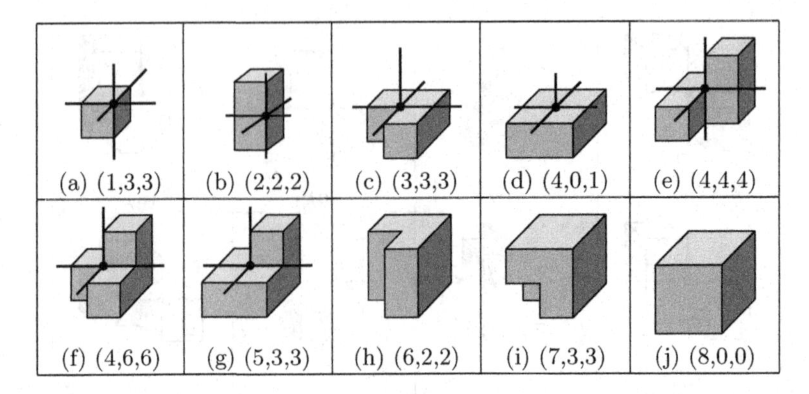

Fig. 9. Configurations of orthogonal polyhedra w.r.t. a vertex v excluding the cases of singular points or singular line segments. For each vertex, the configuration comprises the 3-tuple (# incident UGCs, # incident edges, # incident faces).

Observation 4. *A singular edge cannot be embedded in a plane.*

It is evident from the observations that self intersection at a vertex or an edge cannot be embedded in a plane. Let us consider the orthogonal polyhedra representing the different configurations containing at most eight UGCs incident at a vertex v [1]. The configurations containing singular points and singular line segments are shown in Fig. 8. Note that the configurations in Fig. 8 (a) and (b) contain singular point and those in Fig. 8(c)–(k) contain singular points as the end points of singular line segments. If a singular point v is embedded on a plane, then at least one of its incident edges intersects another edge. Since, each point on a singular line segment is a singular point, embedding a singular line segment also results in edge intersections.

4 Planar Graph Drawing

In order to propose a planar graph drawing algorithm, Fig. 9 shows all possible configurations w.r.t. v that represent an orthogonal polyhedron without singular point and without singular line segment. Considering at most eight UGCs may be incident at a vertex, the configurations are based on the 3-tuple (# incident UGCs, # incident edges, # incident faces) w.r.t the vertex. A vertex with degree 3, 4, or 6 may serve as a proper vertex to be embedded on a plane. Hence, configurations shown in Fig. 9 (a,c,e,f,g, and i) are considered for planar graph drawing. The graph corresponding to a configuration is considered as planar if there exists as least one way of embedding it on a plane. Hence, the orthogonal polyhedron is viewed either from the top or from the front to ensure clarity of the planar graph. The infinite face in the graph represents the face of the orthogonal polyhedron which is parallel to and invisible from the xy-plane (in case of front view) or from the zx-plane (in case of top view). Thus, the graph represents the orthogonal polyhedron back to front or bottom to top. In Fig. 10, the planar graphs corresponding to the configurations in Fig. 9 (a,c,e,f,g, and i) are displayed. Note that the face adjacency of the orthogonal polyhedron is maintained in the corresponding graph. That is why configurations like Fig. 8(e) do not admit a planar graph drawing.

The planar graph drawing of such an orthogonal polyhedron follows certain graphic standards. The incidence structure of vertices, edges, and faces in the polyhedron are maintained in the graph. Although the edge lengths and face areas are not in accordance with those of the orthogonal polyhedron, they can be calculated from the information stored in the graph. Vertices are represented by filled discs. Edge lengths are proportional to the edges in the orthogonal polyhedron. Straight line drawing standard is adopted for drawing edges, i.e., edges are straight line segments with no bend in edges. Size of the faces increases in the direction farther from view. Also, faces at the same level of view are equal in size. Thus, symmetry is maintained. A scheme to decide the angular resolution of the angles incident at a vertex may be decided in future. Also, minimum distance between vertices may be specified in future for further accuracy in maintaining graphic standards.

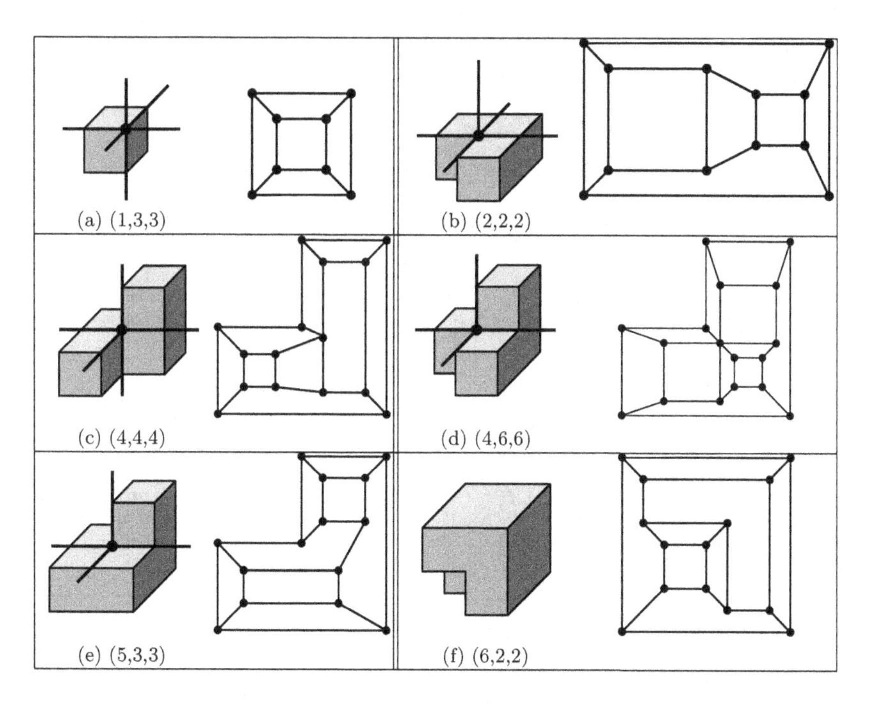

Fig. 10. Planar graphs of orthogonal polyhedra having permissible configurations w.r.t. a vertex v.

5 Algorithm

Given an orthogonal polyhedron $\mathcal{P}$ of genus 0, its graph drawing algorithm (Fig. 11) will be characterized by the following features:

i. If there exists any singular point or singular line segment in $\mathcal{P}$, then its planar graph cannot be drawn.
ii. Otherwise all vertices, edges and faces of $\mathcal{P}$ represent the vertices, edges and faces of the planar graph respectively.
iii. Depending on the structural complexity of the given orthogonal polyhedron, either the top view or the front view is considered for representing the planar graph.
iv. Accordingly the face invisible from view is represented as the infinite face in the graph.
v. If there exists a concave region in the polyhedron, then it must be adjacent to a hole polygon on a face of the polyhedron. The edges connecting the concave region to the hole polygon are represented by dotted lines in the planar graph. All other edges are represented by solid lines.
vi. The resultant planar graph is represented by an adjacency list G. Each value of $G[i,j]$ contains the coordinates of the vertices i and j, the face number to which these vertices belong and a marker to state whether the edge (i,j) is to be represented by a dotted line or solid line.

```
Algorithm PLANAR_GRAPHDRAW(V, E, F, view)

01. for each face f ∈ F
02.   for each edge e = (v1, v2) ∈ f    ▷ v1 = (x1, y1, z1),
                                          v2 = (x2, y2, z2)
03.     if (POINTTYPE(v1) = FALSE
             or POINTTYPE(v2) = FALSE)
04.       print: Graph is not planar.
05.       exit
06.     if view = top
07.       ai = xi, bi = zi, ci = yi          ▷ i ∈ {1, 2}
08.     else if view = front
09.       ai = xi, bi = yi, ci = zi          ▷ i ∈ {1, 2}
10.     if ((a1 ≠ a2) or (b1 ≠ b2)
          or ((c1 ≠ c2) and LINETYPE(E, f) = False))
11.       G[v1, v2] ← {(x1, y1, z1), (x2, y2, z2), 1, f}
12.     else if ((c1 ≠ c2) and LINETYPE(E, f) = True)
13.       G[v1, v2] ← {(x1, y1, z1), (x2, y2, z2), 2, f}
14. return G
```

```
Procedure LINETYPE(E, f)

01. for each edge e ∈ f in E
02.   e' = pair(e)
03.   f' = face(e')
04.   if (f'.flag > 1)
           ▷ f'.flag = 1 ⟹ f' is face,
             f'.flag > 1 ⟹ f' is hole on the face.
05.     return TRUE
06. return FALSE
```

```
Procedure POINTTYPE(v)

01. if v = (1, 3, 3) or v = (3, 3, 3) or v = (4, 4, 4)
      or v = (4, 6, 6) or v = (5, 3, 3) or v = (7, 3, 3)
02.   return TRUE
03. return FALSE
```

Fig. 11. The algorithm for planar graph drawing and its related procedures.

5.1 Reconstruction

The input orthogonal polyhedron $\mathcal{P}$ is represented as a Doubly Connected Edge List (DCEL) [6] containing the vertex list, edge list, and face list with the following fields.

- Vertex list V: vertex id, coordinates
- Edge list E: half-edge id, source vertex, destination vertex, face number, flag, pair, next, previous
- Face list F: face id, source vertex

While traversing the resultant adjacency matrix G, $G[i,j]$ represents a half-edge (i,j) between vertices i and j. This half-edge can be inserted in the edge list with an assigned half-edge id, source vertex i, destination vertex j, and face number f stored in $G[i,j]$. Consequently, vertices i and j are inserted in the vertex list with assigned vertex ids and coordinates stored in $G[i,j]$. Also, the face id f stored in $G[i,j]$ is entered in the face list along with the source vertex i. The half-edge id of (i,j) may be stored separately and used to populate the pair field when $G[j,i]$ is traversed. The value of the marker in $G[i,j]$, that indicates whether the face f is a hole or not, is used to populate the flag field. Thus the vertex list, face list, and the first six fields of the edge list are populated by traversing the adjacency list G once. A separate list of faces may be maintained during the traversal of G, which contains the set of half-edges belonging to each face. A single traversal of this list of faces will be enough to identify the order of the half-edges comprising each face which may be used to populate the next and previous fields. Thus, the resultant adjacency matrix G suffices to reconstruct the orthogonal polyhedron (represented as DCEL) leading to the efficiency of the algorithm.

5.2 Time Complexity

Let $|E|$ be the number of edges in the edge list representing the orthogonal polyhedron. In the algorithm PLANAR_GRAPHDRAW, w.r.t. a face f, its edges are traversed (Steps 1 and 2). For each edge $e \in f$, the type of two end vertices are checked to decide whether or not the graph is planar (Steps 3–5 and Procedure POINTTYPE). In case of a planar graph, the edge e is embedded as solid line (denoted by 1 in G) (Steps 10 and 11) unless the pairing edge e' belongs to a hole polygon on some face f' (Procedure LINETYPE). Then edge e is represented by dotted line (denoted by 2 in G) (Steps 12 and 13). The total procedure can be done by traversing the edge list exactly once. Hence time complexity of the algorithm is given by $O(|E|)$. The graph can be embedded on the plane while populating the adjacency matrix G. Hence the time complexity remains $O(|E|)$.

During the reconstruction procedure as described in Sect. 5.1, each edge of the graph is traversed exactly once to populate the face list, vertex list and the first six fields of the DCEL using vertex coordinates and the face *id* stored in $G[i,j]$. The separate list of faces (mentioned in Sect. 5.1) is traversed once to populate its next and previous pointer fields. Hence, the time complexity of the reconstruction procedure will be $O(|E|) + O(|F|) \approx O(|E|)$.

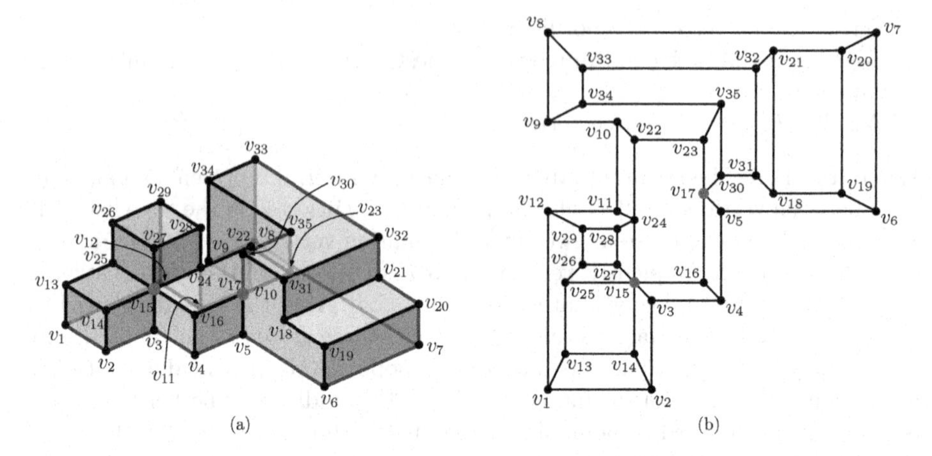

Fig. 12. (a) A non self-intersecting orthogonal polyhedron of genus 0, (b) Planar graph corresponding to the polyhedron. (Color figure online)

6 Discussion

A few orthogonal polyhedra and their corresponding planar graphs are shown in Figs. 12–14. The orthogonal polyhedra are of genus 0 and do not contain singular points or singular line segments. Note that in Fig. 12(a), the vertices v_{15} and v_{17} (shown in red) belong to the types of vertices (Fig. 9(f) and (e) respectively).

In Fig. 14(a), the polyhedron contains a number of convex (marked as C_1, C_3, and C_4) and concave (marked as C_2 and C_5) regions (cavities) that touch the polyhedron at a single face but do not touch its boundary edges. Such a concave or convex region touches the polyhedron face along the edges of a hole polygon. For the sake of clarity each component is marked by different colors instead of numbering all the vertices. Such a portion is shown as a separate component in a planar graph. Hence, the graph in Fig. 14(b) consists of five components apart from the polyhedron component. C_1 lies on the front face of the polyhedron in the top view. C_2 lies on the front face of the polyhedron where the open face $v_1v_2v_3v_4v_5v_6$ touches the front face but the lower face $v_7v_8v_9v_{10}v_{11}v_{12}$ does not touch any face of the polyhedron. A small nested concavity exists on the lower face. C_3 portrays a nested convex region on a vertical face of the polyhedron and C_4 is a convex portion on a face of the polyhedron with a small cavity on one of its faces. C_5 depicts the reverse scenario of C_4, i.e., a convex region is nested in a concavity at the back face of the polyhedron.

While drawing the graphs in Figs. 12(b), 13(b), and 14(b), the top view of the polyhedron is considered. The following conventions are followed.

i. The more distant a polytope face is, the larger is its size in the graph,
ii. The edges connecting a concave region to the hole polygon on the face of the polyhedron are shown as dotted lines.
For instance, in Fig. 14(b), the polygon representing the back face of the

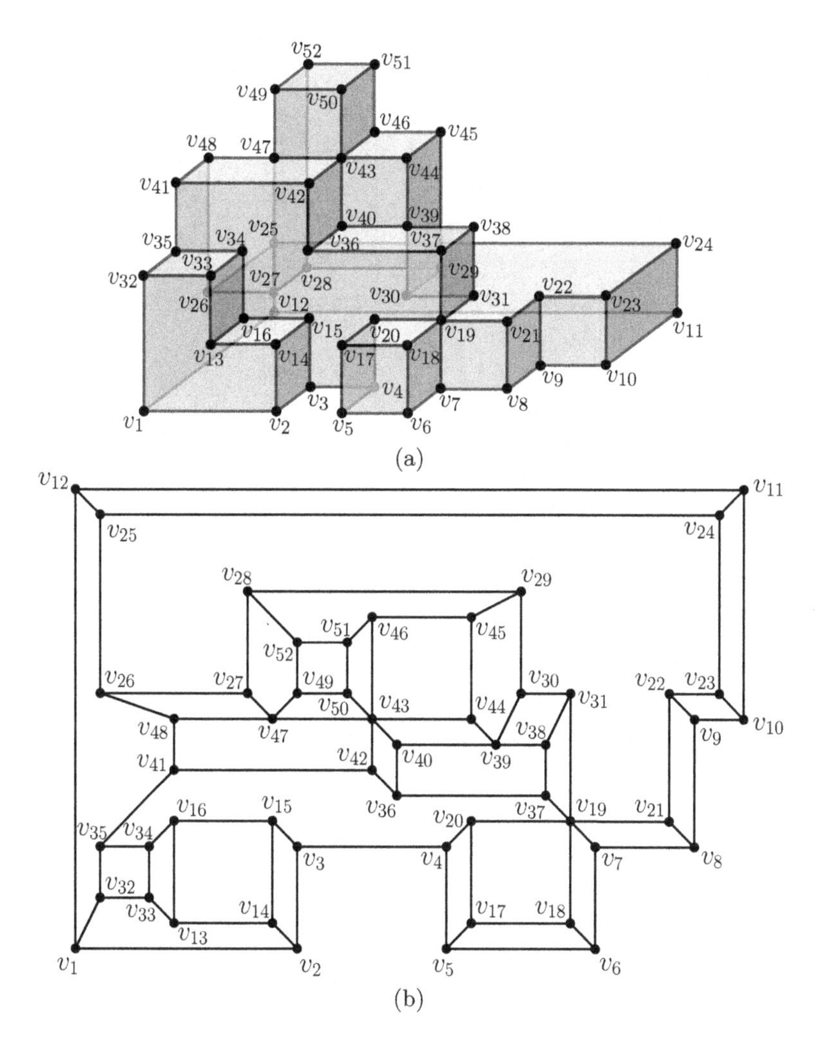

Fig. 13. (a) A non self-intersecting orthogonal polyhedron of genus 0, (b) Planar graph corresponding to the polyhedron.

component C_2 is connected by dotted lines to the polygon representing the front face.

iii. The back face of a polyhedron is represented by the infinite face outside the graph representing the polyhedron.

iv. If a depressed concave region lies on the back face of a polyhedron, then the component of the graph representing the region is placed in the infinite face outside the graph representing the polyhedron.
For instance, in Fig. 14(a), C_5 lies on the back face of the polyhedron. Its corresponding component in the graph (Fig. 14(b)) is shown in the infinite face outside the polyhedron component. Note that the region projected out

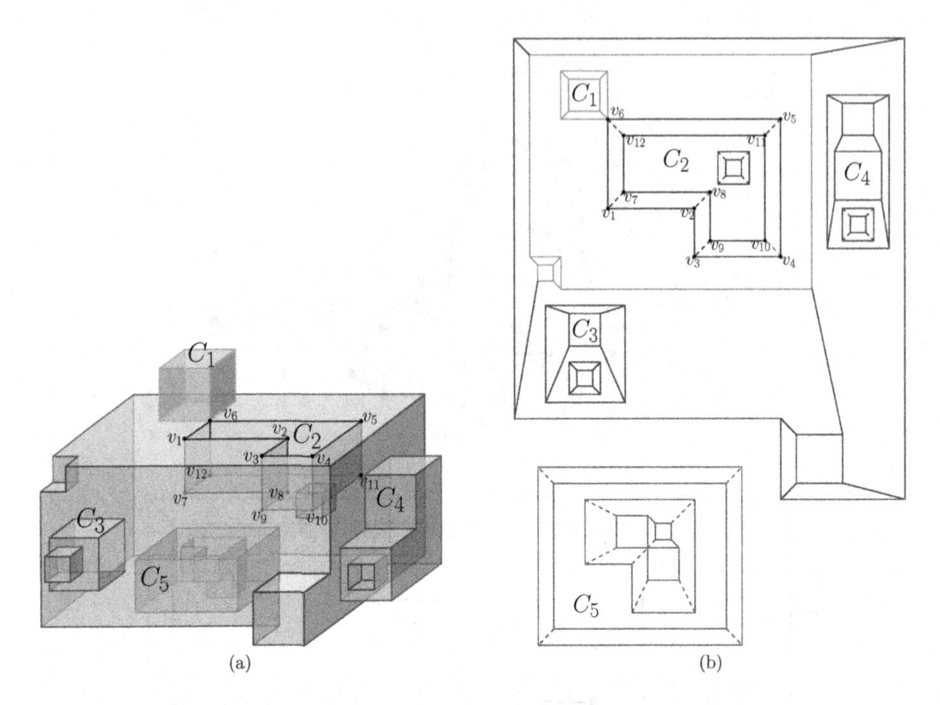

Fig. 14. (a) A non self-intersecting orthogonal polyhedron of genus 0, (b) Planar graph corresponding to the polyhedron (Color figure online).

from C_5 is shown partially by dotted lines due to the convention followed in ii.

v. If a protruded convex region lies on the back face of a polyhedron, then it is treated according to convention i.

It is evident from the conventions that corresponding to each face of $\mathcal{P}$ there exists a face in the graph G. Hence, corresponding to the vertices and edges of a face in $\mathcal{P}$ there exists vertices and edges of the face in G. Hence, the number of vertices, edges, and faces (including the infinite face) in G are equal to those in P. The incidence structure of the vertices, edges, and faces of $\mathcal{P}$ are maintained in the corresponding graph. Though the facial angles and edge lengths are not maintained, enough information is maintained in G to calculate those and hence to reconstruct $\mathcal{P}$, as explained in Sect. 5.1.

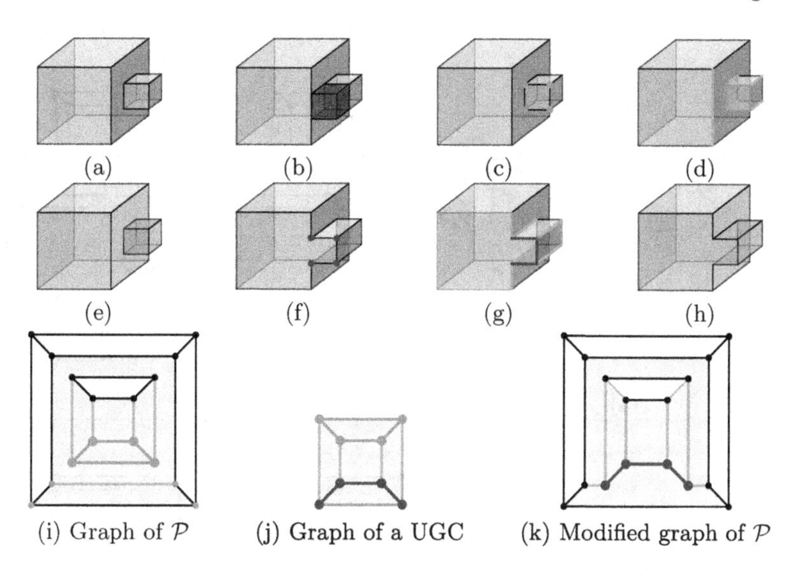

Fig. 15. (a) An orthogonal polyhedron $\mathcal{P}$ is augmented with (b) a UGC 2-adjacent to $\mathcal{P}$ along 2 UGC-faces. In the process, (c) vertices to be deleted (shown in orange), (d) edges to be deleted (orange) and extended or reduced (turquoise), (e) faces to be deleted (orange) and extended or reduced (turquoise), (f) vertices added (magenta), (g) edges added (magenta) and extended or reduced (turquoise), and (h) faces extended or reduced (turquoise) are shown. In this case, none of the vertices have undergone a type change and no face has been added. The graphs in (i) and (j) are planar and their combination (k) is also planar. The vertices, edges, and faces in (i,j,k) are marked in accordance with those in (c–h). (Color figure online)

Dynamic Planar Graph Drawing

The planar graph drawing algorithm explained above can adapt itself to dynamic changes in the orthogonal polyhedron. Figure 15 illustrates a situation where an orthogonal polyhedron $\mathcal{P}$ is augmented with a UGC 2-adjacent to $\mathcal{P}$ along 2 UGC-faces. The resultant changes in the vertices, edges, and faces of $\mathcal{P}$ along with the corresponding changes in the planar graph are shown in Fig. 15(a-k). Another example in Fig. 16 demonstrates the step by step changes in the planar graph as the orthogonal polyhedron $\mathcal{P}$ is augmented with subsequent UGCs.

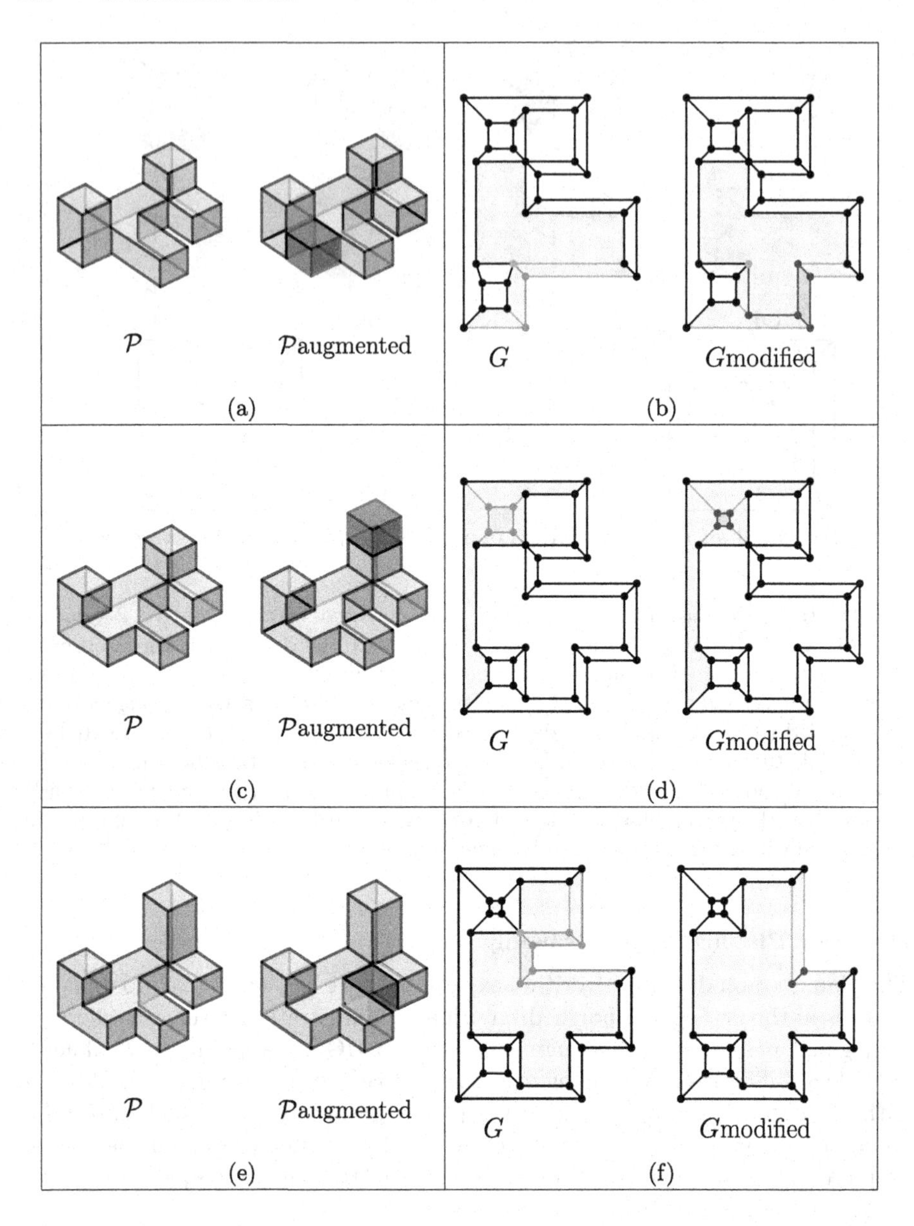

Fig. 16. (a,c,e)$\mathcal{P}$ is augmented with a new UGC step by step. (b,d,f)The corresponding changes in the planar graph. In each case, only a section of the graph is affected. The inserted (magenta), deleted (orange), and modified (turquoise) vertices, edges, and faces are marked. (Color figure online)

7 Conclusion

In this paper, we have described a planar graph-drawing algorithm for a certain class of orthogonal polyhedra based on digital-geometric properties. Many operations on such 3D objects can thus be envisaged as algorithms for planar graphs. The application areas for this work include computer vision, robot motion planning, and 3D integrated circuits. The planar embedding may be useful for the characterization of the polyhedron and subsequent treatise for purposes like reachability andor shortest traversal from a given face to another and it may also be useful for topological analysis.

References

1. Aguilera, A.: Orthogonal Polyhedra: Study and Application. Ph.D. thesis, Universitat Politécnica de Catalunya (1998)
2. Alexa, M.: Merging polyhedral shapes with scattered features. In: Proceedings of the International Conference on Shape Modeling and Applications, SMI 1999, p. 202 (1999)
3. Alexa, M.: Merging polyhedral shapes with scattered features. Vis. Comput. **16**(1), 26–37 (2000)
4. Batini, C., Nardelli, E., Talamo, M., Tamassia, R.: A Grap-theoretic approach to aesthetic layout of information systems diagrams. In: 10th International Workshop on Graph-theoretic Concepts in Computer Science, Trauner Verlag, Berlin, pp. 9–18 (1984)
5. Battista, G.D., Eades, P., Tamassia, R., Tollis, I.G.: Algorithms for drawing graphs: an annotated bibliography. Comput. Geom. **4**(5), 235–282 (1994)
6. Berg, M.D., Cheong, O., Kreveld, M.V., Overmars, M.: Computational Geometry-Algorithms and Applications, 3rd edn. Springer, Heidelberg (1997)
7. Biedl, T.C., Genc, B.: When can a graph form an orthogonal polyhedron? In: Canadian Conference On Computational Geometry, pp. 53–56 (2004)
8. Biedl, T., Genc, B.: Cauchy's theorem for orthogonal polyhedra of genus 0. In: Fiat, A., Sanders, P. (eds.) ESA 2009. LNCS, vol. 5757, pp. 71–82. Springer, Heidelberg (2009). https://doi.org/10.1007/978-3-642-04128-0_7
9. Cruz, I.F., Tamassia, R.: Graph Drawing Tutorial. http://cs.brown.edu/people/rtamassi/gd-tutorial.html
10. Deo, N.: Graph Theory with Application to Engineering and Computer Science. PHI Learning Private Limited, New Delhi (2009)
11. Duijvestijn, A.J.W.: The number of polyhedral (3-connected planar) graphs. Math. Comput. **65**, 1289–1293 (1996)
12. Eades, P.: A heuristic for graph drawing. Congr. Numer. **42**, 149–160 (1984)
13. Eades, P., Garvan, P.: Drawing stressed planar graphs in three dimensions. In: Brandenburg, F.J. (ed.) GD 1995. LNCS, vol. 1027, pp. 212–223. Springer, Heidelberg (1996). https://doi.org/10.1007/BFb0021805
14. Eppstein, D.: The topology of bendless three-dimensional orthogonal graph drawing. In: Tollis, I.G., Patrignani, M. (eds.) GD 2008. LNCS, vol. 5417, pp. 78–89. Springer, Heidelberg (2009). https://doi.org/10.1007/978-3-642-00219-9_9
15. Eppstein, D., Mumford, E.: Steinitz theorems for orthogonal polyhedra. In: Proceeedings 2010 Annual Symposium on Computational Geometry, SoCG 2010, ACM, New York, USA, pp. 429–438 (2010)

16. Genc, B.: Reconstruction of Orthogonal Polyhedra. Ph.D. thesis, University of Waterloo (2008)
17. Grünbaum, B.: Graphs of polyhedra. Polyhedra Graphs. Discrete Math. **307**(3–5), 445–463 (2007)
18. Henk, M., Richter-Gebert, J., Ziegler, G.M.: Basic properties of convex polytopes, second edn. In: Goodman, J.E., O'Rourke, J. (eds.) Handbook of Discrete and Computational Geometry, chap. 15, pp. 243–270. CRC Press LLC, Boca Raton, FL, USA (2004)
19. Hong, S.H., Nagamochi, H.: Extending Steinitz's theorem to upward star-shaped polyhedra and spherical polyhedra. Algorithmica **61**(4), 1022–1076 (2011)
20. Hopcroft, J., Tarjan, R.E.: Efficient planarity testing. J. ACM **21**(4), 549–568 (1974)
21. Karmakar, N., Biswas, A., Bhowmick, P., Bhattacharya, B.B.: Construction of 3D orthogonal cover of a digital object. In: Aggarwal, J.K., Barneva, R.P., Brimkov, V.E., Koroutchev, K.N., Korutcheva, E.R. (eds.) IWCIA 2011. LNCS, vol. 6636, pp. 70–83. Springer, Heidelberg (2011). https://doi.org/10.1007/978-3-642-21073-0_9
22. Karmakar, N., Biswas, A., Bhowmick, P., Bhattacharya, B.B.: A combinatorial algorithm to construct 3D isothetic covers. Int. J. Comput. Math. **90**(8), 1571–1606 (2013)
23. Klette, R., Rosenfeld, A.: Digital Geometry: Geometric Methods for Digital Picture Analysis. Morgan Kaufmann, San Francisco (2004)
24. Lempel, A., Even, S., Cederbaum, I.: An algorithm for planarity testing of graphs. In: International Symposium on Theory of Graphs, Gordon and Breach, New York, pp. 215–232 (1967)
25. Lipton, R., North, S., Sandberg, J.: A method for drawing graphs. In: ACM Symposium on Computational Geometry, pp. 153–160 (1985)
26. Orbanić, A., Boben, M., Jaklič, G., Pisanski, T.: Algorithms for drawing polyhedra from 3-connected planar graphs. Spec. Issue: Theor. Comput. Sci. Guest Editors: Boštjan Vilfan **28**, 239–243 (2004)
27. O'Rourke, J.: Unfolding orthogonal polyhedra. Contemp. Math. **453**, 307 (2008)
28. Richter-Gebert, J.: Realization Spaces of Polytopes. Lecture Notes in Mathematics, vol. 164. Springer-Verlag, Berlin (1996)
29. Schnyder, W.: Embedding planar graphs on the grid. In: Proceedings 1st Annual ACM-SIAM Symposium on Discrete Algorithms, SODA 1990, pp. 138–148 (1990)
30. Tamassia, R.: Planar orthogonal drawings of graphs. In: IEEE International Symposium on Circuits and Systems, vol. 1, pp. 319–322 (1990)
31. Tamassia, R., Battista, G.D., Batini, C.: Automatic graph drawing and readability of diagrams. IEEE Trans. Syst. Man Cybern. SMC-**18**(1), 61–79 (1988)
32. Tarjan, R.E.: Algorithm design. Commun. ACM **30**(3), 205–212 (1987)
33. Ziegler, G.M.: Convex Polytopes: Extremal Constructions and f -Vector Shapes, vol. 14. IAS/Park City Mathematics, Salt Lake City (2004)

Extractive Text Summarization Using Topological Features

Ankit Kumar and Apurba Sarkar(✉)

Department of Computer Science and Technology,
Indian Institute of Engineering Science and Technology, Howrah, Shibpur, India
as.besu@gmail.com

Abstract. The amount of text data generated these days is increasing exponentially, and it is becoming a very tedious process to extract meaningful information from the huge amounts of text data. In this work, we propose two methods to summarize the texts using topological features that capture the information over the topological structure, such as connected components and holes in the text data. The first method uses the concept of minimum dominating set to group the sentences into multiple clusters and to find the similarities between clusters using topological features (connected components and tunnels). Sentence scoring and extraction of key sentences from each cluster are done by the existing method of TextRank. The second method uses the pretrained *GloVe* (global vectors to represent the words) and to find the similarities between sentences using topological features. A classical set cover based algorithm has been used to extract the key sentences for the summary. Both methods are compared on the basis of rouge scores with the existing method, i.e., TextRank, and the results are satisfactory.

Keywords: Extractive text summarization · Minimum dominating set · Persistent homology · Topological features

1 Introduction

Text summarization reduces the text of a document to provide concise representation retaining the semantic of the entire document. It is categorized into Extractive and Abstractive summarizations [8]. In extractive summarization, the summary is created by using the words, sentences, or keyphrases of the original document. It includes only those sentences or words which are in the source documents and do not include new sentences or words. Whereas in abstractive summarization, summary represents the idea behind the source document by including sentences or words which may or may not be present in the source document. The paper presents an extractive summarization approach using topological features.

Various document summarization methods have been proposed based on different graph features such as betweenness centrality, degree centrality, etc. However, these measures do not consider the structure of data, therefore they may

R. P. Barneva et al. (Eds.): IWCIA 2022, LNCS 13348, pp. 105–121, 2023.
https://doi.org/10.1007/978-3-031-23612-9_7

lose the information over topological structure such as the connected components and holes in data. The proposed methods extract these information from the topological structure of text data.

It has been observed that topological data analysis is used in different scenarios like, Almgren et al. proposed a method [2] for mining social media data to cluster the images based on their popularity using topological data analysis and it outperforms the traditional K-means, hierarchical clustering algorithms. Chuan-Shen et al. proposed a method [6] for video summarization using topological data analysis that generates video summaries without any training procedure. Hui-Guan et al. proposed a method [4] Docollapse, a topological collapse-based unsupervised method for document summarization. It outperforms the graph-based keyphrase extraction methods such as TF-IDF, TextRank, and TopicRank. Gholizadeh et al. [3] proposed a method that uses persistent homology to predict the author based on the graph of the main character extracted from the novel. Similarly, Proposed methods extract the topological features from the text to summarize the document using the above-mentioned features thus obtained.

2 Background

To give a brief overview of topology. It is the study of the shape of objects that does not change on any transformation except gluing and tearing, because gluing and tearing change the neighborhood relation. Topology tries to describe an object based on connected components or cavities (Fig. 1).

2.1 Simplicial Complex

Definition 1. *A p-simplex [12] σ is the convex hull of p+1 affinely independent points $\{x_0, x_1, \ldots, x_p\} \in \mathbb{R}^d$. We denote $\sigma = conv\{x_0, x_1, \ldots, x_p\}$ and the $dim\{\sigma\}$ is p.*

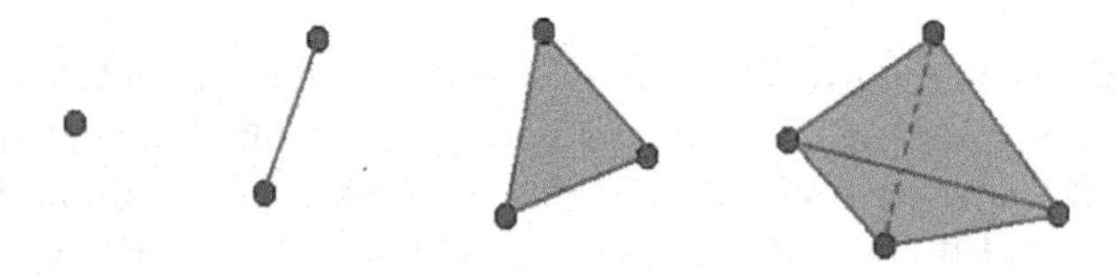

Fig. 1. *0-d* simplex: Vertex, *1-d* simplex: edge, *2-d* simplex: triangle, *3-d* simplex: tetrahedron.

Definition 2. *A simplicial complex [12] K is a finite set of simplices such that $\sigma \in K$ and τ being a face of σ implies $\tau \in K$, and $\sigma, \sigma' \in K$ implies $\sigma \cap \sigma'$ is either empty or a face of both σ and σ'.*

The above definition implies that if a simplex (σ) is in simplicial complex (K), all its faces need to be in K, too and the simplices have to be glued together along with whole faces or be separate. The proposed methods use Vietoris-Rips complex for sentences or words.

A Vietoris-Rips complex [12] of diameter ϵ is the simplicial complex

$$VR(\epsilon) = \{\sigma | dia(\sigma) \leq \epsilon\}$$

where, dia(σ) is the largest distance between two points in σ. For a way to form a Vietoris-Rips complex, some threshold value needs to be defined. As threshold value increases, the lower dimensional simplex gets merged to form higher-dimensional simplex and a Vietoris-Rips complex is formed. In the simplicial complexes a tool, called Persistent Homology can be used for the analysis of connected components, holes, and voids.

2.2 Persistent Homology

Persistent homology is a mathematical topological data analysis tool which has the advantage to capture the structural features of documents [12]. It identifies the holes in each dimension known as betti numbers [3], as well as it captures at what value of ϵ these holes appear and how long they persist. Betti numbers in each dimension are described as β_0 (*0-d* connected components), β_1 (*1-d* holes), β_2 (*2-d* void) and β_k (*k-d* holes) in simplicial complex. The lifespan (birth, death) of each homological feature is called persistence interval at different values of ϵ. These persistence intervals can be easily visualized using the persistence diagram and persistence barcode.

Persistence diagram or persistence barcode are used for the comparison between simplicial complexes. The proposed methods use the Gudhi [5] python library to create persistence diagrams for VR-complexes. Two persistence diagrams P_i and P_j can be compared to obtain the similarity between simplicial complexes K_i and K_j respectively where, $i, j \in [n]$ and $[n]$ is the set of number of simplicial complexes. Such comparison can be done by measuring the distance between matching points in the persistence diagram P_i and P_j. There are two distance metrics: Bottleneck distance and q-Wasserstein distance defined to measure the similarity between the persistence diagram P_i and P_j.

Definition 3. *Let P and Q be two persistence diagrams. The bottleneck distance [1] between P and Q is defined as*

$$d_B(P, Q) = \inf_{\gamma} \sup_{x \in P} \parallel x - \gamma(x) \parallel_\infty$$

where, γ ranges over all matchings from P to Q and $\parallel p - q \parallel_\infty = max(|p_1 - q_1|, |p_2 - q_2|)$ for $p = (p_1, p_2), q = (q_1, q_2) \in \bar{\mathbb{R}}^2$ with $|\infty - \infty| = 0$.

Basically, bottleneck distance b is the shortest maximum distance between two points in matching from P to Q for which any matched point is at maximum distance b. While q-Wasserstein distance is calculated as the total distance between all matched pairs of points.

Definition 4. *Let P and Q be two persistence diagrams. The q-Wasserstein distance [1] between P and Q is defined as*

$$d_{W_q}(P,Q) = \inf_{\gamma} \left(\sum_{x \in P} \| x - \gamma(x) \|_q \right)^{\frac{1}{q}}$$

where, γ ranges over all matchings from P to Q and $\| p - q \|_q = (|p_1 - q_1|^q + |p_2 - q_2|^q)^{\frac{1}{q}}$ for $p = (p_1, p_2), q = (q_1, q_2) \in \bar{\mathbb{R}}^2$ with $|\infty - \infty| = 0$.

2.3 Minimum Dominating Set

Definition 5. *Let $G(V, E)$ a graph, the dominating set of G is a subset S of vertices V with the properties that each $v \in V$ is either in S or $adj(v^\star)$. $adj(v^\star)$ implies adjacent to $v^\star$ where, $v^\star \in S$.*

Minimum Dominating Set is the dominating set with a minimum number of vertices of G. Shen et al. [11] used minimum dominating set properties for multi-document text summarization, where each vertex v_i is the representative of sentence S_i in document. An edge $e_{i,j}$ gets added on the basis of cosine similarity between v_i and v_j. The vertices in the minimum dominating set represent that they cover all the information about the adjacent vertices to it. Finding the minimum dominating set is a NP-hard problem, so it uses an approximation algorithm that uses a greedy approach.

2.4 Cosine Similarity

Cosine similarity $CS(A, B)$ used to find the similarity between two vectors A and B that is calculated as

$$CS(A,B) = \frac{\sum_{i=1}^{n} A_i B_i}{\sqrt{\sum_{i=1}^{n} {A_i}^2} \sqrt{\sum_{i=1}^{n} {B_i}^2}} \tag{1}$$

Rest of the paper is organized as follows. Section 3 presents two methodologies for text summerization where algorithms for respective methods are also given. Section 4 presents experimental results with comparison of both methods with TextRank based on Rouge score. Finally, the paper concludes in Sect. 5.

3 Proposed Methodology

We propose two methods for extractive summarization with the help of topological features. As an initial step, the input document needs to be preprocessed to remove stop words, extra white space, and newline characters. After that, proposed methods are applied to the preprocessed text as described below.

3.1 Proposed Method (I)

This method uses the minimum dominating set to cluster the preprocessed text and persistence interval to find the similarity between clusters. Initially, we need to tokenize the preprocessed text into sentences and construct the similarity graph $G(V, E)$. In the graph, V is the set of vertices where each $v \in V$ corresponds to a sentence represented by a TF-IDF vector. E is the set of edges where edge $e(i, j)$ gets added if the cosine similarity between the vertices v_i and v_j is greater than T_{th}. The threshold, T_{th} is defined as mean of the cosine similarities between all the vertices. Isolated vertices in the graph denote that they do not contain relevant information related to the document.

The Algorithm 1 is used for the sentence clustering which, takes a similarity graph $G(V, E)$ as input and returns a dictionary in the form of key, value pair. A key of the dictionary is the $v^\star \in M$ and values are $v \in adj(v^\star) - \{M \cup Adj\}$ that avoids the selection of a single vertex into multiple pairs. Where, $adj(v^\star)$ is a set of adjacent vertices to $v^\star$ and M is the set of dominating vertices. Consider the sentences of each key, values pair into a cluster. In this, the vertex $v^\star$ which has $max\{|adj(v^\star)|\}$ gets added first and so on. So Algorithm 1 groups the sentences into ones that have more similarities as compared to the sentences of other clusters.

Algorithm 1. Algorithm for sentence clustering

1: **procedure** SENTENCE_CLUSTERING($G(V, E)$)
2: $M \leftarrow \phi$
3: $Adj \leftarrow \phi$
4: $dict = \{\}$
5: **while** $|V| \geq (|M| + |Adj|)$ **do**
6: **for** $v \in (V - M)$ **do**
7: $s(v) = |adj(v) - \{M \cup Adj\}|$
8: $v^\star \leftarrow arg\ max_v s(v)$
9: $M \leftarrow M \cup \{v^\star\}$
10: $dict[v^\star] = adj(v^\star) - \{M \cup Adj\}$
11: $Adj \leftarrow Adj \cup adj(v^\star)$
12: **return** $dict$, M

The sentences of the document have been divided into multiple clusters $[C]$ where, $|[C]|$ euqal to $|M|$. A $VR(\epsilon)$ complex has been constructed for each cluster c_i where, $c_i \in [C]$. The construction of the $VR(\epsilon)$ complex needs a pairwise distance matrix, so the proposed method uses the cosine distance matrix. The cosine distance matrix has been calculated between each pair of the sentences in a cluster. For the proposed method, $\epsilon = 1$ has been taken because cosine distance has the range [0,1]. Hence, the $VR(1)$ complex captures the homological features till the maximum value of ϵ. For the proposed method, *2*-simplex is considered as maximum dimensional simplex. Next, persistence diagrams P for each cluster are drawn using the Gudhi [5] package in python. From the persistence diagram,

0-dimensional persistence interval (I^0) and *1*-dimensional persistence interval (I^1) have been extracted for each cluster.

Two VR-complex $VR_i(\epsilon)$ & $VR_j(\epsilon)$ have the same persistence interval shows that $VR_i(\epsilon)$ & $VR_j(\epsilon)$ are topologically equivalent due to the number of connected components, *1*-d holes generated and collapsed at same value of ϵ. So, the both $VR_i(\epsilon)$ & $VR_j(\epsilon)$ have been merged into one, that results in only unique VR-complexes with distinct persistence intervals. The Algorithm 2 is used for merging the complexes. It takes C (contain all clusters c_i) as input and results in clusters $C^\star$ (contain left clusters) that form VR-complex with distinct persistence intervals.

Algorithm 2. Algorithm for merging the cluster

```
1: procedure CLUSTER_COLLAPSING(C)
2:     D ← Array(|C|,|C|)
3:     for c_i ∈ C do
4:         Calculate cosine distance for c_i
5:         Build VR-complex VR_i(1)
6:         Extract I_i^0 and I_i^1
7:     for i ∈ [|C|] do
8:         for j ∈ [|C|] do
9:             D[i][j] = Dis(VR_i(1), VR_j(1))
10:    thr ← mean.D
11:    min ← arg min.D
12:    i,j ← Index(min.D)
13:    if (min ≤ thr) then
14:        merge(c_i, c_j)
15:        goto Step(2)
16:    else
17:        C* ← C
18:        return C*
```

Let persistence diagram P_i has been drawn and found from the persistence interval of I_i^0 and I_i^1 for each $VR_i(\epsilon)$. A $VR_i(\epsilon)$ is the representative of each cluster c_i. The similarity between $VR_i(\epsilon)$ and $VR_j(\epsilon)$, $\{i, j \in [|M|]\}$ is defined by the distance calculated using Eq. 2. Where d_{W_1} is the q-Wasserstein distance at q=1 as defined in Defnition 4 . If q-Wasserstein distance between $VR_i(\epsilon)$ and $VR_j(\epsilon)$ is less than *thr* then, merge the cluster c_i and c_j. The threshold, *thr* is defined as the mean for all Dis($VR_i(\epsilon)$,$VR_j(\epsilon)$) where, $i, j \in [|M|]$.

$$Dis(VR_i(\epsilon), VR_j(\epsilon)) = \{d_{W_1}(I_i^0, I_j^0) + d_{W_1}(I_i^1, I_j^1)\} \tag{2}$$

After the merging, some $C^\star$ clusters remain. These clusters contain the sentences of the document. Now, the sentences of each cluster $c_i \in C^\star$ are scored using traditional TextRank [9] for weighted graphs as,

$$S(v_i) = (1-d) + d * \sum_{v_j \in adj(v_i)} \frac{w_{j,i}}{\sum_{v_k \in adj(v_j)} w_{j,k}} S(v_j) \tag{3}$$

where $S(v_i)$ and $S(v_j)$ are the TextRank score of sentence s_i and s_j represented as node v_i and v_j such that $v_j \in adj(v_i)$. The weight $w_{j,i}$ on the edge $e_{j,i}$ is the cosine similarity between v_j and v_i. The damping factor d is set to 0.85 same as Mihalcea et al. [9]. Next, sentences are sorted on the basis of their scores in descending order, and the n top ranked sentences are extracted from each cluster c_i. After that, the indexes of extracted n sentences from each c_i are rearranged.

3.2 Proposed Method (II)

In the previous proposed method, VR-complexes were constructed for each cluster, but now VR-complexes are constructed for each sentence of the document with the maximum *2*-dimensional simplices. The words of the documents are represented by pretrained *GloVe* (Global vector) [10]. Cosine distance has been calculated between the words of a sentence to generate the VR-complex for each sentence.

The proposed method calculates the bottleneck distance between persistence diagrams as defined in Defnition 3 to find the similarity between them. We know that persistence diagrams are drawn for each VR-complex, which is representative of the sentences. So basically, the proposed method tried to find the similarity between sentences using the persistence diagram and the bottleneck distance. The sum of the distance from P_i to P_j, where j = *0,1,2,....n* and $j \neq i$ describes the score of the sentence S_i. The Algorithm 3 describes the process of selecting the sentences for the summary that is based on the classical set cover algorithm.

Algorithm 3. Sentence Cover Algorithm

```
1: procedure SENTENCE_COVER(S,B,n,Score,K)
2:     Repr_Sent ← φ
3:     while K > |Repr_sent| do
4:         W ← Score.keys
5:         for i ∈ min.Score_i do
6:             if K > |Repr_Sent| then
7:                 if i ∈ W then
8:                     Repr_Sent ← Repr_Sent ∪ {S_i}
9:                     Remove Score_i
10:                    T_r = (1/n) Σ^n_{j∈0} B_ij
11:                    for j ∈ 0 to n do
12:                        if B_ij < T_r then
13:                            W ← W − {j}
14:            else
15:                Break
16:    return Repr_Sent
```

In the algorithm, *Repr_Sent* is an empty set initially, which will later contain the indexes of the sentences that will appear in the summary. K is the number of sentences to be extracted for the summary. B describes the bottleneck distance among the persistence diagrams. S are the sentences, and n describes the total number of sentences in the document. The algorithm works in pass (similar to bubble sort) until the given number of sentences *(K)* for the summary are extracted. It starts with P_i having the least score and for every P_i, the algorithm discards P_j having the bottleneck distance less than the threshold $T_r = \frac{1}{n}\sum_{j\in 0}^{n} B_{ij}$. The sentence S_i representing P_i is included in *Repr_Sent*. So Algorithm 3 tries to cover the more dissimilar information from the input documents.

4 Experimental Results

The performance of proposed methods is evaluated on 100 documents of the DUC-2002 dataset using the ROUGE (Recall-Oriented Understudy for Gisting Evaluation) [7] score. Specifically, ROUGE-1, ROUGE-2, and ROUGE-L are considered for performance metrics. ROUGE-N is the N-gram overlapping between system-generated summaries and reference summaries, computed as

$$Rouge\_N = \frac{\sum_{s\in ref\_summary}\sum_{gram_n\in s} C_{match}(gram_n)}{\sum_{s\in ref\_summary}\sum_{gram_n\in s} C(gram_n)} \tag{4}$$

where n stands for the length of the n-gram. The $C(gram_n)$ and $C_{match}(gram_n)$ are the maximum number of n-gram co-occurring in the system-generated summary and set of reference summaries, respectively. ROUGH-L refers to the longest common sub-sequence between the system-generated summaries and reference summaries. LCS based f-measures to estimate the similarity between two summaries X of length m and Y of length n, are computed using Eqs. 5, 6, and 7. Let X be a reference summary sentence and Y be a system-generated summary sentence.

$$R_{lcs} = \frac{LCS(X,Y)}{m} \tag{5}$$

$$P_{lcs} = \frac{LCS(X,Y)}{n} \tag{6}$$

$$F_{lcs} = \frac{(1+\beta^2)R_{lcs}P_{lcs}}{R_{lcs}+\beta^2 P_{lcs}} \tag{7}$$

where $LCS(X,Y)$ denotes the length of the longest common subsequence of X and Y, and β is equal to P_{lcs}/R_{lcs}. The Rouge score of both proposed methods has been compared with TextRank and the results are mentioned in further sections.

4.1 Proposed Methodology (I) and TextRank

In method 3.1, 10% of sentences have been extracted from each cluster $c_i \in C^\star$ and rearranged based on their index for the summary. For TextRank, 10% of sentences were extracted from the entire document. To compute the rouge scores, the summary generated by Method 3.1 is compared with the summary generated by TextRank. The evaluated results are shown in Table 1 and Figs. 2, 3, and 4.

Table 1. ROUGE Score of 10% summary extracted from the proposed method (I) and TextRank.

	Proposed method (I)			TextRank		
	F-Score	Precision	Recall	F-Score	Precision	Recall
Rough-1	0.10619 to 0.56818	0.15860 to 0.82222	0.05941 to 0.56731	0.14359 to 0.57292	0.14583 to 0.84000	0.10784 to 0.57000
Rouge-2	0.00000 to 0.38095	0.00000 to 0.63636	0.00000 to 0.32673	0.00000 to 0.38411	0.00000 to 0.63636	0.00000 to 0.30693
Rouge-L	0.10526 to 0.58333	0.13122 to 0.75676	0.06024 to 0.50649	0.12214 to 0.52893	0.11739 to 0.75758	0.09722 to 0.46753

In 38 documents out of 100 documents, the proposed method gives better Rough-1 F-Score. In 36 documents out of 100 documensts, the proposed method gives better Rough-2 F-Score. In 40 documents out of 100 documents, the proposed method gives better Rough-L F-Score, as compared to TextRank.

4.2 Proposed Methodology (II) and TextRank

In Method 3.2, 10% of the sentences have been extracted and rearranged based on their index for the summary. Similarly, for TextRank, 10% of sentences have been taken from the same document. To compute the rouge scores, the summary generated by Method 3.2 is compared with the summary generated by TextRank. The evaluated results are shown in Table 2 and Figs. 5 ,6, and 7.

Table 2. ROUGE Score of 10% summary extracted from the proposed method (II) and TextRank.

	Proposed method (II)			TextRank		
	F-Score	Precision	Recall	F-Score	Precision	Recall
Rough-1	0.15254 to 0.60714	0.13183 to 0.84000	0.08108 to 0.52703	0.14359 to 0.57292	0.14583 to 0.84000	0.10784 to 0.57000
Rouge-2	0.00000 to 0.48193	0.00000 to 0.70455	0.00000 to 0.40404	0.00000 to 0.38411	0.00000 to 0.63636	0.00000 to 0.30693
Rouge-L	0.10769 to 0.60937	0.09722 to 0.82927	0.08108 to 0.52703	0.12214 to 0.52893	0.11739 to 0.75758	0.09722 to 0.46753

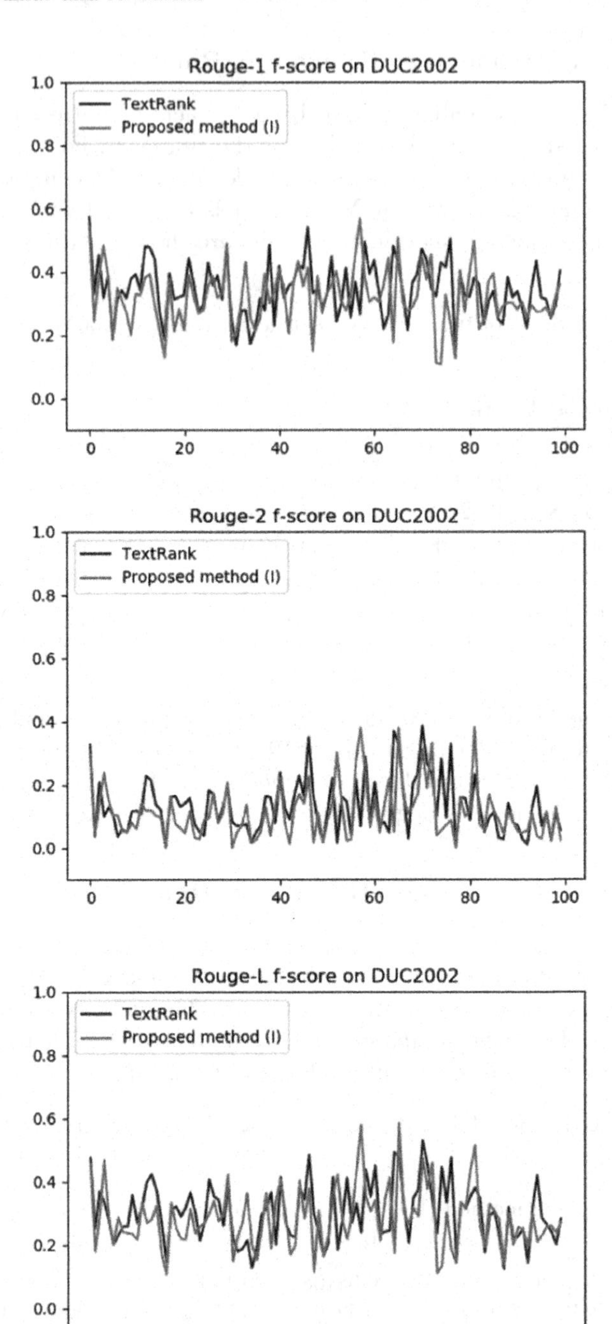

Fig. 2. F-Score comparison between Method 3.1 and TextRank w.r.t. i^{th} document. Where X-axis represents the ith document out of 100 and the Y-axis represents the respective Rough-1 F-Score, Rough-2 F-Score, and Rough-L F-Score.

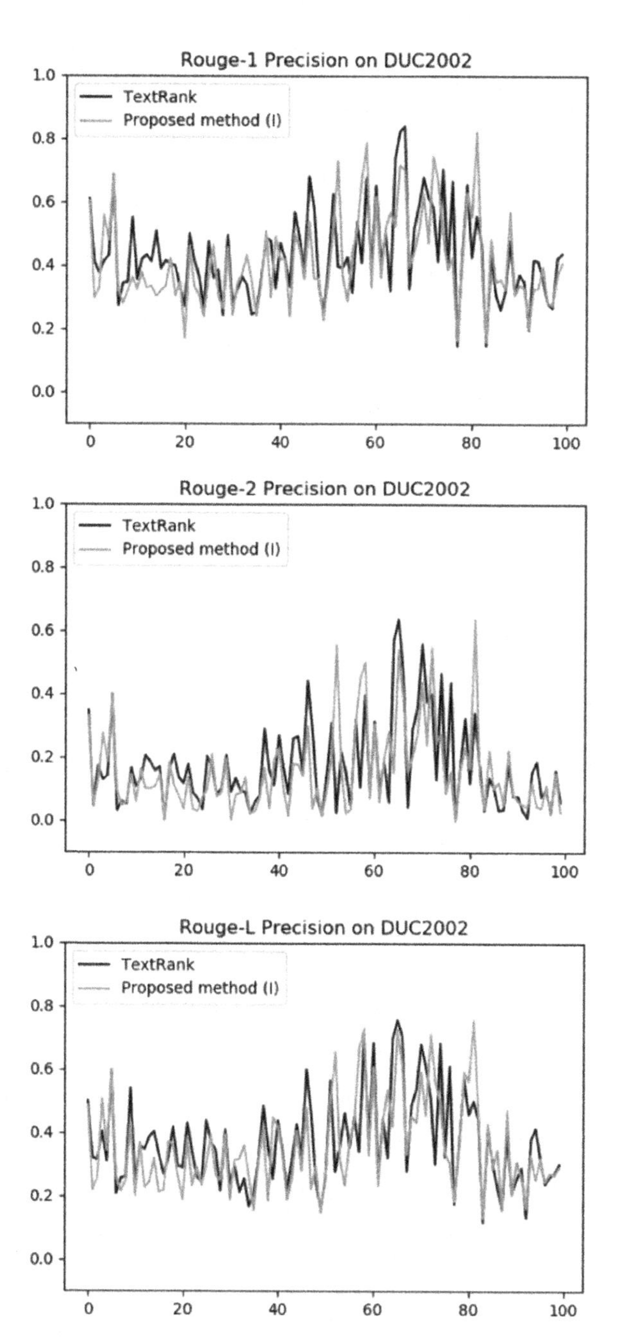

Fig. 3. Precision comparison between Method 3.1 and TextRank w.r.t. i^{th} document. Where X-axis represents the ith document out of 100 and the Y-axis represents the respective Rough-1, Rough-2, and Rough-L Precision.

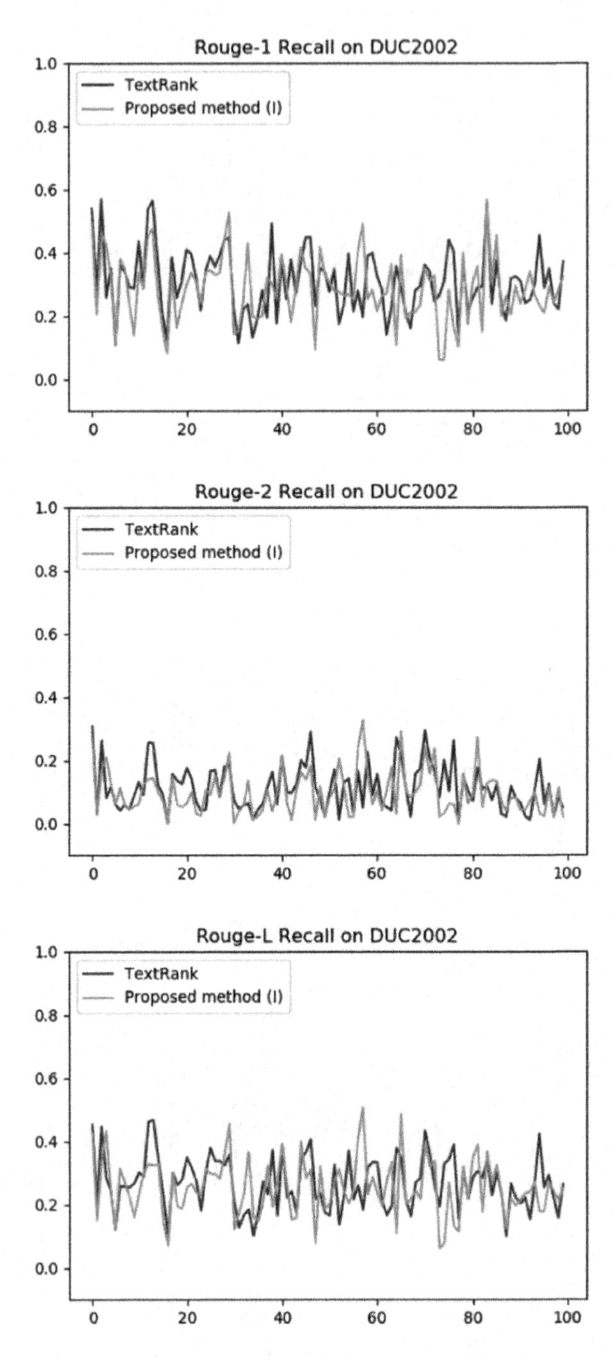

Fig. 4. Recall comparison between Method 3.1 and TextRank w.r.t. i^{th} document. Where X-axis represents the ith document out of 100 and the Y-axis represents the respective Rough-1, Rough-2, and Rough-L Recall.

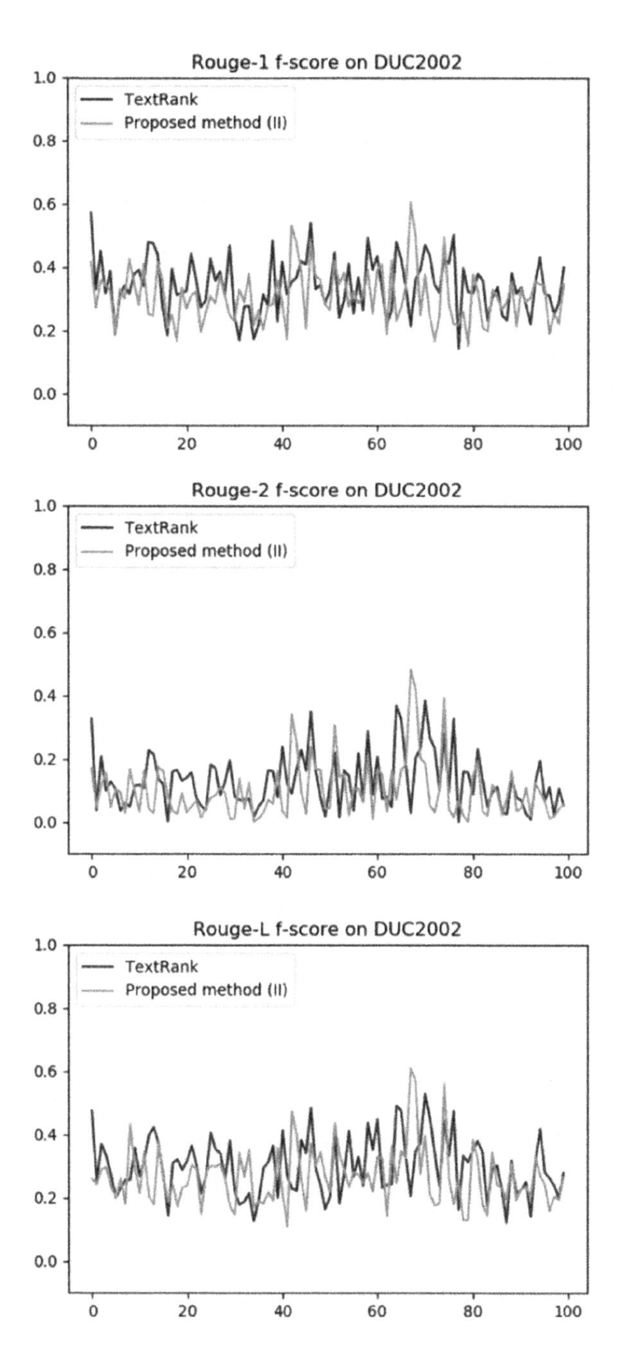

Fig. 5. *F*-Score comparison between Method 3.2 and TextRank w.r.t. i^{th} document. Where *X*-axis represents the ith document out of 100 and the *Y*-axis represents the respective Rough-1 *F*-Score, Rough-2 *F*-Score, and Rough-L *F*-Score.

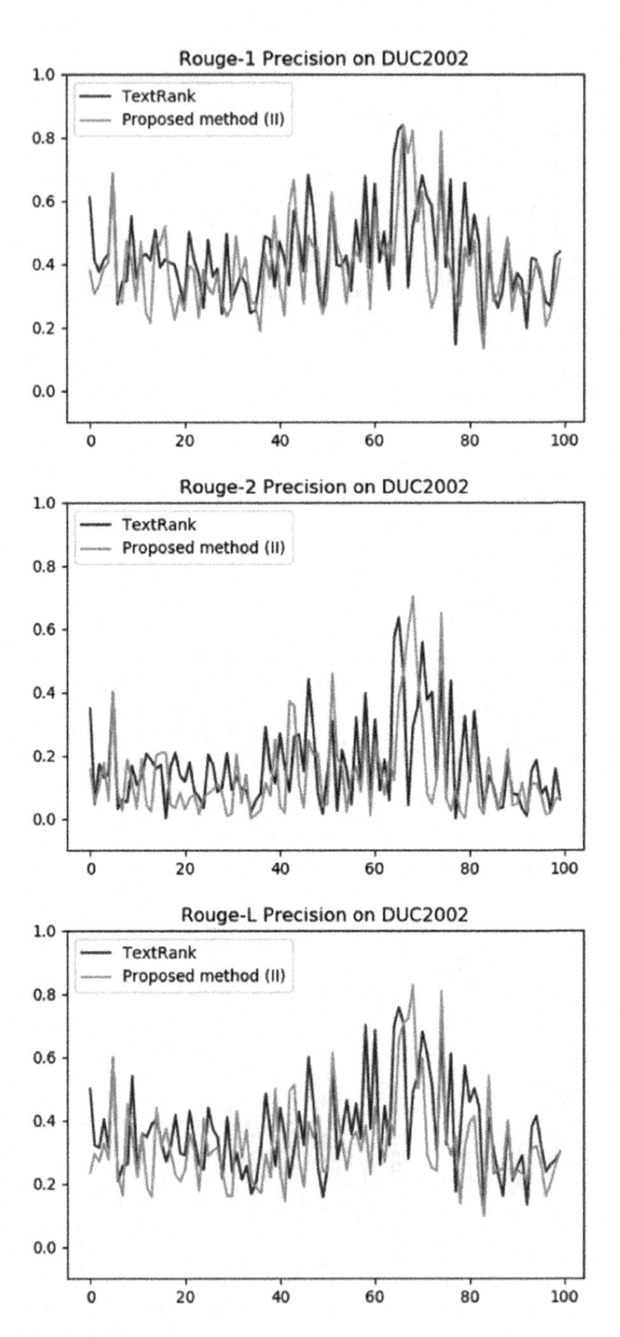

Fig. 6. Precision comparison between Method 3.2 and TextRank w.r.t. i^{th} document. Where X-axis represents the ith document out of 100 and the Y-axis represents the respective Rough-1, Rough-2, and Rough-L Precision.

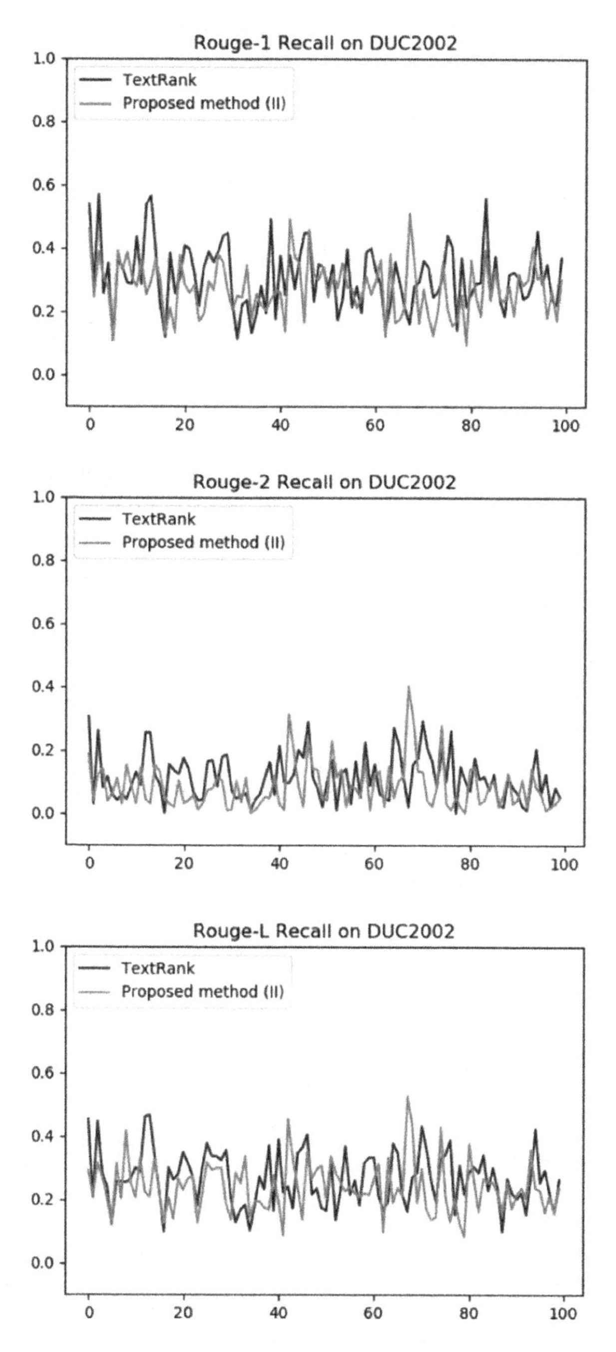

Fig. 7. Recall comparison between Method 3.2 and TextRank w.r.t. i^{th} document. Where X-axis represents the ith document out of 100 and the Y-axis represents the respective Rough-1, Rough-2, and Rough-L Recall.

In 39 documents out of 100 documents, the proposed method gives better Rough-1 F-Score. In 36 documents out of 100 documensts, the proposed method gives better Rough-2 F-Score. In 38 documents out of 100 documents, the proposed method gives better Rough-L F-Score, as compared to TextRank.

5 Conclusion

In this paper, we proposed two methods for extractive text summarization using topological features. The unique characteristics of topology are the n-dimensional holes that can help to distinguish objects. We tried to distinguish the clusters and sentences based on the *0-d* and *1-d* persistence intervals at which these holes appear and disappear. Moreover, these n-dimensional holes in data can be used as important features for network analysis and classification. Method 3.1 uses Algorithm 1 and the minimum dominating set for clustering the sentences and Algorithm 2 for merging the simplicial complexes into one, based on their topological features. We used a *1*-Wasserstein distance to find the similarity between VR-complexes, and TextRank is used for scoring the sentences in each cluster. In Method 3.2, VR-Complexes are formed for each sentence, and sentences are scored based on the bottleneck distance. Algorithm 3 can be used for selection of sentences in other Extractive Text Summarization methods also.

References

1. Aktas, M.E., Akbas, E., Fatmaoui, A.E.: Persistence homology of networks: methods and applications. Appl. Netw. Sci. **4**(1), 1–28 (2019). https://doi.org/10.1007/s41109-019-0179-3
2. Almgren, K., Kim, M., Lee, J.: Mining social media data using topological data analysis. In: 2017 IEEE International Conference on Information Reuse and Integration (IRI), pp. 144–153. IEEE (2017)
3. Gholizadeh, S., Seyeditabari, A., Zadrozny, W.: Topological signature of 19th century novelists: persistent homology in text mining. Big Data Cognit. Comput. **2**(4), 33 (2018)
4. Guan, H., Tang, W., Krim, H., Keiser, J., Rindos, A., Sazdanovic, R.: A topological collapse for document summarization. In: 2016 IEEE 17th International Workshop on Signal Processing Advances in Wireless Communications (SPAWC), pp. 1–5. IEEE (2016)
5. Gudhi python documentation. https://gudhi.inria.fr/python/latest/
6. Hu, C.S., Yeh, M.C.: A topological data analysis approach to video summarization. In: 2019 IEEE International Conference on Image Processing (ICIP), pp. 1815–1819. IEEE (2019)
7. Lin, C.Y.: Rouge: a package for automatic evaluation of summaries. In: Text summarization Branches Out, pp. 74–81 (2004)
8. Madhuri, J.N., Kumar, R.G.: Extractive text summarization using sentence ranking. In: The 2019 International Conference on Data Science and Communication (IconDSC), pp. 1–3. IEEE (2019)
9. Mihalcea, R., Tarau, P.: Textrank: bringing order into text. In: Proceedings of the 2004 Conference on Empirical Methods in Natural Language Processing, pp. 404–411 (2004)

10. Pretrained GloVe. https://nlp.stanford.edu/projects/glove/
11. Shen, C., Li, T.: Multi-document summarization via the minimum dominating set. In: Proceedings of the 23rd International Conference on Computational Linguistics (Coling 2010), pp. 984–992 (2010)
12. Zhu, X.: Persistent homology: an introduction and a new text representation for natural language processing. In: IJCAI, pp. 1953–1959 (2013)

Largest Area Parallelogram Inside a Digital Object in a Triangular Grid

Md Abdul Aziz Al Aman[1], Raina Paul[1], Apurba Sarkar[1(✉)], and Arindam Biswas[1,2]

[1] Department of Computer Science and Technology, Indian Institute of Engineering Science and Technology, Botanical Garden Road, Howrah 711103, West Bengal, India
as.besu@gmail.com

[2] Department of Information Technology, Indian Institute of Engineering Science and Technology, Botanical Garden Road, Howrah 711103, West Bengal, India

Abstract. A combinatorial algorithm to construct Largest Area Parallelogram (*LAPT*) inside a digital object lying on a Triangular grid is proposed in this work. An inner triangular cover (ITC) is first constructed, where the sides of ITC lies on the grid line and within the object. After the ITC is constructed, the proposed algorithm maintain few lists and a set of rules to find the *LAPT*. It is observed that the algorithm runs in $O(k \cdot \frac{n}{g} \lg \frac{n}{g})$ time where n number pixel on the boundary of the digital object, g is grid size, and k is the number of convexities.

Keywords: Triangular grid · Triangular cover · Largest parallelogram · Largest rectangle

1 Introduction

A number of work has been done to compute different geometric figures of maximum area inscribed inside a polygon. The application of this type of work can be found in internal approximation of polygon, textile industry [1,2], metal sheet cutting with minimum wastage etc. Apart from the industrial application these works are also of theoretical interest in computational geometry. To give a brief overview of the works done so far, Mackenn et al. [3] proposed an algorithm of finding maximum empty rectangle inside an orthogonal polygon in the 1985. There are other similar kind of work after that. To mention a few of them, algorithm to find largest equilateral triangles and squares was proposed by DePano et al. [4] that runs in $O(n^2)$ time. An algorithm to find largest box inside a convex polygon proposed by N. Amenta [5]. S. Fekete [6] proposed an algorithm to find the inscribed squares in a convex polygon with one vertex of the square lying on one vertex of polygon in $O(n \log^2 n)$ time. Alt et al. [7] and Daniels et al. [8] proposed an algorithm to find largest empty axis parallel rectangle inside a polygon. Finding largest empty rectangles from a point set was proposed by Chaudhuri et al. [9]. Ahn et al. [10] gave an approximation algorithm

R. P. Barneva et al. (Eds.): IWCIA 2022, LNCS 13348, pp. 122–135, 2023.
https://doi.org/10.1007/978-3-031-23612-9_8

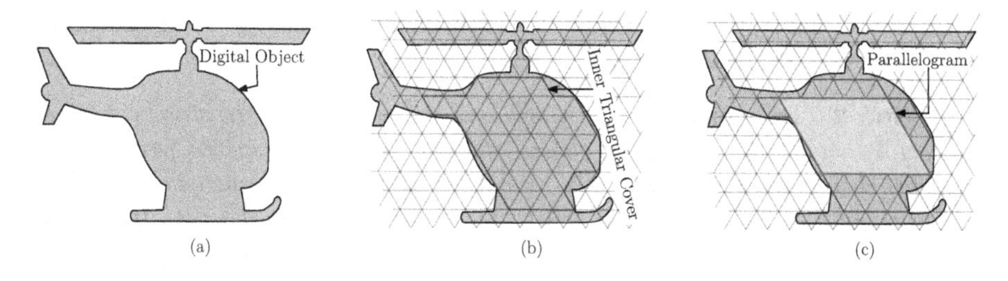

Fig. 1. (a) Object (b) Object with ITC (c) Largest Area Parallelogram

to find axially symmetrical polygon inscribed inside convex sets. Optimization problem to find the longest line segment (stick) or maximum area triangle or convex body (potato) was given by Hall et al. [11]. Largest inscribed rectangle inside a given polygon with n vertices is proposed by Knauer et al. [12]. They have provided an algorithm, assuming that the vertices are provided in order. Apurba et al. [13,14] proposed a combinatorial algorithm to find largest rectangle (LR) inside a given digital object, which runs in $O(n \log n)$ time. To obtain the largest rectangle inside the inner triangular cover of the digital object, they first construct a histogram polygon based on the convex edge and then within the histogram polygon, the LR is determined. However, there are no reported work to find $LAPT$ within a digital object in the literature.

A combinatorial algorithm has been proposed to find the largest area parallelogram in triangular grid which we call $LAPT$ inside a digital object. The digital object and inner triangular cover are shown in Fig. 1 (a), (b) respectively and the largest area parallelogram for the object is shown in Fig. 1(c).

The rest of the paper is organized as follows. All the required definitions and preliminaries are presented in Sect. 2. The procedure to compute sub-polygon and finding parallelogram within sub-polygon are explained in Sect. 3 and 3.1 respectively. The reduction rules are explained in Sect. 3.2. The proposed algorithm with explanation and time complexity are given in Sect. 4 and Sect. 4.1 respectively. Section 5 contains some output of the proposed algorithm. Finally, Sect. 6, concludes with future work.

2 Definitions and Preliminaries

This section defines few terminologies those are required to explain the proposed algorithm.

Definition 1 (Digital object): *A digital object, Q, is a finite subset of $\mathbb{Z}^2$ consisting of one or more $k(= 4$ or $8)$-connected components [15].*

A discrete triangular coordinate system [16] (as shown in Fig. 2) consists of three sets of lines, namely, $\mathbb{L}_0$, $\mathbb{L}_{60}$, and $\mathbb{L}_{120}$, where θ is the set of parallel lines inclined at an angle θ with respect to the X-axis, separated by unit distance, measured along a line inclined at $(\theta + 60)$ or $(\theta + 120)$. Note that the unit distance is equal to the side of the equilateral triangles constituting the discrete triangular plane in the discrete coordinate system.

Definition 2 (Triangular Grid): *A triangular grid (henceforth simply referred as grid)* $\mathbb{T} := (\mathbb{L}_0, \mathbb{L}_{60}, \mathbb{L}_{120})$ *consists of three sets of equispaced parallel grid lines, which are inclined at* $0°$, $60°$, *and* $120°$ *w.r.t. x-axis, such that one line from each of* $\mathbb{L}_0$, $\mathbb{L}_{60}$, *and* $\mathbb{L}_{120}$ *meet at the origin of the discrete triangular coordinate system.*

The grid lines in $\mathbb{L}_0$, $\mathbb{L}_{60}$, and $\mathbb{L}_{120}$ correspond to three distinct coordinate axes, namely α, β, and γ. A grid point is the intersection between three grid lines $\mathbb{L}_0$, $\mathbb{L}_{60}$, and $\mathbb{L}_{120}$. A grid size, g is the distance between two consecutive grid points along a grid line. A grid edge is the grid line segment of length g and connecting two consecutive grid points. An unit grid triangle (UGT) is the smallest-area triangle formed by three grid edges (marked by grey in Fig. 2). A grid point, p consists of six neighboring UGT's, given by $\{T_i\colon i = 0, 1, 2, ...5\}$ as shown in Fig. 2.

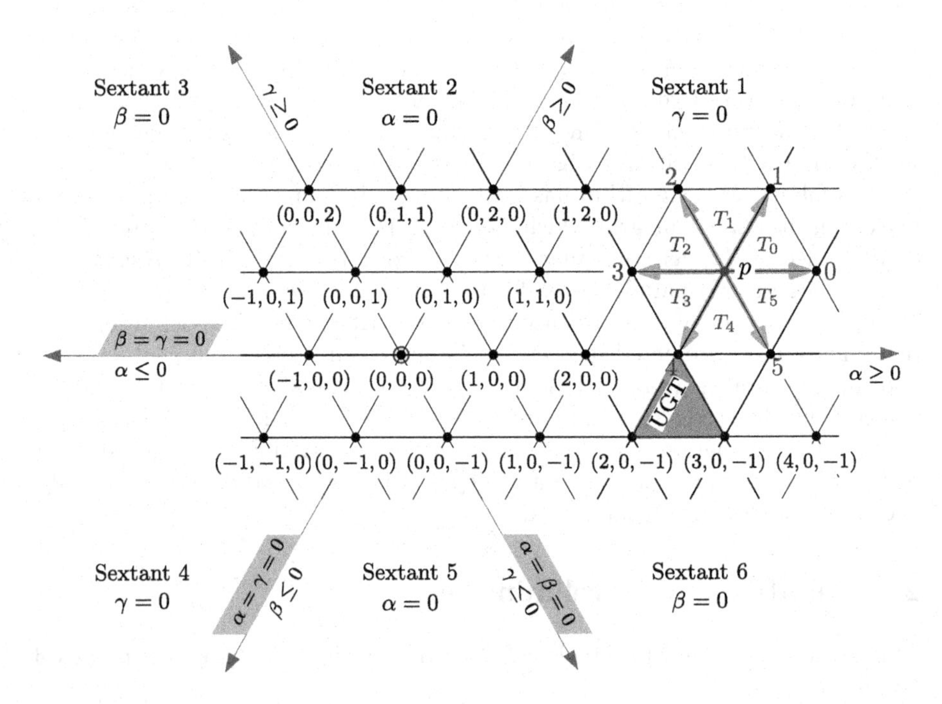

Fig. 2. A triangular canvas with UGT's $T_0, T_1, \ldots, T_5$ incident at a vertex p, and the direction codes $0, 1, ..., 5$ of neighboring grid points of p

Definition 3 (Digital object in triangular grid): *A digital object in the triangular grid is a finite subset of triangles of the discrete triangular plane. It may be one or more connected components (edge adjacent or vertex adjacent).*

Definition 4 (Triangular distance): *The triangular distance (d_t) between two points $p(\alpha_p, \beta_p, \gamma_p)$ and $q(\alpha_q, \beta_q, \gamma_q)$ is defined by $d_t(p, q) = max(| \alpha_p - \alpha_q |, | \beta_p - \beta_q |, | \gamma_p - \gamma_q |)$.*

Definition 5 (Triangular Polygon): *A (finite) polygon P imposed on the grid $\mathbb{T}$ is termed as a triangular polygon if its sides are collinear with lines in $\mathbb{L}_0$, $\mathbb{L}_{60}$, and $\mathbb{L}_{120}$.*

Definition 6 (ITC): *The inner triangular cover of a digital object($\underline{Q}$) is the maximum area triangular polygon inscribing the digital object Q in $\mathbb{T}$.*

The inner triangular cover of a digital object($\underline{Q}$) consists of different vertex types. A vertex q is said to be type i, $i \in \{1, 2, 3, 4, 5\}$ if the internal angle at a vertex q is $i \times \frac{\pi}{3}$. A digital object (Fig. 1(a)) with its inner triangular cover is shown in Fig. 1(b).

Definition 7 (Concavity): *A type* 5 *vertex or two consecutive vertices of types* 44, 45, 54, *and* 55 *create concavity in* $\underline{Q}$.

Different types of concavities are shown in Fig. 4.

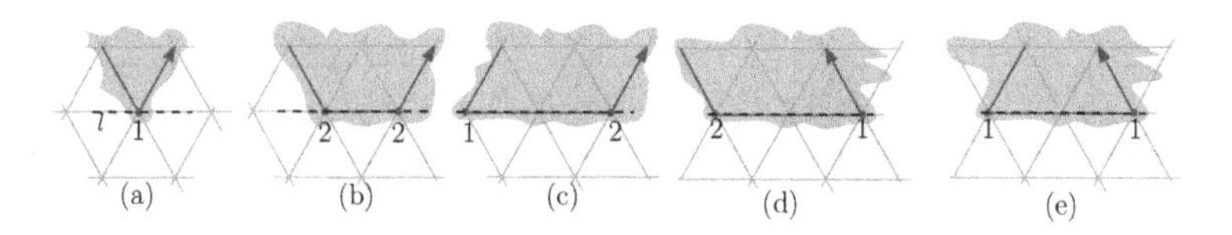

Fig. 3. (a) Convex portion due to single vertex, (b), (c), (d), (e) convex portion due to pattern xy where $x, y \in \{1, 2\}$

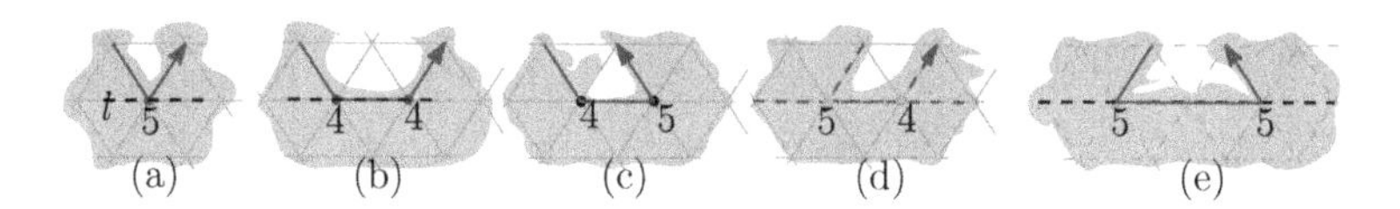

Fig. 4. (a) Concave portion due to single vertex, (b), (c), (d), (e) concave portion due to pattern xy where $x, y \in \{4, 5\}$

2.1 Deriving the Inner Triangular Cover (ITC)

To construct the inner triangular cover of a digital object the combinatorial algorithm proposed in [17] is used. The grid points of the triangular canvas can be divided into three categories after as follows: (i) none of the UGTs of a grid point completely occupies the object pixel as shown in the Fig. 5 (a); (ii) all UGTs of a grid point completely occupies the object pixel as shown in the Fig. 5 (b) and (iii) some UGTs of a grid point entirely occupies the object pixel as shown in the Fig. 5 (c), (d), and (e) which is taking part to construct the inner triangular cover. These grid points or vertex are further divided into five different types based on the object occupancy as shown in Fig. 6. Another type of vertices that does not part of a simple polygon are shown in Fig. 7. During the construction of $\underline{Q}$, the type of a vertex q is determined based on the incoming direction (d_{in}) and the outgoing direction (d_{out}) at q. An object occupancy vector $A_q = \langle a_0, a_1, \ldots a_5 \rangle$ ($\langle T_i : i = 0, 1, \ldots 5 \rangle$) incident at q is maintained throughout the process. If the incoming direction is d_{in} then $a_j = 1$ and $a_{(j+1)} mod\ 6 = 1$ where $j = (d_{in} + 2)\ mod\ 6$. Now d_{out} determined by incrementing j until the next 1-bit in A_q. If, say at j', $A_q[j'] = 1$, then $d_{out} = j'$. A type 3 vertex is considered as an edge point. Type 1, 2 vertices are convex vertices and type 4, 5 vertices are concave vertices. The construction of $\underline{Q}$ keeps the background of A to the right during the traversal. After determining the type outgoing direction of q, the polygon is traced to the next grid point q_n and type of q_n and the direction of traversal from q_n is determined. The traversal continues until the start vertex, v_s, is reached. During the construction of $\underline{Q}$, 4 lists are maintained L, L_α, L_β, and L_γ, where, L is a list of vertices of $\underline{Q}$ and L_α, L_β, and L_γ simultaneously contain vertices as well as edge points of $\underline{Q}$ in lexicographically sorted order with their respective primary and secondary keys. The primary key for L_α is α and secondary key is β, similarly the keys for L_β and L_γ can be defined. An index (in increasing order) is assigned to each vertex in order of their occurrences in $\underline{Q}$.

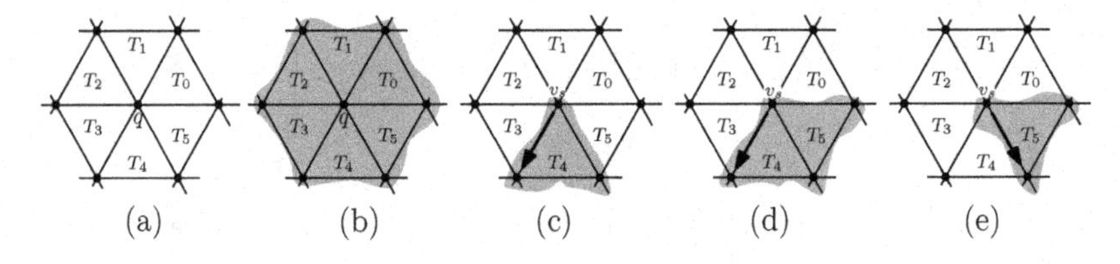

Fig. 5. (a) Strictly outside and (b) strictly inside of a digital object. (c), (d) and (e) Start vertex from the top-left. Object-occupied UGT's are shown in light grey, and start directions shown by arrow

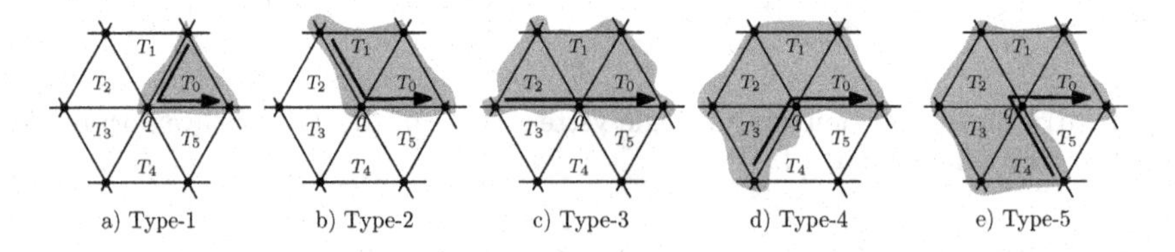

Fig. 6. Different vertex type based on object occupancy

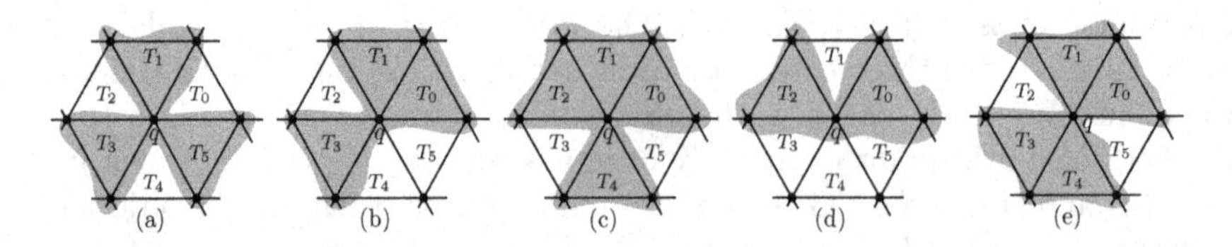

Fig. 7. Different vertex type that does not belong to simple polygon

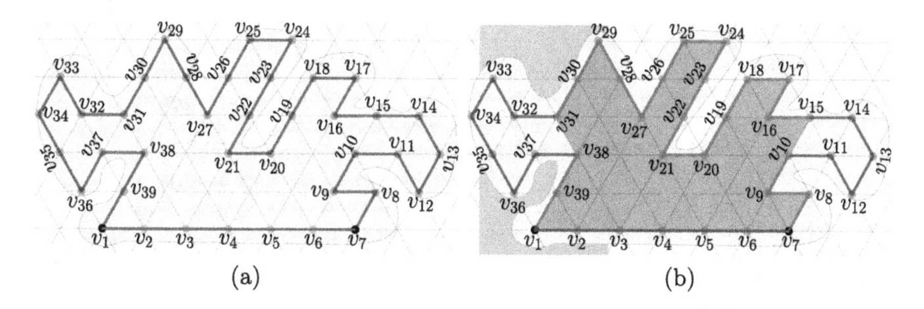

Fig. 8. (a) inner triangular cover and (b) sub-polygon(shaded) based on convex edge(v_1v_7)

3 Procedure to Compute *LAPT*

To find the largest area parallelogram, the inner triangular cover is constructed using the method described in [17]. The ITC thus obtained contains many convex and concave edges. The proposed algorithm runs in two passes, in the first pass it traverses in anti-cock wise direction from the top-left corner of the ITC and three extreme points reachable(that lies on the ITC and which we reach without going outside the object) along the three axes (α, β and γ) from each vertex are found out and stored with that vertex. For example the extreme points of v_1 along α and β are v_7 and v_{28} respectively. There is no extreme vertex of v_1 along γ direction. The extreme points of v_4 along α, β, and γ are respectively v_7, v_{18}, and v_{30}. Similarly extreme points of all other vertices are found and stored. During the second pass, again the traversal starts from the top left corner of the ITC and whenever a convex edge is encountered a sub-polygon is constructed assuming the convex edge as the base of the sub-polygon. To construct the sub-polygon for an convex edge say (u, v) an anti-clock wise traversal is made starting from vertex v until it reaches u. Only the vertices which has extreme points on the convex edge (u, v) either in α, β or γ are made part of the sub-polygon. For example, Fig. 8 shows the sub-polygon corresponding to the convex edge (v_1, v_7). After the sub-polygon is constructed, the *LAPT* inscribed within it is found out using the procedure explained in Sect. 3.1. This way, *LAPT* corresponding to each convex edge is found out and maximum of all of them gives the *LAPT* inscribed in the digital object.

3.1 Finding *LAPT* Within Sub-polygon

To find the largest area parallelogram inscribed within the sub-polygon corresponding to a convex edge say (u, v) a traversal is made in anti-clock wise direction from v. Whenever a convex edge say (u_i, v_i) is encountered the extreme points of u_i say u_{iex} and extreme points of v_i say v_{iex} determined. It is to be noted that these information of extreme points are already kept with u_i and v_i. The area of the parallelogram consisting of the points $u_, v_i, v_{iex}$ and u_{iex} is calculated and stored in a temporary variable. The convex edge u_i, v_i is then reduced

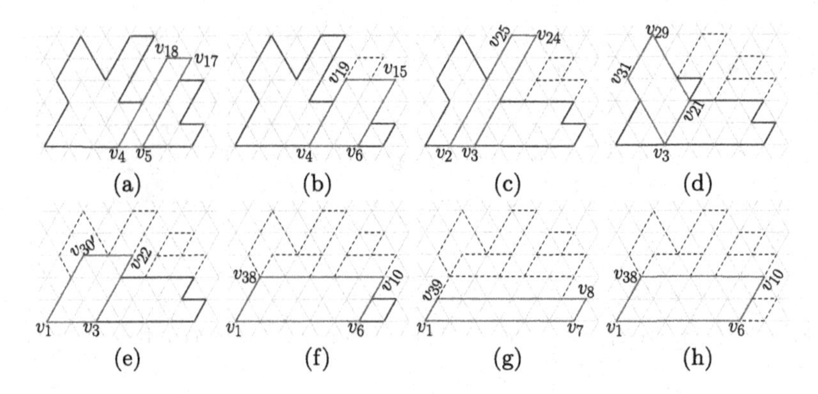

Fig. 9. Largest area parallelogram inside sub-polygon (Color figure online)

and the traversal proceeds to the next point on the sub-polygon. This process continues till it reaches the vertex u_i. The procedure is explained in steps in Fig. 9. In this figure, the intermediate parallelogram is shown in red. The maximum among all these red parallelogram is the largest area parallelogram for this sub-polygon.

3.2 Reduction Rules

During the process of finding $LAPT$, the convex portion (Fig. 3) of polygon and sub-polygons were reduced in different phases. For that, various combinatorial rules are formulated and used through out. The different reduction rules used are explained in this section. A convex region may occur due to a single convex vertex (type 1) or two or more consecutive convex vertices (type 1, 2). Different convexities created by single type 1 vertex is shown in Fig. 3(a). Similarly different convexities created by two convex vertices are shown in Fig. 3(b-e). All possible patterns that create convexity are 1, 11, 12, 21, and 22. For the convexity formed by a single type 1 vertex, four most recently visited vertices v_0, v_1, v_2, and v_3, are considered, where v_1 is type 1 and v_3 is the most recently visited vertex. Rule 1 and its sub cases are used to remove this type of convexity. On the other hand, if the pattern is 11, 12, 21 or 22, then five most recently visited vertex will be considered. Rule 2 with its possible subcases are used to remove this type of convexity. Both the rules i.e. Rule 1 and Rule 2 with their possible subcases are explained bellow. Figure 10 and 11 shows the result of both the rules respectively.

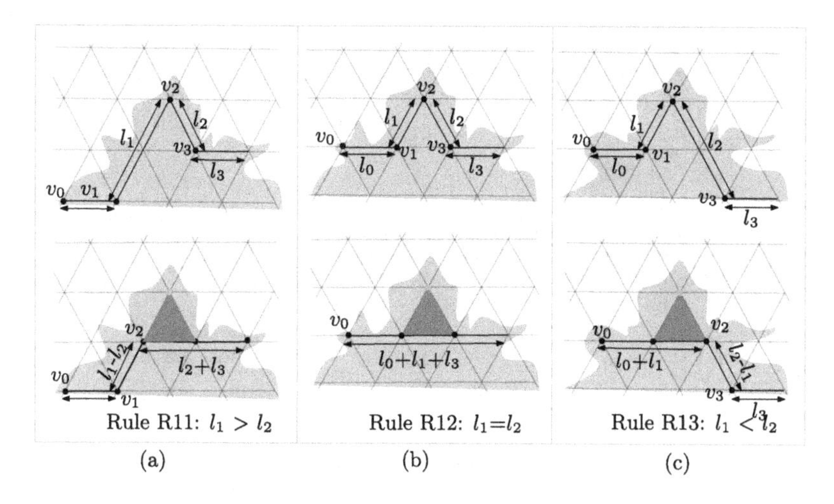

Fig. 10. Illustration of Rule 1

Rule R1: The pattern $t_1 1 t_2$, where $t_1, t_2 \in \{4, 5\}$ implies a convex region of the polygon. In this pattern, a type 1 vertex is preceded and followed by either type 4 or type 5 vertex. The shape of the convex region for this pattern is always an equilateral triangle. The rules are shown in Fig. 10.

R11 ($l_1 > l_2$): $< v_0(t_0, l_0), v_1(t_1, l_1), v_2(t_2, l_2), v_3(t_3, l_3) > \rightarrow$

$$< v_0(t_0, l_0), v_1(t_1, l_1 - l_2), v_2(t_2 + 1, l_2), v_3(t_3 - 1, l_3) >$$

R12 ($l_1 = l_2$): $< v_0(t_0, l_0), v_1(t_1, l_1), v_2(t_2, l_2), v_3(t_3, l_3) > \rightarrow$

$$< v_0(t_0, l_0), v_1(t_1 - 1, l_1), v_3(t_3 - 1, l_3) >$$

R13 ($l_1 < l_2$): $< v_0(t_0, l_0), v_1(t_1, l_1), v_2(t_2, l_2), v_3(t_3, l_3) > \rightarrow$

$$< v_0(t_0, l_0), v_1(t_1 - 1, l_1), v_2(t_2 + 1, l_2 - l_1), v_3(t_3, l_3) >$$

where, v_0, v_1, v_2, and v_3 are the most recently traversed vertices.

Rule R2: This pattern represents a vertex $t_1 \in \{4, 5\}$ followed by two consecutive vertices t_2, $t_3 \in \{1, 2\}$ and another vertex $t_4 \in \{4, 5\}$. The rules for the elimination of the convex portion is shown in Fig. 11.

R21 ($l_1 > l_2$): $< v_1(t_1, l_1), v_2(t_2, l_2), v_3(t_3, l_3), v_4(t_4, l_4) > \rightarrow$

$$< v_1(t_1, l_1 - l_3), v_2(t_2, l_1 + l_2), v_4(t_4 - 1, l_4) >$$

R22 ($l_1 = l_2$): $< v_1(t_1, l_1), v_2(t_2, l_2), v_3(t_3, l_3), v_4(t_4, l_4) > \rightarrow$

$$< v_1(t_1 - 1, l_1 + l_2), v_4(t_4 - 1, l_4) >$$

R23 ($l_1 < l_2$): $< v_1(t_1, l_1), v_2(t_2, l_2), v_3(t_3, l_3), v_4(t_4, l_4) > \rightarrow$

$$< v_1(t_1 - 1, l_1 + l_2), v_3(t_3, l_3 - l_1), v_4(t_4, l_4) >$$

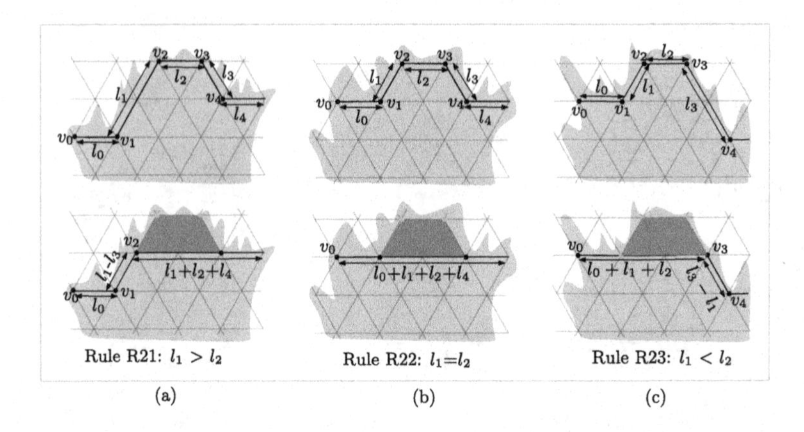

Fig. 11. Illustration of Rule 2

4 Algorithm

The algorithm starts from the top-left corner of the ITC of the digital object Q and continues till it reaches the start vertex (steps $3-23$). While traversing when a convex edge is encountered each vertex of the convex edge is appended to the list L_{base} which forms the base of the sub-polygon (Steps 5–13). Starting from vertex(v_2) at step 12 another traversal is made and all the vertices which satisfies the criteria mentioned in Sect. 3 is added to the list L_{sp}. Procedure 1 in step 14 finds the sub-polygon and returns it. Procedure FIND_LAP() in step 15 returns the largest empty parallelogram within the sub-polygon just computed. Steps 16–17 compares the area of the parallelogram just computed with the maximum area parallelogram computed so far and updates the variable LAP which stores the global maximum. In step 19, the convex edge is reduced following the rules explained in Sect. 3.2 and the process is repeated till the sub-polygon can be reduced no more.

In procedure FIND_SUBPOLY(Procedure 1), the traversal starts from the end point v_2 of the base. The direction of the base is determined in steps 3, 6, and 9. For each vertex during the anti-clockwise traversal from v_2 to v_1, the procedure FIND_INTERSECTION() appends all the vertex encountered such that vertex has a projection on the base (i.e. extreme vertex). For example, in Fig. 8(b), the extreme vertex of v_{14} is v_{11} that does not have a projection with the base so it is discarded, on the other hand the vertex v_{15} whose extreme point v_6 meets with base. So, the vertex v_{15} is appended to the list L_{sp}. This procedure continues till the traversal reaches to v_1.

The procedure FIND_PARALLELOGRAM()(Procedure 2) constructs all possible parallelogram within the sub-polygon and find the largest area parallelogram among them. The traversal starts from v_2 and when a convex edge appears (in step 4), the procedure COMPUTE_PARALLELOGRAM() constructs the parallelogram and it is saved (step 7). In step 6, the procedure CAL_AREA() calculates the area of the parallelogram and compare it with the previously stored LAP and updates the global variable for LAP in step 7. After that, the reduction rules are applied on the convex edge of the sub-polygon (step 9). The traversal will continue till it reaches the vertex v_1.

4.1 Complexity Analysis

The proposed algorithm finds all possible parallelograms inscribed inside the ITC of the digital object and reports the largest one among them. Let n be the number of contour points of the digital object and g be the grid size of $\mathbb{T}$. Initially, the inner triangular cover of the object is constructed which takes $O(\frac{n}{g})$ time. To constructs all possible parallelograms and find the $LAPT$, the algorithm traverses the ITC, and whenever a convex edge encountered, the procedure FIND_SUBPOLY() is called. To compute the sub-polygon, we need to find extreme vertex of all vertices of ITC using the search L_α, L_β, or L_γ list. Time taken to compute the sub-polygon is $O(\frac{n}{g} \lg \frac{n}{g})$. The procedure like APPLY_RULE(), CAL_AREA(), consists of only some arithmetic operations like addition, subtraction or multiplication, and which takes $O(1)$. During the whole procedure the

Algorithm 1. FIND_LAP()

Require: Input Image Q
Ensure: Largest area Parallelogram

```
 1: L, Lα, Lβ, Lγ = CONSTRUCT_ITC(Q);
 2: vs, curr ← Lstart
 3: while curr → next ≠ vs do
 4:     v1 ← curr
 5:     if v1.type ∈ {1,2} then
 6:         APPEND(Lbase, v1)
 7:         while curr.next ∈ {3} do
 8:             APPEND(Lbase, curr.next)
 9:             curr ← curr.next
10:         end while
11:         v2 ← curr.next
12:         if v2.type ∈ {1,2} then
13:             APPEND(Lbase, v2)
14:             SP ← FIND_SUBPOLY(L, Lbase, v1, v2)
15:             curr_LAP ← FIND_LAP(SP)
16:             if CAL_AREA(LAP)≤CAL_AREA(curr_LAP) then
17:                 LAP ← curr_LAP
18:             end if
19:             APPLY_RULE(L, v1, v2)
20:         end if
21:     end if
22:     curr ← curr.next
23: end while
      return LAP
```

Procedure 1. FIND_SUBPOLY(L, L_{base}, v_1, v_2)

```
1: L_sp ← Append(v1)
2: while v1 ≠ v2.next do
3:     if L_base.dir ∈ {0,3} then
4:         v ← FIND_INTERSECTION(L, L_base.β, L_base.γ)
5:         L_sp ← APPEND(v)
6:     end if
7:     if L_base.dir ∈ {2,5} then
8:         v ← FIND_INTERSECTION(L, L_base.α, L_base.β)
9:         L_sp ← APPEND(v)
10:    end if
11:    if L_base.dir ∈ {1,4} then
12:        v ← FIND_INTERSECTION(L, L_base.α, L_base.γ)
13:        L_sp ←APPEND(v)
14:    end if
15:    v2 ← v2.next
16: end while
       return L_sp
```

Procedure 2. FIND_PARALLELOGRAM(L_{sp}, v_1, v_2)

```
1: v ← v1
2: while v2.next ≠ v do
3:     v1 ← v2.next; v2 ← v1.next
4:     if (v1.type ∈ {1,2} and v2.type ∈ {1,2}) or v1.type ∈ {1} then
5:         temp_Parallelogram ← COMPUTE_PARALLELOGRAM(v1, v2, L_sp)
6:         if CAL_AREA (L_parallelogram) ≤ CAL_AREA (temp_Parallelogram) then
7:             L_parallelogram ← temp_Parallelogram
8:         end if
9:         APPLY_RULE(v1, v2)
10:    end if
11:    v2 ← v2.next
12: end while
       return L_parallelogram
```

ITC is traversed only once and each convexity is also considered only once. So, if there are k ($k << n$) such convexities encountered then the total time to compute $LAPT$ is $O(k \cdot \frac{n}{g} \lg \frac{n}{g}) + O(\frac{n}{g}) + O(1)) \simeq O(k \cdot \frac{n}{g} \lg \frac{n}{g})$.

5 Experimental Results

The proposed algorithm is implemented in python 3.9 in Ubuntu 16.04, 64-bit kernel version, the processor being Intel®Core^{TM}i7-6700 CPU @ 3.40GHz × 8. To check the correctness of the proposed algorithm, it is tested exhaustively on several digital objects. The result on a few different objects are shown in the Fig. 12. The blue line shows the triangular inner cover, and the red line indicates the largest parallelogram inscribed in the corresponding digital object.

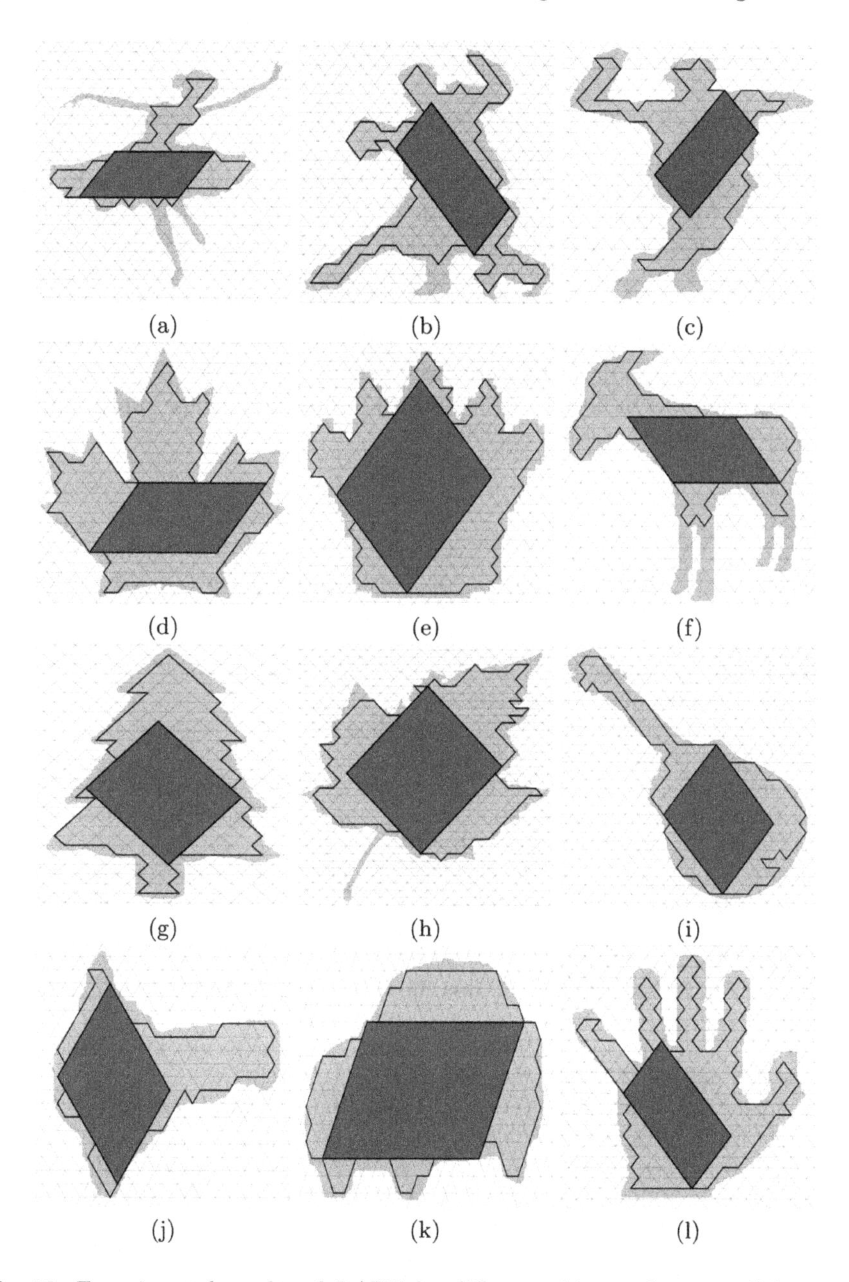

Fig. 12. Experimental results of *LAPT* for different objects where parallelogram is marked by red, ITC is marked by blue for grid size $g = 8$ (Color figure online)

6 Conclusions

A combinatorial algorithm to find and locate the largest area parallelogram inside a digital object that lies on a triangular grid has been proposed. The proposed algorithm maintains some lists and uses few reduction rules to derive the desired output. It take $O(\frac{n}{g} \lg \frac{n}{g})$ time where n is the number of pixel on the boundary of the digital object and g is grid size. In this paper, the algorithm tested on digital objects only but it can be applied to any simple triangular polygon. Though the work is theoretical in nature but can be useful in areas like industrial metal sheet cutting or VLSI design. The algorithm can also be useful for shape analysis of objects by iteratively generating the largest parallelogram up to a certain limit. In future, the algorithm can be modified to apply on digital objects or polygons that contains holes.

References

1. Milenkovic, V., Daniels, K., Li, Z.: Automatic marker making. In: Proceedings of the Third Canadian Conference on Computational Geometry, pp. 243–246. Simon Fraser University (1991)
2. Milenkovic, V., Daniels, K., Li, Z.: Placement and Compaction of Nonconvex Polygons for Clothing Manufacture (1992)
3. MacKenna, M., O'Rourke, J., Suri, S.: Finding the largest rectangle in an orthogonal polygon (1985)
4. DePano, A., Ke, Y., O'Rourke, J.: Finding largest inscribed equilateral triangles and squares. In: Proceedings of the 25th Allerton Conference Communication Control Computing, pp. 869–878 (1987)
5. Amenta, N.: Largest volume box is convex programming. Pers. Commun. (1992)
6. Fekete, S.P.: Finding all anchored squares in a convex polygon in subquadratic time. In: Proceedings of the Fourth Canadian Conference on Computational Geometry, pp. 71–76 (1992)
7. Alt, H., Hsu, D., Snoeyink, J.: Computing the largest inscribed isothetic rectangle. In: CCCG, pp. 67–72 (1995)
8. Daniels, K., Milenkovic, V., Roth, D.: Finding the largest area axis-parallel rectangle in a polygon. Comput. Geom. **7**(1–2), 125–148 (1997)
9. Chaudhuri, J., Nandy, S.C., Das, S.: Largest empty rectangle among a point set. J. Algorithms **46**(1), 54–78 (2003)
10. Ahn, H.-K., Brass, P., Cheong, O., Na, H.-S., Shin, C.-S., Vigneron, A.: Inscribing an axially symmetric polygon and other approximation algorithms for planar convex sets. Comput. Geom. **33**(3), 152–164 (2006)
11. Hall-Holt, O., Katz, M.J., Kumar, P., Mitchell, J.S., Sityon, A.: Finding large sticks and potatoes in polygons. In: SODA, vol. 6, pp. 474–483 (2006)
12. Knauer, C., Schlipf, L., Schmidt, J.M., Tiwary, H.R.: Largest inscribed rectangles in convex polygons. J. Discrete Algorithms **13**, 78–85 (2012)
13. Sarkar, A., Biswas, A., Dutt, M., Bhattacharya, A.: Finding a largest rectangle inside a digital object and rectangularization. J. Comput. Syst. Sci. **95**, 204–217 (2018)
14. Sarkar, A., Biswas, A., Dutt, M., Bhattacharya, A.: Finding largest rectangle inside a digital object. In: Computational Topology in Image Context, pp. 157–169 (2016)

15. Klette, R., Rosenfeld, A.: Digital Geometry: Geometric Methods for Digital Picture Analysis. Morgan Kaufmann Publishers Inc., San Francisco, CA, USA (2004)
16. Nagy, B., Abuhmaidan, K.: A continuous coordinate system for the plane by triangular symmetry. Symmetry **11**(2) (2019)
17. Biswas, A., Bhowmick, P., Bhattacharya, B.B., Das, B., Dutt, M., Sarkar, A.: Triangular covers of a digital object. Appl. Math. Comput. **58**(1), 667–691 (2018)

Picture Languages

Weighted Three Directions OTA and Weighted Hexapolic Picture Automata

Meenakshi Paramasivan[1(✉)] and D. G. Thomas[2]

[1] FB IV - Informatikwissenschaften, Universität Trier, 54286 Trier, Germany
meena_maths@yahoo.com

[2] Department of Mathematics, Madras Christian College, Chennai, India

Abstract. Two-dimensional hexagonal arrays seen on a triangular grid can be treated as two-dimensional representations of three-dimensional rectangular parallelepipeds. We are introducing weighted 3 directions on-line tessellation automata (W3OTA) and investigating formal power series on hexagonal pictures. These are functions that map hexagonal pictures to elements of a semiring and provide an extension of two-dimensional hexagonal picture languages to a quantitative setting.

Keywords: Picture series · Two-dimensional languages · Hexagonal pictures · Recognizable hexagonal picture languages · Weighted 3OTA

1 Introduction

Picture languages generated by grammar models and recognized by automata models have been investigated since the 1970s for their complications raised in the framework of pattern recognition and image analysis [10,17,18,20]. These two-dimensional picture languages have a connection with the generation of Kolam patterns [19,22], which are traditional pieces of the South Indian style of painting. Dora Giammarresi and her co-authors investigated two-dimensional picture languages and their connection to tiling systems [2,8] through local and recognizable picture languages. In [12], characterizations of recognizable picture series were investigated. In [7], weighted two-dimensional on-line tessellation automata (W2OTA) were introduced and proven that the picture series is recognizable by some weighted two-dimensional on-line tessellation automaton if and only if it has the behavior of a weighted picture automaton. A Nivat theorem on W2OTA has been proved; (see [3]).

Siromoney has defined an arrowhead catenation for the two-dimensional hexagonal arrays. These arrays on a triangular grid can be viewed or treated as two-dimensional representations of three-dimensional rectangular parallelepipeds [16]. Hexagonal cellular automata (HCA) were introduced as a variation of the rectangular two-dimensional cellular automata (RCA). The equivalence of HCA and RCA was shown in [11]. Hexagonal array patterns are found

Mathematics Subject Classification: 68Q45, 68Q70, 68R01

R. P. Barneva et al. (Eds.): IWCIA 2022, LNCS 13348, pp. 139–153, 2023.
https://doi.org/10.1007/978-3-031-23612-9_9

in the literature on picture processing and scene analysis. Hexagonal Image Processing (HIP) provides an introduction to the processing of hexagonally sampled images. The utility of the HIP framework is demonstrated by implementing several basic image processing techniques. The HIP framework serves as a tool for comparing the processing of images defined on a square versus a hexagonal grid [13].

In [5,6], two classes, namely (i) local hexagonal picture languages (HLOC) and (ii) reognizable hexagonal picture languages (HREC), were introduced, and also hexagonal wang systems (HWS) and hexagonal tiling systems (HTS) were used to study these picture languages. In [21], to recognize HREC, 3 directions on-line tessellation automata were introduced. Jaya Abraham et al. [1] studied characterizations of hexagonal recognizable picture series through weighted hexapolic picture automata (WHPA).

A preliminary version of this paper and some of its properties were presented in [14,15]. Now the paper is organized as follows: In Sect. 2, we recall some basics. In Sect. 3, we introduce weighted three directions on-line tessellation automata (W3OTA) and investigate formal power series on hexagonal pictures. In Sect. 4, we show that the W3OTA recognizable series are WHPA recognizable. In Sect. 5, we conclude with a further connection to MSO logic over hexagonal pictures.

2 Pictures and Hexagonal Pictures

In this section, we shall briefly recall some of the required standard notations and definitions of two-dimensional hexagonal pictures and languages [6,21].

2.1 Two-Dimensional Hexagonal Pictures and Languages

A *hexagonal picture* p over the finite alphabet Σ is a hexagonal array of symbols from Σ. The set of all non-empty hexagonal pictures over Σ is denoted by Σ^{++H}. Let $p \in \Sigma^{++H}$. We get the bordered version of p denoted by $\hat{p}$ when the special symbol $\# \notin \Sigma$ is added as boundary to p.

Example 1. A hexagonal picture $p \in \Sigma^{++H}$ over $\Sigma = \{\mathtt{a}, \mathtt{b}, \mathtt{c}\}$ and its bordered version $\hat{p}$ is given in Fig. 1.

With respect to a triad (axes x, y, z) of triangular axes x, y, z the coordinates of each element of the hexagonal pictures can be fixed. For a similar hexagonal picture in Example 1 we depict the coordinates in Fig. 2.

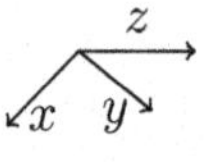

Given a hexagonal picture $p \in \Sigma^{++H}$ let ℓ, m, n denote the number of elements in the direction of x, y, z respectively. The directions are fixed with origin of reference as the upper left vertex, having coordinates $(1, 1, 1)$. The triple (ℓ, m, n) denotes the size of the hexagonal picture p. Hexagonal pictures of size $(0, m, n)$, $(\ell, 0, n)$, or $(\ell, m, 0)$ where $\ell, m, n > 0$ are not defined. Furthermore, let p_{ijk} denote the symbol in p with coordinates (i, j, k), where $1 \le i \le \ell$, $1 \le j \le m$,

$$\begin{array}{ccccc} & \mathtt{b} & & \mathtt{b} & \\ \mathtt{a} & & \mathtt{c} & & \mathtt{a} \\ & \mathtt{b} & & \mathtt{b} & \end{array} \qquad \text{and} \qquad \begin{array}{ccccccccc} & & \# & & \# & & \# & & \\ & \# & & \mathtt{b} & & \mathtt{b} & & \# & \\ \# & & \mathtt{a} & & \mathtt{c} & & \mathtt{a} & & \# \\ & \# & & \mathtt{b} & & \mathtt{b} & & \# & \\ & & \# & & \# & & \# & & \end{array}$$

Fig. 1. A hexagonal picture over $\Sigma = \{\mathtt{a}, \mathtt{b}, \mathtt{c}\}$ and its bordered version.

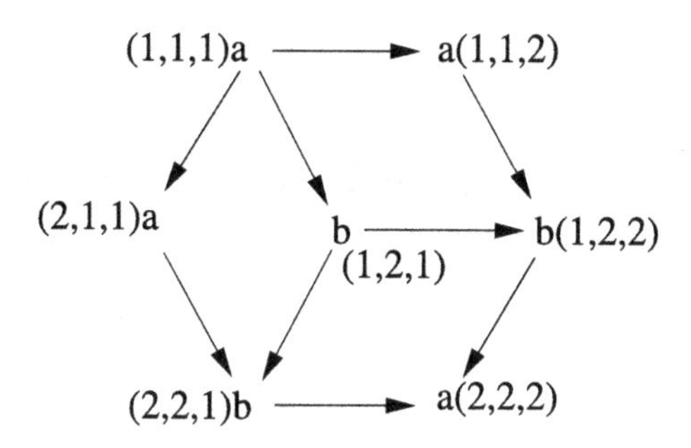

Fig. 2. Coordinates of a hexagonal picture p.

$1 \leq k \leq n$. For instance, p in Example 1 is of size $(2, 2, 2)$ and $p_{111} = b$, $p_{221} = c$, and so on. Every subset $L \subseteq \Sigma^{++H}$ is a hexagonal picture language. Given a hexagonal picture p of size (ℓ, m, n), if $g \leq \ell$, $h \leq m$ and $k \leq n$ then we denote by $B_{g,h,k}(p)$ the set of all hexagonal blocks (or hexagonal sub-pictures) of p of size (g, h, k). In fact a *hexagonal tile* is a hexagonal picture of size $(2, 2, 2)$.

Definition 1 [6]. *A* hexagonal tiling system *(in short, HTS) is a 4-tuple* $\mathcal{T} = (\Sigma, \Gamma, \pi, \Theta)$*, where* Σ *and* Γ *are two finite alphabets,* $\pi : \Gamma \to \Sigma$ *is a projection and* Θ *is a finite set of hexagonal tiles over the alphabet* $\Gamma \cup \{\#\}$*.*

Note that HREC [6,21] is exactly the family of all hexagonal picture languages recognizable by hexagonal tiling systems $\mathcal{L}(HTS)$.

Example 2. Let $\Sigma = \{1, 2, 3\}$;

$$\Theta = \left\{ \begin{array}{ccccc} & \# & & \# & \\ \# & & 1 & & 1 \\ & 2 & & 2 & \end{array}, \quad \begin{array}{ccccc} & \# & & \# & \\ 1 & & 1 & & 1 \\ & 2 & & 2 & \end{array}, \quad \begin{array}{ccccc} & \# & & \# & \\ 1 & & 1 & & \# \\ & 2 & & 2 & \end{array}, \right.$$

$$\begin{array}{ccccc} & \# & & 1 & \\ \# & & 2 & & 2 \\ & \# & & 3 & \end{array}, \quad \begin{array}{ccccc} & 1 & & 1 & \\ 2 & & 2 & & 2 \\ & 3 & & 3 & \end{array}, \quad \begin{array}{ccccc} & 1 & & \# & \\ 2 & & 2 & & \# \\ & 3 & & \# & \end{array},$$

$$\left. \begin{array}{ccccc} & 2 & & 2 & \\ \# & & 3 & & 3 \\ & \# & & \# & \end{array}, \quad \begin{array}{ccccc} & 2 & & 2 & \\ 3 & & 3 & & 3 \\ & \# & & \# & \end{array}, \quad \begin{array}{ccccc} & 2 & & 2 & \\ 3 & & 3 & & \# \\ & \# & & \# & \end{array} \right\}.$$

Then

$$L_1 = \pi(L(\Theta)) = \left\{ \begin{array}{ccccc} & 1 & & 1 & \\ 2 & & 2 & & 2 \\ & 3 & & 3 & \end{array}, \begin{array}{ccccccc} & 1 & & 1 & & 1 & \\ 2 & & 2 & & 2 & & 2 \\ & 3 & & 3 & & 3 & \end{array}, \begin{array}{ccccccc} & 1 & & 1 & & 1 & \\ 2 & & 2 & & 2 & & 2 \\ & 3 & & 3 & & 3 & \end{array}, \ldots \right\}.$$

L_1 is the set of all hexagons of sizes $(2, 2, k) (k \geq 2)$ with z direction elements respectively at the top are 1, at the middle are 2 and at the bottom are 3.

3 Weighted Automata over Hexagonal Pictures

In this section, we shall briefly recall some of the required standard notations and definitions with respect to picture series and hexagonal picture series.

3.1 Series on Pictures

A *semiring* $(K, +, \cdot, 0, 1)$ is a structure K such that $(K, +, 0)$ is a commutative monoid, $(K, \cdot, 1)$ is a monoid, multiplication distributes over addition, and $x \cdot 0 = 0 = 0 \cdot x$ for all elements $x \in K$. If multiplication is commutative, K is called *commutative*. Examples of semirings useful to model problems in operations research and carrying quantitative properties for many devices include e. g. the *Boolean* semiring $\mathbb{B} = (0, 1, \vee, \wedge, 0, 1)$, the natural numbers $\mathbb{N} = (\mathbb{N}, +, \cdot, 0, 1)$, the *tropical* semiring $\mathbb{T} = (\mathbb{R} \cup \{\infty\}, \min, +, \infty, 0)$, the *artical* (or *max-plus*) semiring $\mathbb{A} = (\mathbb{N} \cup \{-\infty\}, \max, +, -\infty, 0)$, the language-semiring $(\mathscr{P}(\Sigma^*), \cup, \cap, \emptyset, \Sigma^*)$ and $([0, 1], \max, \cdot, 0, 1)$ (to capture probabilities).

Subsequently, K will always denote a commutative semiring. Let Σ, Γ be alphabets. We will now assign weights to hexagonal pictures. This provides a generalization of the theory of hexagonal pictures to formal power series over hexagonal pictures. In [4, 12] this generalization was done to rectangular pictures.

A *picture series* is a mapping $S : \Sigma^{++} \to K$. We let $K\langle\langle \Sigma^{++} \rangle\rangle$ comprise all picture series. We write (S, p) for $S(p)$, then a picture series S often is written as a formal sum $S = \sum_{p \in \Sigma^{++}} (S, p) \cdot p$. The set $\text{supp}(S) = \{p \in \Sigma^{++} \mid (S, p) \neq 0\}$ is the *support* of S. For a picture language $L \subseteq \Sigma^{++}$, the *characteristic series* $\mathbb{1}_L : \Sigma^{++} \to K$ is defined by setting $(\mathbb{1}_L, p) = 1$ if $p \in L$, and $(\mathbb{1}_L, p) = 0$ otherwise. For $K = \mathbb{B}$, the mapping $L \mapsto \mathbb{1}_L$ gives a natural bijection between languages over Σ and series in $\mathbb{B}\langle\langle \Sigma^{++} \rangle\rangle$.

We define *rational* operations $\oplus$ and $\odot$, referred to as *sum* and *Hadamard product*, respectively, and also $\cdot : K \times K\langle\langle \Sigma^{++} \rangle\rangle \to K\langle\langle \Sigma^{++} \rangle\rangle$, the *scalar multiplications* with elements of the semiring, in the following way:

For two series $S, T \in K\langle\langle \Sigma^{++} \rangle\rangle$, $k \in K$ and $p \in \Sigma^{++}$, we set

$$(S \oplus T, p) := (S, p) + (T, p),$$
$$(S \odot T, p) := (S, p) \cdot (T, p) \quad \text{and} \quad (k \cdot S, p) := k \cdot (S, p).$$

Note that $k \cdot S = (k \cdot \mathbb{1}_{\Sigma^{++}}) \odot S$. Now, defining projections and inverse projections for series, given additionally $\pi : \Gamma \to \Sigma$, $R \in K\langle\langle \Gamma^{++} \rangle\rangle$ and $q \in \Gamma^{++}$, we put

$$(\pi(R), p) := \sum_{\pi(p')=p} (R, p') \quad \text{and}(\pi^{-1}(S), q) := (S, \pi(q)).$$

We will call the series, $\pi(R) \in K\langle\langle\Sigma^{++}\rangle\rangle$ *projection* of R by π and $\pi^{-1}(S) \in K\langle\langle\Gamma^{++}\rangle\rangle$ *inverse projection* of S by π, respectively. In the boolean case we get for languages $L \subseteq \Sigma^{++} : \pi^{-1}(L) = \{p \in \Gamma^{++} \mid \pi(p) \in L\}$. There are further rational operations on picture series like horizontal/vertical multiplication and horizontal/vertical star. The closure of the class of series having finite support (polynomials) under rational operations and projections defines the family of projections of rational picture series which coincides with the family of series that are behaviours of weighted picture automata (WPA) [12]. In [7] the equivalence of weighted 2-dimensional on-line tessellation automata and WPA is proved.

3.2 Series on Hexagonal Pictures

A *hexagonal picture series* [1] is a mapping $S : \Sigma^{++H} \to K$. We let $K\langle\langle\Sigma^{++H}\rangle\rangle$ contains all hexagonal picture series over Σ. We write (S, p) for $S(p)$, then a hexagonal picture series S is written as $S = \Sigma_{p\in\Sigma^{++H}}(S, p)\cdot p$. The set $\text{supp}(S) = \{p \in \Sigma^{++H} \mid (S, p) \neq 0\}$ is the *support* of S. For a language $L \subseteq \Sigma^{++H}$, the *characteristic series* $\mathbb{1}_L : \Sigma^{++H} \to K$ is defined by setting $(\mathbb{1}_L, p) = 1$ if $p \in L$, and $(\mathbb{1}_L, p) = 0$ otherwise. For $K = \mathbb{B}$, the mapping $L \mapsto \mathbb{1}_L$ gives a natural bijection between languages over Σ and series in $\mathbb{B}\langle\langle\Sigma^{++H}\rangle\rangle$.

We recall *rational* operations on hexagonal picture series $\oplus$, $\odot$, $\oslash$, $\otimes$ and $\ominus$ referred to as *sum*, *Hadamard product*, *x-directional multiplication*, *y-directional multiplication* and *z-directional multiplication* respectively, and also $\cdot : K \times K\langle\langle\Sigma^{++H}\rangle\rangle \to K\langle\langle\Sigma^{++H}\rangle\rangle$, the *scalar multiplications* with elements of the semiring, in the following way:

For two series $S, T \in K\langle\langle\Sigma^{++H}\rangle\rangle$, $k \in K$ and $p \in \Sigma^{++H}$, we set

$$(S \oplus T, p) := (S, p) + (T, p)\,,$$
$$(S \odot T, p) := (S, p) \cdot (T, p) \quad \text{and} \quad (k \cdot S, p) := k \cdot (S, p).$$

Note that $k \cdot S = (k \cdot \mathbb{1}_{\Sigma^{++H}}) \odot S$. Now, defining projections and inverse projections for series, given additionally $\pi : \Gamma \to \Sigma$, $R \in K\langle\langle\Gamma^{++H}\rangle\rangle$ and $q \in \Gamma^{++H}$, we put

$$(\pi(R), p) := \sum_{\pi(p')=p} (R, p') \quad \text{and } (\pi^{-1}(S), q) := (S, \pi(q))\,.$$

We will call the series, $\pi(R) \in K\langle\langle\Sigma^{++H}\rangle\rangle$ *projection* of R by π and $\pi^{-1}(S) \in K\langle\langle\Gamma^{++H}\rangle\rangle$ *inverse projection* of S by π, respectively. In the boolean case we get for languages $L \subseteq \Sigma^{++H} : \pi^{-1}(L) = \{p \in \Gamma^{++H} \mid \pi(p) \in L\}$. There are further rational operations on hexagonal picture series (similar to horizontal/vertical multiplication/star on picture series) which we define in the following:

$$(S \oslash T, p) = \sum_{p=p_1 \oslash p_2} (S, p_1) \cdot (T, p_2).$$

$$(S \obslash T, p) = \sum_{p=p_1 \obslash p_2} (S, p_1) \cdot (T, p_2).$$

$$(S \ominus T, p) = \sum_{p=p_1 \ominus p_2} (S, p_1) \cdot (T, p_2).$$

$$(S^{\oslash +}, p) = \sum_{p=p_1 \oslash p_2 \oslash \cdots \oslash p_n} (S, p_1) \cdot (S, p_2) \cdots (S, p_n).$$

$$(S^{\obslash +}, p) = \sum_{p=p_1 \obslash p_2 \obslash \cdots \obslash p_n} (S, p_1) \cdot (S, p_2) \cdots (S, p_n).$$

$$(S^{\ominus +}, p) = \sum_{p=p_1 \ominus p_2 \ominus \cdots \ominus p_n} (S, p_1) \cdot (S, p_2) \cdots (S, p_n).$$

The closure of the class of series having finite support (polynomials) under rational operations and projections defines the family of projections of rational hexagonal picture series which coincides with the family of series that are behaviours of weighted hexapolic picture automata (WHPA) [1] (See Definition 4 for WHPA).

In this paper, we introduce W3OTA and in order to prove the equivalence of W3OTA and WHPA, we consider two types of devices for our study on quantitative setting:

1. 3 directions on-line tessellation automata (3OTA) [21].
2. weighted hexapolic picture automata (WHPA) [1] (See Definition 4).

Definition 2 [21]. *A non-deterministic (deterministic) 3 directions on-line tessellation automaton (3OTA) is defined by a 5-tuple* $\mathcal{A} = (\Sigma, Q, q_0, F, \delta)$ *where*

- Σ *is the input alphabet*
- Q *is a finite set of states*
- $q_0 \in Q$ *is the initial state*
- $F \subseteq Q$ *is the set of final states*
- $\delta : Q \times Q \times Q \times \Sigma \to 2^Q$ $(\delta : Q \times Q \times Q \times \Sigma \to Q)$ *is the transition function.*

A run of $\mathcal{A}$ *on a hexagonal picture* $p \in \Sigma^{++H}$ *consists of associating a state (from the set* Q*) to each position* (i, j, k) *of* p*. Such state is given by the transition function* δ *and depends on the states already associated. For* p*, consider* $\hat{p}$ *and let all the border letters # in* p *be associated with state* q_0*. The computation of the automaton starts at time* $t = 1$*, by reading* p_{111} *and associating the state* $\delta(q_0, q_0, q_0, p_{111})$ *to position* $(1, 1, 1)$*. In general, we view* $\delta(q_1, q_2, q_3, p_{ijk})$ *as*

$$\begin{array}{c} q_2 \quad q_3 \\ \swarrow \; \searrow p_{ijk} \\ q_1 \end{array}$$

At time $t = 2$, *states are simultaneously associated to positions* p_{211} *and* p_{112}. *This process continues until a state is associated to position* $(\ell_1(p), \ell_2(p), \ell_3(p))$. *A 3OTA* $\mathcal{A}$ *recognizes a hexagonal picture* p *if there exists a run of* $\mathcal{A}$ *on* $\hat{p}$ *such that the state associated to position* $(\ell_1(p), \ell_2(p), \ell_3(p))$ *is a final state. The set of all hexagonal pictures recognized by* $\mathcal{A}$ *is denoted by* $\mathcal{L}(A)$. *Let* $\mathcal{L}(3OTA)$ *be the set of hexagonal picture languages recognized by 3OTAs.*

Example 3. A 3OTA for L_1 in Example 2 is a 5-tuple $\mathcal{A}_1 = (\Sigma_1, Q_1, q_0, F_1, \delta_1)$ where $\Sigma_1 = \{1, 2, 3\}$, $Q_1 = \{q_0, q_1, q_2, q_3\}$, $\delta_1(q_0, q_0, q_0, 1) = q_1$, $\delta_1(q_0, q_0, q_1, 2) = q_2$, $\delta_1(q_1, q_0, q_0, 1) = q_1$, $\delta_1(q_2, q_1, q_1, 2) = q_2$, $\delta_1(q_0, q_2, q_2, 3) = q_3$, $\delta_1(q_3, q_2, q_2, 3) = q_3$, $\delta_1(q_2, q_1, q_0, 2) = q_2$, and $F_1 = \{q_3\}$.

We now present the detailed definition of a weighted 3 directions on-line tessellation automata. It generalizes in a straightforward way the automata-theoretic definition of recognizability for hexagonal picture languages in terms of 3OTA.

Definition 3. *A* weighted 3 directions on-line tessellation automata (in short, W3OTA) *over* Σ *is a tuple* $\mathcal{H} = (Q, E, I, F)$, *consisting of a finite set* Q *of states, a finite set of transitions* $E \subseteq Q \times Q \times Q \times \Sigma \times K \times Q$ *and sets of initial and final states* $I, F \subseteq Q$, *respectively.*

For a transition $e = (q_x, q_y, q_z, a, w, q) \in E$, *we set* $\sigma_x(e) = q_x$, $\sigma_y(e) = q_y$, $\sigma_z(e) = q_z$ *and* $\sigma(e) = q$. *We denote by* label(e) *its label* a *and by* weight(e) *its weight* w. *We extend these both functions to hexagonal pictures by setting, for* $c = (c_{i,j,k}) \in E^{\ell \times m \times n}$:

$$\text{label}(c)(i,j,k) := \text{label}(c_{i,j,k}), \qquad \text{weight}(c) = \prod_{i,j,k} \text{weight}(c_{i,j,k}).$$

It defines functions label : $E^{++H} \rightarrow \Sigma^{++H}$ *and* weight : $E^{++H} \rightarrow K$. *We call* label(c) *the* label *and* weight(c) *the* weight *of* c. *A* run *(or* computation*) in* $\mathcal{H}$ *is an element in* $E^{\ell \times m \times n}$ *satisfying natural compatibility properties, more precisely, for* $c = (c_{i,j,k}) \in E^{\ell \times m \times n}$ *we have* $\forall 1 \le i \le \ell, 1 \le j \le m, 1 \le k \le n$:

$$\sigma_x(c_{i,j,k}) = \sigma(c_{i-1,j,k}), \quad \sigma_y(c_{i,j,k}) = \sigma(c_{i,j-1,k}), \quad \sigma_z(c_{i,j,k}) = \sigma(c_{i,j,k-1}).$$

A run $c \in E^{\ell \times m \times n}$ *is* successful *if for all* $1 \le i \le \ell$, $1 \le j \le m$ *and* $1 \le k \le n$, *we have* $\sigma_x(c_{1,j,k}), \sigma_y(c_{i,1,k}), \sigma_z(c_{i,j,1}) \in I$ *and* $\sigma(c_{\ell,m,n}) \in F$. *The set of all successful runs labelled with a hexagonal picture* p *is denoted by* $I \overset{p}{\rightsquigarrow} F$.

We define a hexagonal picture series $||\mathcal{H}||$ as follows. If $p \in \Sigma^{++H}$ has no successful run in $\mathcal{H}$, $||\mathcal{H}||$ sends p to 0. Otherwise, we define

$$(||\mathcal{H}||, p) = \sum_{c \in I \overset{p}{\rightsquigarrow} F} \text{weight}(c).$$

Intuitively, the weight of a hexagonal picture p is the sum of the weights of all successful runs in $\mathcal{H}$ that read p. We call $||\mathcal{H}||$ the *behaviour* of $\mathcal{H}$ and say that the automaton $\mathcal{H}$ *computes* (or *recognizes*) the hexagonal picture series $||\mathcal{H}|| : \Sigma^{++H} \to K$. We write $K^{\text{rec}}\langle\langle\Sigma^{++H}, W3OTA\rangle\rangle$ for the family of hexagonal series that are computable by W3OTA over Σ, elements of which are referred to as *W3OTA-recognizable* series.

Considering Definition 3 above, where K equals $\mathbb{B}$. We get precisely the definition of a *3 directions OTA (3OTA).* Here, instead of E, one could also define a *transition function* $\delta : Q \times Q \times Q \times \Sigma \to 2^Q$. If $I = 1$ and $\delta : Q \times Q \times Q \times \Sigma \to Q$, we call $\mathcal{H}$ *deterministic.* For an alphabet Σ, devices of 3OTA over Σ define hexagonal picture languages and were shown to compute precisely the family HREC [21]. We shall have an example of hexagonal picture series $S : \Sigma^{++H} \to \mathbb{R} \cup \{\infty\}$.

Similar to common constructions on picture automata and using ideas in [1,4], we have the following.

Proposition 1. *Let K be a commutative semiring. W3OTA-recognizable hexagonal picture series over K are closed under $\odot$, $\oplus$, scalar multiplications with elements of K, projections and inverse projections. For languages, inverse projections of languages that are deterministically 3OTA-recognizable are again recognizable by some deterministic 3OTA. If L is deterministically 3OTA-recognizable then $\mathbb{1}_L$ is W3OTA-recognizable.*

Proof. As usual, for the operations $\odot$ and $\oplus$ we use the direct product and the union of automata, respectively. Let $w \in K$. The W3OTA $\mathcal{H} = (\{0,1\}, E, \{0\}, \{1\})$ defined by

$$E = \bigcup_{a \in \Sigma} \{(0,0,0,a,w,1), (1,0,0,a,1,1), (0,0,1,a,1,1), (0,1,1,a,1,1), \\ (1,1,0,a,1,1), (1,1,1,a,1,1)\}$$

computes $||\mathcal{H}|| : w \cdot \mathbb{1}_{\Sigma^{++H}}$. Now, since in genereal, for a series $S \in K\langle\langle\Sigma^{++H}\rangle\rangle$, we have $w \cdot S = (w \cdot \mathbb{1}_{\Sigma^{++H}}) \odot S$, we get the assertion for scalar multiplications. Let $\pi : \Gamma \to \Sigma$. If $E \subseteq Q \times Q \times Q \times \Sigma \times K \times Q$ denotes the weighted transition set of a W3OTA $\mathcal{H}$ over Γ, then we define the transition set for an automaton computing $\pi(||\mathcal{H}||)$, as

$$\{(q_x, q_y, q_z, \sigma, w, q) \mid a \in \Sigma, w = \sum_{(q_x, q_y, q_z, \gamma, w', q) \in E \;\; \pi(\gamma)=\sigma} w'\}.$$

For the inverse projection let $\mathcal{H} = (Q, E, I, F)$ be a W3OTA computing $S : \Sigma^{++H} \to K$ with $E \subseteq Q \times Q \times Q \times \Sigma \times K \times Q$. We obtain a W3OTA $\mathcal{H}' = (Q, E', I, F)$ on Γ for $\pi^{-1}(S)$ by putting

$$E' := \{(q_x, q_y, q_z, a, w, q) \mid (q_x, q_y, q_z, \pi(a), w, q) \in E\}.$$

This construction also works for the language case of 3OTA and then preserves deterministic devices. For the last claim, let A be a deterministic 3OTA recognizing L. Assigning 1 to every transition in A, and hence extending the transitions to weighted transitions, will result in a W3OTA recognizing $\mathbb{1}_L$. □

Next we define weighted hexapolic picture automata. These devices were introduced by A. Jaya et al. in [1].

Definition 4 [1]. *A* weighted hexapolic picture automaton (WHPA) *is a 8-tuple* $\mathcal{H} = (Q, R, F_n, F_s, F_{nw}, F_{sw}, F_{ne}, F_{se})$ *consisting of finite set* Q *of states, a finite set of rules* $R \subseteq \Sigma \times K \times Q^6$, *as well as six* poles of acceptance $F_n, F_s, F_{nw}, F_{sw}, F_{ne}, F_{se} \subseteq Q$.

Precisely as with W3OTA in Definition 3, for $r = (a, k, q_n, q_s, q_{nw}, q_{sw}, q_{ne}, q_{se}) \in R$, we denote by label$(r)$ its *(input) label* a (extended then to hexagonal pictures), by weight(r) its weight w and corresponding to the six poles $\sigma_n(r) := q_n$, $\sigma_s(r) := q_s$, $\sigma_{nw}(r) := q_{nw}$, $\sigma_{ne}(r) := q_{ne}$, $\sigma_{sw}(r) := q_{sw}$, $\sigma_{se}(r) := q_{se}$. We extend the functions label and weight to hexagonal pictures by setting for a hexagonal picture $c = (c_{i,j,k}) \in R^{++H}$ over the set of rules, label$(c)_{i,j,k}$ = label$(c_{i,j,k})$ and call label(c), the label of c. A *run or (computation)* is an element $c = (c_{i,j,k}) \in R^{\ell\times m\times n}$ satisfying

$$\begin{aligned}
\forall i \leq \ell-1,\ j \leq m,\ k \leq n: &\quad \sigma_n(c_{i,j,k}) = \sigma_s(c_{i+1,j,k}),\\
\forall i \leq \ell,\ j \leq m-1,\ k \leq n: &\quad \sigma_{nw}(c_{i,j,k}) = \sigma_{se}(c_{i,j+1,k}),\\
\forall i \leq \ell,\ j \leq m,\ k \leq n-1: &\quad \sigma_{sw}(c_{i,j,k}) = \sigma_{ne}(c_{i,j,k+1}).
\end{aligned}$$

We put weight$(c) = \prod_{i,j,k}$ weight$(c_{i,j,k})$ and call weight(c) the *weight* of c. A run c is *successful* if it has its (outer) pole-states in the respective poles of acceptance, that is to say:

$$\begin{aligned}
\forall i \leq \ell, j \leq m, k \leq n: \quad & \sigma_n(c_{1,j,k}) \in F_n \quad \sigma_s(c_{\ell,j,k}) \in F_s,\\
& \sigma_{nw}(c_{i,1,k}) \in F_{nw} \quad \sigma_{se}(c_{i,m,k}) \in F_{se},\\
& \sigma_{sw}(c_{i,j,1}) \in F_{sw} \quad \sigma_{ne}(c_{i,j,n}) \in F_{ne}.
\end{aligned}$$

For a successful run c with label$(c) = p$ we shortly write $c \in$ Succ(p). The automaton *computes* a picture series $||\mathcal{H}|| : \Sigma^{++H} \to K$ such that

$$(||\mathcal{H}||, p) = \sum_{c\in \mathrm{Succ}(p)} \mathrm{weight}(c)\,,$$

called the *behaviour* of $\mathcal{H}$. The *weight* of a hexagonal picture p is the sum of the weights of all successful runs with label p. The family of hexagonal picture series computed by WHPA over Σ will be denoted by $K^{\mathrm{rec}}\langle\langle \Sigma^{++H}, \mathit{WHPA}\rangle\rangle$. We call the elements of this family *WHPA-recognizable*.

4 W3OTA-Recognizable Series are WHPA-Recognizable

We shall now convert a weighted 3 directions on-line tessellation automaton into a weighted hexapolic picture automaton. This inclusion is by defining some intermediate "hexagonal tiling" device, describing the context of pixels within their computation. Here these hexagonal tiles are encoded into the states of the new automaton.

Let K be a commutative semiring. For the proof of Theorem 1 we will first convert a given W3OTA into some "deterministic" device of a certain type via a projection similar to a construction in [1] where a Kleene-Schützenberger Theorem for hexagonal picture series is proved. However, in the present paper we apply this contruction to W3OTA rather than to WHPA. The behaviour of the constructed deterministic automaton will then be proved to be WHPA-recognizable.

Definition 5. *A weighted 3 directions on-line tessellation automaton is called* rule deterministic *if for every input label a of the underlying alphabet there is at most one transition with label a.*

Given a rule determinsitic W3OTA with transition set E, for $(q_x, q_y, q_z, a, w, q) \in E$ as a transition with label a we abbreviate (q_x, q_y, q_z, a, w, q) by $r(a)$.

Proposition 2. *Let $\mathcal{H}$ be a W3OTA over Σ. There exists a rule deterministic W3OTA $\mathcal{H}'$ over an alphabet Γ and a projection $\pi : \Gamma \to \Sigma$ satisfying $||\mathcal{H}|| = \pi(||\mathcal{H}'||)$.*

Proof. Let $\mathcal{H} = (Q, E, I, F)$ be a W3OTA over Σ and K computing S. We set $\Gamma := E$ and define a rule deterministic W3OTA over Γ by letting $\mathcal{H}' = (Q, E', I, F)$ such that

$$E' := \{(q_x, q_y, q_z, (q_x, q_y, q_z, a, w, q), w, q) \mid (q_x, q_y, q_z, a, w, q) \in E\}.$$

Clearly, for every input label$(q_x, q_y, q_z, a, w, q) \in \Gamma$ there is at most one transition with label (q_x, q_y, q_z, a, w, q) in E'. We define a projection $\pi : \Gamma \to \Sigma$ by letting

$$\pi(q_x, q_y, q_z, a, w, q) \mapsto a.$$

We have to prove $||\mathcal{H}|| = \pi(||\mathcal{H}'||)$ (*).

Let $p \in \Sigma^{\ell \times m \times n}$. If there was no successful run of p in $\mathcal{H}$, then there is no hexagonal picture in E^{++H} with a successful run in $\mathcal{H}'$, which is mapped to p by π, so (*) holds. For the other case, let $\{c_1, c_2, \ldots, c_s\} \subseteq E^{++H}$ be the set of successful computations for p in $\mathcal{H}$. These runs belong to successful runs $\{c'_1, c'_2, \ldots, c'_s\} \subseteq E'^{++H}$ in $\mathcal{H}'$ such that

$$\forall 1 \leq i \leq s : \quad \pi(label(c'_i)) = p, \sum_{1 \leq i \leq s} \text{weight}(c_i) = \sum_{1 \leq i \leq s} \text{weight}(c'_i).$$

Since there cannot be other successful runs in $\mathcal{H}'$ mapped by the projection π to p, we conclude (*):

$$(||\mathcal{H}||, p) = \sum_{1 \leq i \leq s} \text{weight}(c_i) = \sum_{\pi(p'=p)} (||\mathcal{H}'||, p') = (\pi(||\mathcal{H}'||), p).$$

□

Proposition 3. *Every hexagonal picture series that is recognizable by a rule deterministic W3OTA is WHPA-recognizable.*

Proof. Let $\mathcal{H} = (Q, E, I, F)$ be a rule deterministic W3OTA over the alphabet Σ computing a series $||\mathcal{H}|| : \Sigma^{++H} \to K$. We construct a WHPA $\mathcal{B} = (L, R, F_n, F_s, F_{nw}, F_{sw}, F_{ne}, F_{se})$ over Σ computing $||\mathcal{H}||$ by defining L as the largest subset of $(\Sigma \cup \{\#\})^{2\times2\times2}$ satisfying for all letters $a, b, c, d \in \Sigma$ and $p, q, s \in \Sigma \cup \{\#\}$:

$$\text{If } \left(\begin{array}{ccccc} & p & & q & \\ \# & & a & & s \\ & \# & & \# & \end{array} \in L \vee \begin{array}{ccccc} & \# & & \# & \\ p & & a & & \# \\ & q & & s & \end{array} \in L \right) \text{ then } \sigma_x(r(a)) \in I$$

$$\text{If } \left(\begin{array}{ccccc} & \# & & \# & \\ \# & & a & & s \\ & p & & q & \end{array} \in L \vee \begin{array}{ccccc} & q & & s & \\ p & & a & & \# \\ & \# & & \# & \end{array} \in L \right) \text{ then } \sigma_y(r(a)) \in I$$

$$\text{If } \left(\begin{array}{ccccc} & \# & & p & \\ \# & & a & & q \\ & \# & & s & \end{array} \in L \vee \begin{array}{ccccc} & p & & \# & \\ q & & a & & \# \\ & s & & \# & \end{array} \in L \right) \text{ then } \sigma_z(r(a)) \in I$$

$$\text{If } \begin{array}{ccccc} & c & & d & \\ b & & a & & \# \\ & \# & & \# & \end{array} \in L \text{ then } \sigma(r(a)) \in F.$$

We define

- $F_n = \left\{ \begin{array}{ccccc} & \# & & \# & \\ \# & & a & & b \\ & c & & d & \end{array} \mid a \in \Sigma,\ b, c, d \in \Sigma \cup \{\#\} \right\}$
- $F_s = \left\{ \begin{array}{ccccc} & c & & d & \\ b & & a & & \# \\ & \# & & \# & \end{array} \mid a \in \Sigma,\ b, c, d \in \Sigma \cup \{\#\} \right\}$
- $F_{ne} = \left\{ \begin{array}{ccccc} & \# & & \# & \\ b & & a & & \# \\ & c & & d & \end{array} \mid a \in \Sigma,\ b, c, d \in \Sigma \cup \{\#\} \right\}$
- $F_{sw} = \left\{ \begin{array}{ccccc} & b & & c & \\ \# & & a & & d \\ & \# & & \# & \end{array} \mid a \in \Sigma,\ b, c, d \in \Sigma \cup \{\#\} \right\}$
- $F_{se} = \left\{ \begin{array}{ccccc} & b & & \# & \\ c & & a & & \# \\ & d & & \# & \end{array} \mid a \in \Sigma,\ b, c, d \in \Sigma \cup \{\#\} \right\}$

- $F_{nw} = \left\{ \begin{array}{ccccc} & \# & & b & \\ \# & & a & & c \\ & \# & & d & \end{array} \mid a \in \Sigma,\ b, c, d \in \Sigma \cup \{\#\} \right\}$

We set $R = R_{ulc} \cup R_{lc} \cup R_{rc} \cup R_{ue} \cup R_{le} \cup R_m \subseteq \Sigma \times K \times L^6$ (where ulc, lc, rc, ue, le and m stands for upper left corner, left corner, right corner, upper edge, lower edge and middle respectively) with $(a, b, c, d, f, g, h, i, j, k, l, m, n, t, x, y, z \in \Sigma \cup \{\#\})$:

- $R_{ulc} = \left\{ r = \left(\begin{array}{ccccc} & \# & & a & \\ \# & & c & & d \\ & \# & & g & \end{array}, \begin{array}{ccccc} & \# & & \# & \\ \# & & a & & b \\ & c & & d & \end{array}, \begin{array}{ccccc} & \# & & \# & \\ a & & b & & \# \\ & d & & f & \end{array}, a, \text{weight}(r(a)), \begin{array}{ccccc} & a & & b & \\ c & & d & & f \\ & g & & h & \end{array} \right) \mid a \in \Sigma \right\},$

- $R_{lc} = \left\{ r = \left(\begin{array}{ccccc} & y & & z & \\ \# & & a & & b \\ & c & & d & \end{array}, \begin{array}{ccccc} & \# & & a & \\ \# & & c & & d \\ & \# & & g & \end{array}, \begin{array}{ccccc} & c & & d & \\ \# & & g & & h \\ & \# & & x & \end{array}, c, \text{weight}(r(c)), \begin{array}{ccccc} & a & & b & \\ c & & d & & f \\ & g & & h & \end{array} \right) \mid a, c, g \in \Sigma \right\},$

- $R_{rc} = \left\{ r = \left(\begin{array}{ccccc} & c & & g & \\ b & & x & & \# \\ & f & & y & \end{array}, \begin{array}{ccccc} & x & & \# & \\ f & & y & & \# \\ & z & & \# & \end{array}, \begin{array}{ccccc} & f & & y & \\ h & & z & & \# \\ & a & & \# & \end{array}, y, \text{weight}(r(y)), \begin{array}{ccccc} & b & & x & \\ d & & f & & y \\ & h & & z & \end{array} \right) \mid x, y, z \in \Sigma \right\},$

- $R_{ue} = \left\{ r = \left(\begin{array}{ccccc} & a & & b & \\ c & & d & & f \\ & g & & h & \end{array}, \begin{array}{ccccc} & \# & & \# & \\ a & & b & & i \\ & d & & f & \end{array}, \begin{array}{ccccc} & \# & & t & \\ b & & i & & z \\ & f & & j & \end{array}, b, \text{weight}(r(b)), \begin{array}{ccccc} & b & & i & \\ d & & f & & j \\ & h & & k & \end{array} \right) \mid a, b \in \Sigma \right\},$

- $R_{le} = \left\{ r = \left(\begin{array}{ccccc} & a & & b & \\ c & & d & & f \\ & g & & h & \end{array}, \begin{array}{ccccc} & d & & f & \\ g & & h & & k \\ & \# & & \# & \end{array}, \begin{array}{ccccc} & f & & j & \\ h & & k & & z \\ & \# & & t & \end{array}, h, \text{weight}(r(h)), \begin{array}{ccccc} & b & & i & \\ d & & f & & j \\ & h & & k & \end{array} \right) \mid g, h \in \Sigma \right\},$

- $R_m = \left\{ r = \left(\begin{array}{ccccc} & c & & d & \\ l & & g & & a \\ & m & & n & \end{array}, \begin{array}{ccccc} & h & & b & \\ c & & d & & f \\ & g & & a & \end{array}, \begin{array}{ccccc} & b & & i & \\ d & & f & & j \\ & a & & k & \end{array}, a, \text{weight}(r(a)), \begin{array}{ccccc} & d & & f & \\ g & & a & & k \\ & n & & t & \end{array} \right) \mid a, g, d, f \in \Sigma \right\}$

To prove $||\mathcal{H}|| = ||\mathcal{B}||$, we observe the following: Given a picture $p \in \Sigma^{++H}$ with successful computation $c \in E^{++H}$ in $\mathcal{H}$, for weight(c), the weight of the rule of every pixel of p occurs exactly once in the multiplication. On the other hand, the hexagonal tiles of an arbitrary hexagonal picture p are encoded in L. The given construction with its accepting condition defines an unambiguous weighted hexapolic picture automaton which has a unique successful run for every element in Σ^{++H}. Hence for $p \in \Sigma^{++H}$ we have

$$||\mathcal{B}||(p) = \prod_{\substack{1 \leq i \leq \ell+1 \\ 1 \leq j \leq m+1 \\ 1 \leq k \leq n+1}} \text{weight}(r(p_{i,j,k})) = (||\mathcal{H}||, p).$$

□

Similar to Proposition 1 we can prove that the family of WHPA-recognizable series are closed under projection.

Lemma 1 [1]. *Let* $\pi : \Gamma \to \Sigma$ *and* $S \in K^{\text{rec}}\langle\langle \Gamma^{++H}, WHPA \rangle\rangle$. *Then* $\pi(S) \in K^{\text{rec}}\langle\langle \Sigma^{++H}, WHPA \rangle\rangle$

Theorem 1. $K^{\text{rec}}\langle\langle \Sigma^{++H}, W3OTA \rangle\rangle \subseteq K^{\text{rec}}\langle\langle \Sigma^{++H}, WHPA \rangle\rangle$.

Proof. Immediate by Propositions 2 and 3 and Lemma 1. □

5 Conclusions

(i) The difference between the work of K. S. Dersanambika et al. [6] and D. G. Thomas et al. [21] is the following: Dersanambika et al. introduced two interesting classes of hexagonal picture languages, namely, the class of local hexagonal picture languages (HLOC) and the class of recognizable hexagonal picture languages (HREC). It is known that HLOC is a subset of HREC, where as D. G. Thomas et al. developed a recognizing device called 3 directions on-line tessellation automata (3OTA) to recognize these classes of hexagonal picture languages. It was shown [21] that the class of all hexagonal picture languages recognized by 3OTAs is exactly the family of hexagonal picture languages recognized by hexagonal tiling systems (HTS).

(ii) The difference between the work of Jaya Abraham et al. [1] and the work of the present paper is the following: Jaya Abraham et al. introduced the weighted hexapolic picture automata (WHPA) and proved the equivalence of the families of projections of rational hexagonal picture series and the series recognized by WHPA. In this paper we introduced weighted 3 directions on-line tessellation (W3OTA) and proved that every W3OTA recognizable hexagonal picture series is WHPA recognizable.

(iii) The work initiated in this paper leads to yield promising results because of the following connections:
  - Monadic Second Order Logic over hexagonal pictures and recognizability of hexagonal tiling systems (HTS),

- Weighted 3OTA and weighted logics,
- Nivat theorem for W3OTA and weighted MSO logics.

The research work done in [3,8,9] will be helpful to explore the connections of two-dimensional hexagonal picture languages and W3OTA with MSO logics.

Acknowledgements. The authors give a big thanks to QuantLA at the University of Leipzig, Germany for the support provided during 2015 and 2019. We also thank the University of Trier, Germany and Madras Christian College, Chennai, India for the visits in 2019 and 2020. The authors owe a great thanks to Manfred Droste for his fruitful discussions and comments.

References

1. Abraham, J., Dersanambika, K.S.: Characterizations of hexagonal recognizable picture series. J. Glob. Res. Math. Arch. **5**(5), 65–71 (2018)
2. Anselmo, M., Giammarresi, D., Madonia, M.: A computational model for tiling recognizable two-dimensional languages. Theor. Comput. Sci. **410**(37), 3520–3529 (2009)
3. Babari, P., Droste, M.: A Nivat theorem for weighted picture automata and weighted MSO logics. J. Comput. Syst. Sci. **104**, 41–57 (2019)
4. Bozapalidis, S., Grammatikopoulou, A.: Recognizable picture series. In: Droste, M., Vogler, H. (eds.) Special Issue on Weighted Automata, Presented at WATA 2004, Dresden, vol. 10, pp. 159–183. Journal of Automata, Languages and Combinatorics (2005)
5. Dersanambika, K.S., Krithivasan, K., Martin-Vide, C., Subramanian, K.G.: Hexagonal pattern languages. In: Klette, R., Žunić, J. (eds.) IWCIA 2004. LNCS, vol. 3322, pp. 52–64. Springer, Heidelberg (2004). https://doi.org/10.1007/978-3-540-30503-3_4
6. Dersanambika, K.S., Krithivasan, K., Martín-Vide, C., Subramanian, K.G.: Local and recognizable hexagonal picture languages. IJPRAI **19**(7), 853–871 (2005)
7. Fichtner, I.: Weighted picture automata and weighted logics. Theory Comput. Syst. **48**(1), 48–78 (2011)
8. Giammarresi, D., Restivo, A.: Two-dimensional languages. In: Rozenberg, G., Salomaa, A. (eds.) Handbook of Formal Languages, vol. III, pp. 215–267. Springer, Heidelberg (1997). https://doi.org/10.1007/978-3-642-59126-6_4
9. Giammarresi, D., Restivo, A., Seibert, S., Thomas, W.: Monadic second-order logic over rectangular pictures and recognizability by tiling systems. Inf. Comput. **125**(1), 32–45 (1996)
10. Krithivasan, K., Siromoney, R.: Array automata and operations on array languages. Int. J. Comput. Math. **4**(A), 3–40 (1974)
11. Mahajan, M., Krithivasan, K.: Hexagonal cellular automata. In: Narasimhan, R. (ed.) A Perspective in Theoretical Computer Science - Commemorative Volume for Gift Siromoney, World Scientific Series in Computer Science, vol. 16, pp. 134–164. World Scientific (1989)
12. Mäurer, I.: Characterizations of recognizable picture series. Theor. Comput. Sci. **374**(1–3), 214–228 (2007)
13. Middleton, L., Sivaswamy, J.: Hexagonal Image Processing: A Practical Approach. Advances in Pattern Recognition, Springer, London (2005). https://doi.org/10.1007/1-84628-203-9

14. Paramasivan, M., Thomas, D.G., Immanuel, S.J., Lakshmi, M.G.: Weighted hexagonal picture automata. In: Droste, M., Gastin, P., Guillon, P., Monmege, B., Vogler, H. (eds.) 10th International Workshop - Weighted Automata: Theory and Applications. WATA 2020 (2021)
15. Paramasivan, M., Thomas, D.G., Immanuel, S.J., Lakshmi, M.G.: Weighted hexagonal picture automata. In: Maletti, A. (ed.) 31. Theorietag "Automaten und Formale Sprachen", pp. 37–40 (2021)
16. Siromoney, G., Siromoney, R.: Hexagonal arrays and rectangular blocks. Comput. Graph. Image Process. **5**, 353–381 (1976)
17. Siromoney, G., Siromoney, R., Krithivasan, K.: Abstract families of matrices and picture languages. Comput. Graph. Image Process. **1**, 284–307 (1972)
18. Siromoney, G., Siromoney, R., Krithivasan, K.: Picture languages with array rewriting rules. Inf. Control (Now Inf. Comput.) **22**(5), 447–470 (1973)
19. Siromoney, G., Siromoney, R., Krithivasan, K.: Array grammars and kolam. Comput. Graph. Image Process. **3**, 63–82 (1974)
20. Subramanian, K.G., Revathi, L., Siromoney, R.: Siromoney array grammars and applications. Int. J. Pattern Recognit. Artif. Intell. **3**, 333–351 (1989)
21. Thomas, D.G., Begam, M.H., David, N.G., de la Higuera, C.: Hexagonal array acceptors and learning. In: Mukund, M., Rangarajan, K., Subramanian, K.G. (eds.) Formal Models, Languages and Applications [this Volume Commemorates the 75th Birthday of Prof. Rani Siromoney]. Series in Machine Perception and Artificial Intelligence, vol. 66, pp. 364–378. World Scientific (2007)
22. Yanagisawa, K., Nagata, S.: Fundamental study on design system of kolam pattern. Forma **22**, 31–46 (2007)

A Myhill-Nerode Theorem for Finite State Matrix Automata and Finite Matrix Languages

Abhisek Midya[1(✉)] and D. G. Thomas[2]

[1] CMR Institute of Technology, Bengaluru, India
abhisekmidyacse@gmail.com
[2] Department of Mathematics, Madras Christian College, Chennai, India

Abstract. We propose a deterministic version of *finite state matrix automaton (DFSMA)* which recognizes *finite matrix languages (FML)*. Our main result is a generalization of the classical Myhill-Nerode theorem for $DFSMA$. Our generalization requires the use of two relations to capture the additional structure of $DFSMA$. *Vertical equivalence* $\equiv_v$ captures that words sharing the same vertical location, *horizontal equivalence* $\equiv_h$ captures that words sharing the same horizontal location. A finite matrix language is defined to be regular if relations $\equiv_v$ and $\equiv_h$ exist that satisfy certain conditions, in particular, they have finite index. We show that the language associated to a $DFSMA$ is regular, and we construct, for each finite matrix language, a $DFSMA$ that accepts this language. Our result provides a foundation for learning algorithms for $DFSMA$.

Keywords: Myhill-Nerode equivalence · Deterministic finite state matrix automata · Finite matrix languages

1 Introduction

Grammatical inference is the realistic common area of research between machine learning and formal language theory. The concept of Grammatical inference deals with the automatic learning of grammars, automata and other language describing devices. We attempt to satisfy both (machine Learning and formal Language Theory) parts of the potential readership of this paper, as it has been shown that the inter-dependencies between both areas are strong. The basic motivation for investigating the learning of $DFSMA$, is to investigate the connection between matrix languages and *automata learning.*

1.1 Learning Aspects

It has been investigated that the passive learning problem of finding a minimal *deterministic finite automata (DFA)* is NP-hard, and it is compatible with a finite set of positive and negative examples, in [16]. In spite of this, many

R. P. Barneva et al. (Eds.): IWCIA 2022, LNCS 13348, pp. 154–170, 2023.
https://doi.org/10.1007/978-3-031-23612-9_10

DFA identification algorithms have been developed. [2] presented an efficient algorithm for active learning a regular language L, which assumes a minimally adequate teacher (MAT) that answers two types of queries about L, with a membership query, the algorithm asks whether or not a given word w is in L, and with an equivalence query it asks whether or not the language L_H of an hypothesized $DFA\ H$ is equal to L. If L_H and L are not same, a word which is in the symmetric difference of the two languages gets returned. Also there are alternative versions of algorithms for learning regular languages in the MAT model appeared in [7,8,17,30,40]. The limits of the model were investigated in [3,5]. There was an interesting question arose, whether it can be extended to the supersets of regular sets.

In [29] Radhakrishnan and Nagaraja proposed a method for the inference of *even linear languages* from positive examples, also the proposed method can be used in a hierarchical manner to infer grammars for complex pictures. The interesting work of [36] and [34] established the reduction technique of the learning of even linear languages (introduced in [1]) to the learning of regular languages. Also, the usefulness of the concept of control languages (originating from [15]) was shown in the reduction of the learning problem of languages through controlled fixed grammars in [20,21,36,38,39]. In particular, Takada used this concept to develop an efficient learning algorithm, called "even equal matrix languages" [37,39]. Also, in [23,24] polynomial time learning algorithms are proposed for interesting subclasses of contextual array and string languages respectively. Also, in [25], a two dimensional automaton had been defined for array languages. In this way, we realize the importance of learning matrix languages and in this paper we deal with *finite matrix languages.*

In this article, we propose a deterministic version of finite state matrix automata ($DFSMA$) which can recognize finite matrix languages (FML). More importantly, we establish a Myhill-Nerode theorem for $DFSMA$ and FML. We know that the Myhill-Nerode theorem refers to a single equivalence relation on words, and constructs a DFA in which states are equivalence classes, our generalization requires the use of two relations to capture the additional structure of $DFSMA$. The Myhill-Nerode theorem makes the platform to develope a learning algorithm for $DFSMA$ using query learning model [2].

Myhill-Nerode theorems are of pivotal importance for learning algorithms. Angluin's classical L^* algorithm for active learning of regular languages, as well as improvements such as [11,19,30], use an observation table to approximate the Myhill-Nerode congruence. Maler and Steiger [22] established a Myhill-Nerode theorem for ω-languages that serves as a basis for a learning algorithm described in [4]. The SL^* algorithm for active learning of register automata of Cassel et al. [10] is directly based on a generalization of the classical Myhill-Nerode theorem to a setting of data languages and register automata (extended finite state machines).

1.2 Formal Language Aspects

Syntactic approaches, on account of their structure-handling capability, have played an important role in the problem of description of picture patterns considered as connected digitized, finite arrays of symbols. Pioneering work in suggesting and applying a linguistic model for the solution of nontrivial problems in picture processing was presented in [27]. Using the techniques of formal string language theory, various types of picture or array grammars have been introduced and investigated in [9,13,14,31,32]. Most of the array grammars are based on Chomskian string grammars. Some recent results on picture languages can be found in [6,12,26].

A picture can be represented as a $m \times n$ matrix in which each entry is a_{ij} where $1 \leq i \leq m, 1 \leq j \leq n$. By an operation on a digitized picture is meant a function which transforms a given picture matrix into another one. Programming languages have types and a function may have an argument, which is of type matrix, and it is not trivial to handle computationally. For practical purposes it is desirable to work with operations on digitized pictures which can be defined in terms of functions having considerably fewer arguments.

In this paper we deal with a linguistic model for the generation of matrices (rectangular arrays of terminals) by the substitution of *regular sets* [18] into well-known families of formal languages. In formal language theory the substitution operator operates on 'string languages' (languages made up of strings of terminals). Here the substitution operator operates on a 'string language' and the resultant is a 'matrix language' (language whose sentences are matrices, i.e., $m \times n$ arrays of terminals). In particular, we recall finite/regular matrix languages and we propose the corresponding deterministic version of automaton, called deterministic finite state matrix automata ($DFSMA$). Matrix grammar refers to a grammar in which the production rules are applied together in fixed sets. There are several variants where the rewriting rules are regular, context-free or context-sensitive with arrays of terminals in the place of strings of terminals. Furthermore, in order to obtain richer families, restrictions are imposed on the use of production rules in well known families of grammars. Several such studies are available in the literature [33]. In this paper, our focus is on FML where the rewriting rules are regular. Some interesting classes of pictures including certain letters of the alphabet, kolam, (traditional picture patterns used to decorate the floor in South Indian homes) and wall paper designs (repetitive patterns) can be generated by finite matrix grammars.

The remainder of this paper is organized as follows. Section 2 recalls the definition of FML and Subsect. 2.1 presents examples of FML for better understanding. Section 3 proposes the definition of $DFSMA$ and examples are discussed in Subsect. 3.1 Section 4 presents some of the important results about FML. In Sect. 5, we discuss the Myhill-Nerode equivalence and establish the Myhill-Nerode theorem for $DFSMA$ with illustrations with examples in Subsect. 5.1 and 5.2. Section 6 concludes the work and shows a future direction of work.

2 Finite Matrix Language (FML)

We recall the definition of FML [35] based on *right linear grammar* [18].

Definition 1 (Finite Matrix Language (FML)). *A Finite matrix grammar (FMG) is a pair $G = (G_1, G_2)$, where $G_1 = (V_1, I_1, P_1, S)$ is a right linear grammar with V_1, a finite set of horizontal non-terminals, I_1, a finite set of intermediates (i.e., $I_1 = \{S_1, S_2, ..., S_k\}$), P_1 is a finite set of right linear grammar production rules called* horizontal production rules, *and S, the start symbol where $S \in V_1$ and $V_1 \cap I_1 = \phi$. We define $G_2 = (\bigcup_{i=1}^{k} G_{2i})$ where $G_{2i} = (V_{2i}, I_{2i}, P_{2i}, S_i)$ is a right linear grammar, V_{2i} is a finite set of vertical non terminals, I_{2i} is a finite set of vertical terminals, S_i is the start symbol, P_{2i} is a finite set of vertical production rules, $V_{2i} \cap V_{2j} = \phi$, if $i \neq j$. The horizontal derivations and vertical derivations are denoted as $\underset{h}{\Rightarrow}, \underset{v}{\Rightarrow}$ respectively. The derivations are obtained by first applying horizontal production rules and then the vertical production rules. Firstly a horizontal string $S_1 S_2 ... S_k \in {I_1}^*$ has been generated using horizontal production rules P_1 in G_1, i.e., $S \underset{h}{\overset{*G_1}{\Longrightarrow}} S_1 S_2 ... S_n$. A vertical derivation has been defined as follows : if there are rules $S_i \downarrow a_{1i} A_i, A_i \downarrow a_{2i} B_{3i}, B_{ji} \downarrow a_{ji} B_{j+1} i, B_{ri} \downarrow a_{ri}, 3 \leq j \leq r-1$ in G_{2i}, where $i \in \{1, ..., k\}$ for $i = 1, ..., n$, then the matrices will be generated in the following way :*

$$S \underset{h}{\overset{*}{\Rightarrow}} \begin{bmatrix} S_1 \ldots S_n \end{bmatrix} \underset{v}{\Rightarrow} \begin{bmatrix} a_{11} \ldots a_{1n} \\ A_1 \ldots A_n \end{bmatrix} \underset{v}{\Rightarrow} \begin{bmatrix} a_{11} & \ldots & a_{1n} \\ . & \ldots & . \\ a_{(r-1)1} & \ldots & a_{(r-1)n} \\ B_{r_1} & & B_{r_n} \end{bmatrix}$$

$$\underset{v}{\Rightarrow} \begin{bmatrix} a_{11} & \ldots & a_{1n} \\ . & \ldots & . \\ a_{(r-1)1} & \ldots & a_{(r-1)n} \\ a_{r1} & & a_{rn} \end{bmatrix}$$

Here $\underset{v}{\overset{*}{\Rightarrow}}$ is the transitive closure of $\Rightarrow$. The vertical derivation gets terminated if $B_{ri} \rightarrow a_{ri}$ are all terminal rules in G_{2i} where $i = 1, ..., n$.

The set of all matrices is defined as follows:

$$L(G) = \{r \times n \text{ arrays } [a_{ij}] \mid i = 1, ..., r, j = 1, ..., n, r, n \geq 1, S \underset{h}{\overset{*G_1}{\Longrightarrow}} S_1 S_2 ... S_n \underset{v}{\overset{*G_2}{\Longrightarrow}} [a_{ij}]\}$$

Remark 1. A single non terminal is produced in each column as the rules are in the form of $A \rightarrow aB, A \rightarrow a$ where $a \in I_2$.

Remark 2. No cell in any column is blank or empty as a rule from one of G_{2i} where $i = 1, ..., k$ is supposed to be ϵ free.

Remark 3. In the definition of finite matrix grammar, the production rules are applied in a simultaneous fashion. In that sense, the grammars are matrix grammars. Moreover, the definition is more general in that the set of rules applied at one stage is not fixed but restricted by the horizontal string generated at the first stage. The name matrix grammar is retained to refer to this generalization also. Importantly, it should be noted that in this paper, the matrix grammars generate matrix languages whose sentences are matrices ($m \times n$ rectangular arrays). On the other hand, in the formal language theory, the matrix languages are considered to be string languages where sentences are strings-generated by grammars written in the form of a matrix.

Remark 4 (Notation). If L' is the language generated by G_1, and $R_1, ..., R_k$ (the subsets of) the regular sets corresponding to G_{2i} where $i = 1, ..., k$, then we can write $L(G) = (L') : \ : (R_1, ..., R_k)$

2.1 FML - Examples

Example 1. Let $G = (G_1, G_2)$ where $G_1 = (\{S, S'\}, \{S_1, S_2\}, \{S \rightarrow S_1S', S' \rightarrow S_2S', S' \rightarrow S_2\}, S), G_2 = G_{21} \cup G_{22}, G_{21} = (\{S_1, A\}, \{X\}, \{S_1 \rightarrow XA, A \rightarrow XA, A \rightarrow X\}, S_1), G_{22} = (\{S_2, A\}, \{., X\}, \{S_2 \rightarrow .A, A \rightarrow .A, A \rightarrow X\}, S_2)$, then $L = \{S_1S_2^n \mid n \geq 1\}, R_1 = \{X^m \mid m \geq 1\}, R_2 = \{(.)^{m-1}X \mid m \geq 1\}$ and $L(G) = (L') : \ : (R_1, R_2)$. $L(G)$ is a finite matrix language and consists of $m \times n$ arrays $(m > 1, n > 1)$ describing the token L.

G generates $m \times n$ matrices $(m > 1, n > 1)$ which describe the token L. We illustrate by generating a 6×5 matrix from G.

$$S \overset{*G_1}{\underset{h}{\Longrightarrow}} \begin{bmatrix} S_1 & S_2 & S_2 & S_2 & S_2 \end{bmatrix} \overset{G_2}{\underset{v}{\Longrightarrow}} \begin{bmatrix} X & . & . & . & . \\ A & A & A & A & A \end{bmatrix} \overset{G_2}{\underset{v}{\Longrightarrow}} \begin{bmatrix} X & . & . & . & . \\ X & . & . & . & . \\ A & A & A & A & A \end{bmatrix}$$

$$\overset{G_2}{\underset{v}{\Longrightarrow}} \begin{bmatrix} X & . & . & . & . \\ X & . & . & . & . \\ X & . & . & . & . \\ A & A & A & A & A \end{bmatrix} \overset{G_2}{\underset{v}{\Longrightarrow}} \begin{bmatrix} X & . & . & . & . \\ X & . & . & . & . \\ X & . & . & . & . \\ X & . & . & . & . \\ A & A & A & A & A \end{bmatrix} \overset{G_2}{\underset{v}{\Longrightarrow}} \begin{bmatrix} X & . & . & . & . \\ X & . & . & . & . \\ X & . & . & . & . \\ X & . & . & . & . \\ X & . & . & . & . \\ A & A & A & A & A \end{bmatrix}$$

$$\overset{G_2}{\underset{v}{\Longrightarrow}} \begin{bmatrix} X & . & . & . & . \\ X & . & . & . & . \\ X & . & . & . & . \\ X & . & . & . & . \\ X & . & . & . & . \\ X & X & X & X & X \end{bmatrix} \in L(G)$$

Example 2. Let $G = (G_1, G_2)$ where $G_1 = (\{S, S'\}, \{S_1, S_2\}, \{S \rightarrow S'S, S \rightarrow S'S_1, S' \rightarrow S_1S_2S_2S_2\}, S), G_2 = G_{21} \cup G_{22}, G_{21} = (\{S_1, A\}, \{X\}, \{S_1 \rightarrow XA, A \rightarrow XA, A \rightarrow X\}, S_1), G_{22} = (\{S_2, S'_2\}, \{., X\}, \{S_2 \rightarrow S'_2S_2, S_2 \rightarrow S'_2X, S'_2 \rightarrow X..\}, S_2)$, then $L = \{S_1S_2S_2S_2^nS_1 \mid n \geq 1\}, R_1 = \{X^{m_1} \mid m_1 \geq 1\}, R_2 = \{(X..)^{m_2}X \mid m_2 \geq 1\}$ and $L(G) = (L') : \ : (R_1, R_2)$. $L(G)$ is regular and describes rectangular grids made up of $r \times s$ rectangles $(r = 1, 2, \ldots, s = 1, 2, \ldots)$ of the same size.

G generates the 2×4 grid $\in L(G)$.

$$S \overset{*G_1}{\underset{h}{\Longrightarrow}} \begin{bmatrix} S_1 & S_2 & S_2 & S_2 & S_1 & S_2 & S_2 & S_2 & S_1 & S_2 & S_2 & S_2 & S_1 & S_2 & S_2 & S_2 & S_1 \end{bmatrix}$$

$$\xrightarrow[h,v]{G_1,G_2} \begin{bmatrix} X & S'_2 & S'_2 & S'_2 & X & S'_2 & S'_2 & S'_2 & X & S'_2 & S'_2 & S'_2 & X & S'_2 & S'_2 & S'_2 & X \\ A & S_2 & S_2 & S_2 & A & S_2 & S_2 & S_2 & A & S_2 & S_2 & S_2 & A & S_2 & S_2 & S_2 & A \end{bmatrix}$$

$$\xrightarrow[v]{*G_1,*G_2} \begin{bmatrix} X & X & X & X & X & X & X & X & X & X & X & X & X & X & X & X & X \\ X & . & . & . & X & . & . & . & X & . & . & . & X & . & . & . & X \\ X & . & . & . & X & . & . & . & X & . & . & . & X & . & . & . & X \\ X & X & X & X & X & X & X & X & X & X & X & X & X & X & X & X & X \\ X & . & . & . & X & . & . & . & X & . & . & . & X & . & . & . & X \\ X & . & . & . & X & . & . & . & X & . & . & . & X & . & . & . & X \\ X & X & X & X & X & X & X & X & X & X & X & X & X & X & X & X & X \end{bmatrix} \in L(G)$$

In the next section we define *deterministic finite state matrix automata (DFSMA)*, to correspond to families of matrices (FML).

3 Deterministic Finite State Matrix Automata (DFSMA)

We define a deterministic version of finite state matrix automata.

Definition 2 (Deterministic finite state matrix automaton(DFSMA)). *A* deterministic finite state matrix automaton *is defined as a* 9 *tuple* $DFSMA = (Q, I, T, \delta, \delta', S, F', F, \$)$ *where*

- T *: Set of horizontal symbols and* $|T|$ *is the number of horizontal symbols and it denotes the number of horizontal states also.*
- I *: Set of vertical symbols.*
- $Q = (\bigcup_{i=1}^{k} Q_i) \cup Q'$ *is the finite set of states where* Q_i *is the finite set of vertical states corresponding to each horizontal state* $S_i \in Q'$, Q' *is the finite set of horizontal states where* $(\bigcup_{i=1}^{k} Q_i) \cap Q' = \phi$ *and* $|Q'| = |T| = k$.
- *Vertical transition function* $\delta : (Q_i \cup Q') \times I \rightarrow Q_i$. *A* vertical transition *of* $DFSMA$ *is of the form :* $\delta(q_i, x) = q_j$ *where* $q_i, q_j \in (Q_i \cup Q')$, *if* $q_i = S_j \in Q'$ *where* $i = 0$, *then it is the first transition of the automata which starts from the start state* $S_0 \in Q'$, *and if* $i \neq 0$ *then it is the first transition from any other horizontal state* $S_j \in Q'$. *The vertical transition function* δ *can be extended to* $\hat{\delta}$ *that operates on states and strings (as opposed to states and symbols), such that,* $\hat{\delta}(q_i, \epsilon) = q_i, \hat{\delta}(q_i, xa) = \delta(\hat{\delta}(q_i, x), a)$.

$$Q_i = \{q_k \mid (\hat{\delta}(q_j, xa) = q_k) \wedge (q_j \in Q_i \vee q_j = S_j)\}.$$

- *Horizontal transition function* $\delta' : F' \times \{\$\} \rightarrow Q'$, *then the horizontal transition is of the form* $\delta'(f_i, \$) = S_j$. $(F' \cap Q_i)$ *is a singleton set which contains only* f_i, $x_{*,i}$ *denotes the ith column vector.*

$$Q' = \{S_j \mid \delta'(f_i, \$) = S_j \wedge (f_i \in F') : \delta'(\hat{\delta}(S_i, x_{*,i}), \$) = S_j\}$$

- $S_0 \in Q'$ *: Initial state*
- *We define finite set of vertically accepting states* $F' = (\bigcup_{i=1}^{k} f_i)$, *such that,*

$$F' = \{f_i \mid \hat{\delta}(q_i, x_{*,i}) = f_i \wedge \delta'(f_i, \$) = S_j \wedge (1 \leq i \leq k)\}$$

- $F = f_k$ *denotes the final state of DFSMA, such that,* $\hat{\delta}(S_k, x_{*,k}) = f_k$ *where* $f_k \in Q_k$ *and* k *is the number of horizontal states.*

- *\$ is the end marker where* $\$ \notin I$

The (standard) semantics of DFSMA is defined as follows.

A *vertical run of DFSMA over* a vertical word $w_v = x_0 \cdots x_n$, is a sequence of steps of $DFSMA$:

$$S_i \xrightarrow{x_0} q_1 \quad \ldots \quad q_n \xrightarrow{x_n} q_{n+1}$$

We say a vertical run is *accepting* if $q_{n+1} = f_i \in F'$. It is *rejecting* if $q_{n+1} \notin F'$. Vertical word w_v is *accepted (rejected)* if $DFSMA$ has an accepting (rejecting) run over w_v. There must be an accepting vertical run corresponding each intermediate state $S_j \in Q'$ where $1 \leq j \leq k$.

A *horizontal run of DFSMA over* a horizontal word $w_h = x_{*,0}x_{*,1} \cdots x_{*,n}$ where $x_{*,i}$ denotes the *ith* column of the matrix, a horizontal word w_h is a sequence of column vectors where each column vector is followed by the end marker $\$ \notin I$, such that, $w_h = x_{*,0}\$ \; x_{*,1}\$ \cdots x_{*,n}$, is a sequence of steps of $DFSMA$:

$$S_0 \xrightarrow{x_{*,0}} f_0 \xrightarrow{\$} S_1 \quad \ldots \quad S_n \xrightarrow{x_{*,n}} f_n.$$

We say a horizontal run is *accepting* if $f_n = f_k \in F'$ where $|T| = k$. Horizontal word w_h is *accepted (rejected)* if $DFSMA$ has an accepting (rejecting) run over w_h. The *language* of $DFSMA$, notation $L(DFSMA)$, is the set of all horizontal words or images that are accepted by $DFSMA$.

Our proposed DFSMA has single initial state S_0. The automaton starts reading from the first column of the input matrix. All the vertical and horizontal moves are unique. It reaches f_i and then using enmarker $\$_i$ goes to another column corresponding to some S_j. If $i = j$ then it will create a loop, otherwise it will go to new horizontal state. If there exist an input matrix $m \times n$ then the automaton reads till the *nth* column, there will not be any endmarker $\$_n$ followed by the *nth* column, so the last horizontal move is based on the endmarker $\$_{n-1}$ which takes the automaton to S_n. The automaton will finish the reading with the nth set of vertical moves corresponding to S_n, and it ends up with $f_n = F$, it has single final state (Fig. 1).

3.1 DFSMA - Examples

Example 3. We define $DFSMA = (Q, I, T, \delta, \delta', S, F, F', \$)$ which can accept the language of Example 2.1 (L token).

- $Q = \{Q_1 \cup Q_2 \cup Q'\}$ where $Q_1 = \{q_{1_1}\}, Q_2 = \{q_{2_1}, q_{2_2}\}$ and $F' = \{q_{1_1}, q_{2_2}\}$ where $q_{1_1} = f_1, q_{2_2} = f_2$ and $Q' = \{S_1, S_2\}$,
- $I = \{., x\}$,
- $T = \{S_1, S_2\}$,
- $F' = \{f_1 = q_{1_1}, f_2 = q_{2_2}\}$, and $f_2 = q_{2_2}$ is the final state.
- S_1 is the initial state.
- \$ is the end marker where $\$ \notin I$
- Vertical transitions (δ) and horizontal transitions (δ') are given below.

1. $\delta(S_1, x) = q_{1_1}$ where $q_{1_1} = f_1$
2. $\delta(q_{1_1}, x) = q_{1_1}$
3. $\delta'(q_{1_1}, \$) = S_2$
4. $\delta(S_2, .) = q_{2_1}$
5. $\delta(q_{2_1}, .) = q_{2_1}$
6. $\delta(q_{2_1}, x) = q_{2_2}$
7. $\delta'(q_{2_2}, \$) = S_2$

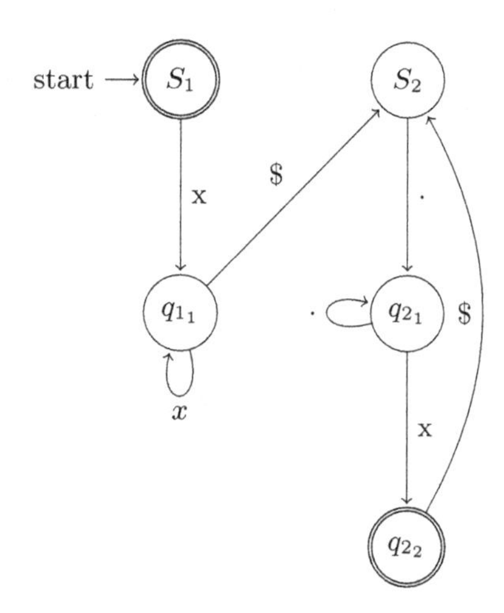

Fig. 1. Deterministic finite state matrix automaton.

Example 4. We define $DFSMA = (Q, I, T, \delta, \delta', S, F, F', \$)$ which can accept the language of Example 2.2 (L token).

- $Q = \{Q_1 \cup Q_2 \cup Q_3 \cup Q_4 \cup Q'\}$ where $Q_1 = \{q_{1_1}\}, Q_2 = \{q^1_{2_1}, q^1_{2_2}, q^1_{2_3}\},, Q_2 = \{q^1_{2_1}, q^1_{2_2}, q^1_{2_3}\}, Q_3 = \{q^2_{2_1}, q^2_{2_2}, q^2_{2_3}\}, Q_4 = \{q^3_{2_1}, q^3_{2_2}, q^3_{2_3}\}$ and $F' = \{q_{1_1}, q^1_{2_3}, q^2_{2_3}, q^3_{2_3}\}$ where $q_{1_1} = f_1, q^1_{2_3} = f_2, q^2_{2_3} = f_3, q^3_{2_3} = f_4$ and $Q' = \{S_1, S^2_1, S^2_2, S^2_3\}$,
- $I = \{., x\}$,
- $T = \{S_1, S^2_1, S^2_2, S^2_3\}$,
- $F' = \{f_1 = q_{1_1}, f_2 = q^2_{2_3}, f_3 = q^2_{2_3}, f_4 = q^3_{2_3}\}$, and $f_1 = q_{1_1}$ is the final state.
- S_1 is the initial state.
- \$ is the end marker where $\$ \notin I$
- Vertical transitions (δ) and horizontal transitions (δ') are given below (Fig. 2).

1. $\delta(S_1, x) = q_{1_1}$ where $q_{1_1} = f_1$
2. $\delta(q_{1_1}, x) = q_{1_1}$
3. $\delta'(q_{1_1}, \$) = S^2_1$
4. $\delta(S^2_1, X) = q^1_{2_1}$

5. $\delta(q^1_{2_1}, .) = q^1_{2_2}$
6. $\delta(q^1_{2_2}, .) = S^2_1$
7. $\delta(q^1_{2_2}, X) = q^1_{2_3}$
8. $\delta'(q^1_{2_3}, \$) = S^2_2$
9. $\delta(S^2_2, X) = q^2_{2_1}$
10. $\delta(q^2_{2_1}, .) = q^2_{2_2}$
11. $\delta(q^2_{2_2}, .) = S^2_2$
12. $\delta(q^2_{2_2}, X) = q^2_{2_3}$
13. $\delta'(q^2_{2_3}, \$) = S^2_3$
14. $\delta(S^2_3, X) = q^3_{2_1}$
15. $\delta(q^3_{2_1}, .) = q^3_{2_2}$
16. $\delta(q^3_{2_2}, .) = S^2_3$
17. $\delta(q^3_{2_2}, X) = q^3_{2_3}$
18. $\delta'(q^3_{2_3}, \$) = S_1$

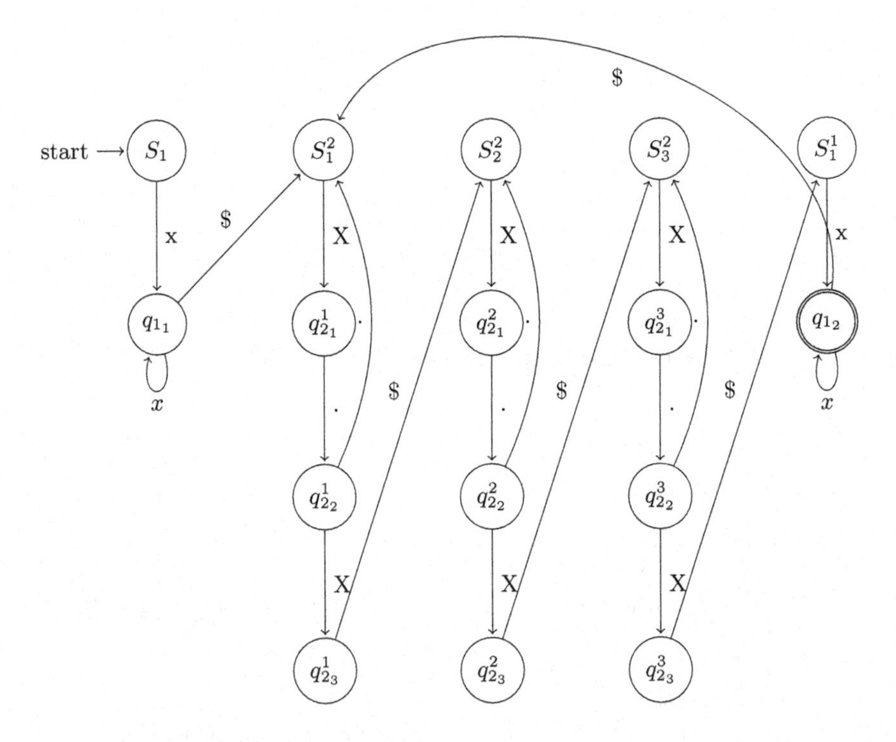

Fig. 2. Deterministic finite state matrix automaton for 2×4 grid

In the next section, we show some of the important results of FML.

4 Properties of Finite Matrix Languages

In this section, we summarize some of the important closure properties of FML in Table 1. Also, we present some of the decidable results of FML in Table 2. The following important results had been eastablished in [28].

Table 1. Closure property results

Union	Closed
Concatenation	Closed
Kleene Closure	Closed
Complementation	Closed
Intersection	Closed

Table 2. Decidability results

$L = \phi$	Decidable
$L = \Sigma^*$	Decidable
$L_1 = L_2$	Decidable
$w \in L$	Decidable

As the membership problem, ($w \in L$), is decidable, it would be possible to apply MAT model to learn $DFSMA$. In order to apply MAT model, very importantly we must establish the important Myhill - Nerode theorem for $DFSMA$ and FML.

In the next section, we discuss the Myhill - Nerode equivalence of $DFSMA$ and FML.

5 Myhill - Nerode Equivalence

The Myhill-Nerode equivalence [2] considers two words w and w' of a language L equivalent if there does not exist a suffix u that distinguishes them, that is, only one of the words wu and $w'u$ is in L. The Myhill-Nerode theorem states that L is regular if and only if this equivalence relation has a finite index, and moreover that the number of states in the smallest deterministic finite automaton (DFA) recognizing L is equal to the number of equivalence classes. In this section, we present a Myhill-Nerode theorem for DFSMA and FML. In string languages, Myhill and Nerode only needs a single equivalence relation on words to capture DFAs, we need two relations $\equiv_v$, $\equiv_h$ on words to capture the richer structure of DFSMA.

Here, first we define *Right invariant vertical equivalence relation* and *Right invariant horizontal equivalence relation* in order to establish the Myhill - Nerode theorem for DFSMA and FML.

Definition 3 (Right invariant vertical equivalence relation). *A vertical equivalence relation $\equiv_v$ on I^* is said to be right invariant if, for $x, y, z \in I^*$, $x \equiv_v y \Longrightarrow \forall z(xz \equiv_v yz)$.*

Example 5. Suppose $L = (L') : \quad : (R_1, ..., R_k)$ be a language over I^{**} where each $R_i, i \geq 1$ be a language over I. If there exist an equivalence relation $\equiv_R$ on

I^* then it is a right invariant equivalence relation on I^*. We define $x \equiv_R y$ if and only if $\forall z(xz \in R \iff yz \in R)$. It can be easily cross-checked that $x \equiv_R y$ is an equivalence relation as it satisfies reflexive, symmetric, and transitive properties. We assume that $x \equiv_R y$ where $x, y \in I^*$ and $z \in I^*$ be an arbitrary, now our claim is $xz \equiv_R yz$, that is, $(\forall w)(xzw \in R \iff yzw \in R)$. For any arbitrary $w \in I^*$, we write $u = zw$, now since $x \equiv_R y$, we have $xu \in R \iff yu \in R$, so $xzw \in R \iff yzw \in R$.

Definition 4 (Right invariant horizontal equivalence relation). *Horizontal equivalence relation $\equiv_h$ on I^{**} is said to be right invariant if, for $x, y, z \in I^{**}$, $x \equiv_h y \Longrightarrow \forall z(xz \equiv_h yz)$.*

Example 6. Suppose $L = (L') : \; : (R_1, ..., R_k)$ be a language over I^{**}. If there exist an equivalence relation $\equiv_{L'}$ on I^{**} then it is a right invariant equivalence relation on I^{**}. We define $x \equiv_{L'} y$ if and only if $\forall z(xz \in L' \iff yz \in L')$ where $x, y, z \in I^{**}$. It can be easily verified that $x \equiv_{L'} y$ is an equivalence relation as it satisfies reflexive, symmetric, and transitive properties. We assume that $x \equiv_{L'} y$ where $x, y \in I^{**}$ and $z \in I^{**}$ be an arbitrary, Now we claim $xz \equiv_{L'} yz$, that is, $(\forall w)(xzw \in L' \iff yzw \in L')$. For any arbitrary $w \in I^{**}$, we write $u = zw$, now since $x \equiv_{L'} y$, we have $xu \in L' \iff yu \in L'$, so $xzw \in L' \iff yzw \in L'$.

Lemma 1. *Suppose $DFSMA = (Q, I, T, \delta, \delta', S, F, F', \$)$. There exist a vertical equivalence $\equiv_{DFSMA_v}$ and it is right invariant.*

Proof. We define $x \equiv_{DFSMA_v} y$ if and only if $\hat{\delta}(s_0, x) = \hat{\delta}(s_0, y)$ where $x, y \in I^*$. It is trivial that $x \equiv_{DFSMA_v} y$ is an equivalence relation as it satisfies reflexive, symmetric and transitive properties. We consider $x \equiv_{DFSMA_v} y$ that is $\hat{\delta}(s_0, x) = \hat{\delta}(s_0, y)$, for $z \in I^*$, $\hat{\delta}(s_0, xz) = \hat{\delta}(\hat{\delta}(s_0, x), z) = \hat{\delta}(\hat{\delta}(s_0, y), z) = \hat{\delta}(s_0, yz)$ as we know already that $\hat{\delta}(s_0, x) = \hat{\delta}(s_0, y)$. (See Definition 3)

Lemma 2. *Suppose $DFSMA = (Q, I, T, \delta, \delta', S, F, F', \$)$. There is a horizontal relation $\equiv_{DFSMA_h}$ on I^{**}, and it is right invariant.*

Proof. Suppose $w = x_{*,0} \cdots x_{*,n}$ and $w' = y_{*,0} \cdots y_{*,n'}$ then $w \equiv_{DFSMA_h} w'$ if and only if -

$$\delta'(\hat{\delta}(s_0, x_{*,0}), \$) = \delta'(\hat{\delta}(s_0, y_{*,0}), \$)$$
$$\delta'(\hat{\delta}(s_1, x_{*,1}), \$) = \delta'(\hat{\delta}(s_1, y_{*,1}), \$)$$
$$\vdots$$
$$\delta'(\hat{\delta}(s_n, x_{*,n}), \$) = \delta'(\hat{\delta}(s_{n'}, y_{*,n'}), \$)$$

Now it can be easily understood that $\equiv_h$ is right invariant equivalence relation if, for all $z \in I^{**}$, wz and $w'z$ leads $DFSMA$ to same state. (See Definition 4)

We can now state and prove the celebrated result of Myhill & Nerode.

Theorem 1. *Suppose there is a $DFSMA$. Then $L(DFSMA) = (L') :: (R_1, ..., R_k)$.*

Proof. *Assume $(L') :: (R_1, ..., R_k)$ is recognized by $DFSMA = (Q, I, T, \delta, \delta', S, F, F', \$)$,*

- *For $(x \in R_i)$, if $\hat{\delta}(s_0, x) = p$, then*

$$[x]_i = \{y \in R_i \mid \hat{\delta}(s_0, y) = p\}$$

(All those strings member of R_i, if we put them in the initial state S_0 and if they reach p, they are equivalent to x).
- *That is given, $(q \in Q_i)$, we define -*

$$C_q = \{x \mid (x \in R_i) \wedge (R_i \subseteq I^*) \wedge \hat{\delta}(S_0, x) = q\}$$

(All those strings if we put them in the initial state S_0, if they reach q, then those strings are in equivalence class C_q. C_q is possibly empty if q is reachable. So corresponding to each state there is an equivalence class of $\equiv_{R_i}$ and it is finite index.)
- *The vertical equivalence classes corresponding to each R_i are completely determined by the vertical states of $DFSMA$. More over the number of vertical equivalence classes of $\equiv_{R_i}$ for each R_i is less than or equal to the number of vertical states of $DFSMA$ for each R_i. As we know that for each i, $|Q_i|$ is finite, we can conclude that the number of vertical equivalence classes of $\equiv_{R_i}$ for each R_i is finite index.*
-

$$\begin{aligned} R_i &= \{x \in I^* \mid \hat{\delta}(S_0, x) \in F'\} \\ &= \bigcup_{p_v \in F'} \{x \in I^* \mid \hat{\delta}(S_0, x) = p_v\} \\ &= \bigcup_{p_v \in F'} C_{p_v} \end{aligned}$$

(C_{p_v} is a vertical equivalence class corresponding to state p_v, R_i is union of all C_{p_v} for $p_v \in F'$. Some of the intermediate final states may not be reachable, in that case the set is empty) (See Lemma 1)
- *Similarly it can be shown that the horizontal equivalence classes are completely determined by the the horizontal states of $DFSMA$, a horizontal word $w \in I^{**}$ is consisting of multiple column vectors, such that, $w = x_{*,0}\ x_{*,1} \cdots x_{*,j}$, we define, $s_0 \xrightarrow{x_{*,0}\$\ x_{*,1}\$\cdots x_{*,j-1}\$} s_j$ using a sequence of steps of $DFSMA$:*

$$s_0 \xrightarrow{x_{*,0}} f_0 \xrightarrow{\$} s_1 \quad \ldots \quad s_{j-1} \xrightarrow{x_{*,j-1}} f_{j-1} \xrightarrow{\$} s_j,$$

We define $[s_j]$, if $s_0 \xrightarrow{x_{*,0}\$\ x_{*,1}\$\cdots x_{*,j-1}\$} s_j$, then,

$$[x_{*,0}\ x_{*,1} \cdots x_{*,j-1}] = \{y_{*,0}\ y_{*,1} \cdots y_{*,k} \in I^{**} \mid s_0 \xrightarrow{y_{*,0}\$\ y_{*,1}\$\cdots y_{*,k}\$} s_j\}$$

- *More over the number of horizontal equivalence classes of* $\equiv_{L'}$ *is less than or equal to the number of horizontal states of* $DFSMA$.

$$C_{s_j} = \{x_{*,0}\ x_{*,1} \cdots x_{*,n'} \in I^{**} \mid s_0 \xrightarrow{x_{*,0}\$\ x_{*,1}\$ \cdots x_{*,n'}\$} s_j\},$$

is an equivalence class of $\equiv_{L'}$ *and finite index.*

-

$$\begin{aligned} L' &= \{x \in I^{**} \mid \delta'(\hat{\delta}(s_0, w), \$) \in S_j\} \\ &= \bigcup_{p_h \in S_j} \{x \in I^{**} \mid \delta'(\hat{\delta}(s_0, w), \$) = p_h\} \\ &= \bigcup_{p_h \in S_j} C_{p_h} \end{aligned}$$

(C_{p_h} is a horizontal equivalence class corresponding to state p_h, L is union of all C_{p_h} for $p_h \in F$. Some of the final states may not be reachable, in that case the set is empty)(See Lemma 2)

Example 7. In the Example 1 of Subsect. 3.1, the vertical equivalence classes are following:

- $C_{q_{11}} = \{X^m \mid (\hat{\delta}(s_1, X^m) = q_{11}, m \geq 1\}$
- $C_{q_{21}} = \{(.)^m \mid (\hat{\delta}(s_2, (.)^m) = q_{21}, m \geq 1\}$
- $C_{q_{22}} = \{(.)^m X \mid (\hat{\delta}(s_2, (.)^m X) = q_{22}, m \geq 1\}$

Horizontal equivalence class is given below.

$$C_{s_1} = \{\epsilon \mid s_1 \xrightarrow{\epsilon} s_1\}$$

S_1 is the start state.

$$C_{s_2} = \{((.)^m X)_0 \cdots ((.)^m X)_n \mid s_2 \xrightarrow{((.)^m X)_0 \cdots ((.)^m X)_n} s_2\}$$

Theorem 2. *Suppose* $L = (L') : \ : (R_1, ..., R_k)$ *is an* FML *over* I^{**}. *Then there exist a* $DFSMA$ *such that* $L = L(DFSMA)$.

Proof. *We define* $DFSMA = (Q, I, T, \delta, \delta', S, F, F', \$)$ *where*

- $Q = Q'' \cup Q'$ *such that* $Q'' = \bigcup_{i=1}^{k} Q_i$ *where* $\forall i\ Q_i = \{[x]_i \mid (x \in R_i \wedge R_i \subseteq I^*)\}$ *is a finite set of vertical states corresponds to* S_i *and* $Q' = \{[x_{*,i}] \mid x_{*,i} \in I^{**}\}$ *where* $x_{*,i}$ *is the ith column vector and it is followed by the ith end marker* $\$_i$, *then it goes to another horizontal state* S_j. $\forall i$ Q_i *is the set of equivalence classes of* $\equiv_{R_i}$ *and* Q' *is the set of equivalence classes of* $\equiv_{L'}$.
(We consider that for each horizontal state $S_i, i \geq 0$, there is a finite set Q_i of vertical states, all these vertical moves are same as DFA. In case of horizontal moves, the automaton needs to read atleast one column, if it reads the ith column vector $x_{*,i}$, then it encounters with the ith end marker $\$_i$, finally it goes to some another horizontal state S_j, then it is called horizontal move.)

- $S_0 = [\epsilon]$
  (Since we assume that L is nonempty, $\epsilon \in Pref(L)$. Here S_0 is the first horizontal state, always the first horizontal state will be considered as the initial state of the automata. Here S_0 will have Q_0 which will contain a finite set of vertical states and using them the automata will read the corresponding 0th column vector $x_{*,0}$, followed by $\$_0$, then it goes to some S_j where $j \geq 0$.)
- $F = \{[w] \mid w \in L\}$ *where* $w = x_{*,0} \cdots x_{*,n}$.
  ($w = x_{*,0} \cdots x_{*,n}$ where $x_{*,0}$ represents the 0th column, $x_{*,1}$ represents 1th column so on. So 0th to nth column vector, that i, an array of n columns is getting accepted. Every column is followed by an end marker \$ apart from the last column $x_{*,n}$, in case of the last column, it reaches f_n which is the final state. There is no more column, so there does not exist more horizontal state.)
- $F' = \{[x_{*,i}] \in Q_i \mid [x_{*,i}] \in R_i \wedge (1 \leq i \leq k)\}$.
  (Each column vector $x_{*,i}$ takes the automata to vertical final state $f_i \in F'$. After reaching the vertical final state, the automaton encounters with end marker \$, then it goes to another horizontal state.)
- *Vertical transition :* $\delta([x], a) = [xa]$ *and there exist* Q_i*, such that,* $[x] \in Q_i$.
- *Horizontal transition :* $\delta'([x_{*,i}], \$) = [S_i S_j]$ *where* S_i, S_j *corresponds to* $x_{*,i}, x_{*,j}$ *respectively,* $[x_{*,i}] = f_i \in F', [S_i S_j] \in Q'$.

Example 8. Example 1 in Sect. 2.1 is suggested to refer. Now, we show that $L(DFSMA) = R_i$.

Corollary 1. $(R_i \subseteq L(DFSMA) \wedge L(DFSMA) \subseteq R_i) \implies (R_i = L(DFSMA))$

Proof. First we will show that $\forall i\ R_i \subseteq L(DFSMA)$*,* $w_v = x_0 \cdots x_n$*, if* $w_v = x_0 \cdots x_n$ *is an arbitrary element, then the vertical run of DFSMA:*

$$S_i \xrightarrow{x_0} q_1 \quad \cdots \quad q_n \xrightarrow{x_n} q_{n+1}$$

Here the run will be accepted if $q_{n+1} \in F'$*, so any* $w_v \in R_i$*, that will be accepted by DFSMA, that is,* $\forall i\ R_i \subseteq L(DFSMA)$.

Now, we need to show that $L(DFSMA) \subseteq R_i$*, using contrapositive we can write* $\neg R_i \nsubseteq \neg L(DFSMA)$*, then the run of DFSMA:*

$$S_i \xrightarrow{x_0} q_1 \quad \cdots \quad q_n \xrightarrow{x_n} q_{n+1}$$

Here the run of DFSMA will be rejected as $q_{n+1} \notin F'$.

Here, we show that $L(DFSMA) = L$.

Corollary 2. $(L \subseteq L(DFSMA) \wedge L(DFSMA) \subseteq L) \implies (L = L(DFSMA))$

Proof. First we will show that $L \subseteq L(DFSMA)$*, if* $w = x_{*,0}\$\ x_{*,1}\$ \cdots x_{*,n}$ *is an arbitrary element, then the run of DFSMA:*

$$S_0 \xrightarrow{x_{*,0}} f_0 \xrightarrow{\$} S_1 \quad \cdots \quad S_n \xrightarrow{x_{*,n}} f_n.$$

Here the run is accepting if $f_n \in F$, so any $w \in L$, that will be accepted by DFSMA, that is, $L \subseteq L(DFSMA)$.

Now, we need to show that $L(DFSMA) \subseteq L$, using contrapositive we can write $\neg L \nsubseteq \neg L(DFSMA)$, then the run of DFSMA:

$$S_0 \xrightarrow{x_{*,0}} f_0 \xrightarrow{\$} S_1 \quad \ldots \quad S_n \xrightarrow{x_{*,n}} f_n.$$

Here the run of DFSMA is will be rejected as $f_n \notin F$.

6 Conclusion and Future Work

In this paper we define deterministic finite state matrix automata, $DFSMA$, which can recognize finite matrix languages, FML. $DFSMA$ has single initial state and single final state. More importantly, we established the Myhill-Nerode theorem for $DFSMA$ and FML. Unlike the classical Myhill-Nerode theorem, here we need two equivalence relations, called vertical equivalence relation $\equiv_v$ and horizontal equivalence relation $\equiv_h$ to capture the behaviour of $DFSMA$. Now as we have Myhill-Nerode theorem for $DFSMA$ and FML, we can come up with a learning algorithm for $DFSMA$.

So, in the form of future work, it could be the immediate step of this work to develop an efficient learning algorithm for $DFSMA$, it could be interesting to explore query learning model [2] in this context.

References

1. Amar, V., Putzolu, G.: On a family of linear grammars. Inf. Control **7**(3), 283–291 (1964)
2. Angluin, D.: Learning regular sets from queries and counterexamples. Inf. Comput. **75**(2), 87–106 (1987)
3. Angluin, D.: Negative results for equivalence queries. Mach. Learn. **5**(2), 121–150 (1990)
4. Angluin, D., Fisman, D.: Learning regular omega languages. Theoret. Comput. Sci. **650**, 57–72 (2016)
5. Angluin, D., Kharitonov, M.: When won't membership queries help? In: Proceedings of the Twenty-Third Annual ACM Symposium on Theory of Computing, pp. 444–454 (1991)
6. Anselmo, M., Giammarresi, D., Madonia, M.: A common framework to recognize two-dimensional languages. Fund. Inform. **171**(1–4), 1–17 (2020)
7. Bergadano, F., Varricchio, S.: Learning behaviors of automata from shortest counterexamples. In: Vitányi, P. (ed.) EuroCOLT 1995. LNCS, vol. 904, pp. 380–391. Springer, Heidelberg (1995). https://doi.org/10.1007/3-540-59119-2_193
8. Birkendorf, A., Böker, A., Simon, H.U.: Learning deterministic finite automata from smallest counterexamples. SIAM J. Discret. Math. **13**(4), 465–491 (2000)
9. Bunke, H., Sanfeliu, A.: Syntactic and Structural Pattern Recognition: Theory and Applications, vol. 7. World Scientific, Singapore (1990)

10. Cassel, S., Howar, F., Jonsson, B., Steffen, B.: Active learning for extended finite state machines. Formal Aspects Comput. **28**(2), 233–263 (2016). https://doi.org/10.1007/s00165-016-0355-5
11. Cavalcanti, A., Dams, D.: FM 2009: Formal Methods: Second World Congress, Eindhoven, The Netherlands, November 2–6, 2009, Proceedings, vol. 5850. Springer (2009). https://doi.org/10.1007/978-3-642-05089-3
12. Fernau, H., Paramasivan, M., Schmid, M.L., et al.: Simple picture processing based on finite automata and regular grammars. J. Comput. Syst. Sci. **95**, 232–258 (2018)
13. Firschein, O.: Syntactic pattern recognition and applications. Proc. IEEE **71**(10), 1231–1231 (1983)
14. Giammarresi, D., Restivo, A.: Two-dimensional languages. In: Rozenberg, G., Salomaa, A. (eds.) Handbook of Formal Languages, pp. 215–267. Springer, Heidelberg (1997). https://doi.org/10.1007/978-3-642-59126-6_4
15. Ginsburg, S., Spanier, E.H.: Control sets on grammars. Math. Syst. Theory **2**(2), 159–177 (1968)
16. Gold, E.M.: Language identification in the limit. Inf. Control **10**(5), 447–474 (1967)
17. Habarra, O., Jiang, T.: Learning regular languages from counterexamples. J. Comput. Syst. Sci. **43**(2), 299–316 (1991)
18. Hopcroft, J.E., Motwani, R., Ullman, J.D.: Introduction to automata theory, languages, and computation. ACM SIGACT News **32**(1), 60–65 (2001)
19. Isberner, M., Howar, F., Steffen, B.: The TTT algorithm: a redundancy-free approach to active automata learning. In: Bonakdarpour, B., Smolka, S.A. (eds.) RV 2014. LNCS, vol. 8734, pp. 307–322. Springer, Cham (2014). https://doi.org/10.1007/978-3-319-11164-3_26
20. Koshiba, T., Mäkinen, E., Takada, Y.: Learning deterministic even linear languages from positive examples. Theoret. Comput. Sci. **185**(1), 63–79 (1997)
21. Mäkinen, E.: A note on the grammatical inference problem for even linear languages. Fund. Inform. **25**(2), 175–182 (1996)
22. Maler, O., Staiger, L.: On syntactic congruences for ω-languages. Theoret. Comput. Sci. **183**(1), 93–112 (1997)
23. Midya, A., Thomas, D., Malik, S., Pani, A.K.: Polynomial time learner for inferring subclasses of internal contextual grammars with local maximum selectors. In: Hung, D., Kapur, D. (eds.) Theoretical Aspects of Computing- ICTAC 2017. Lecture Notes in Computer Science, vol. 10580, pp. 174–191. Springer, Cham (2017). https://doi.org/10.1007/978-3-319-67729-3_11
24. Midya, A., Thomas, D.G., Pani, A.K., Malik, S., Bhatnagar, S.: Polynomial time algorithm for inferring subclasses of parallel internal column contextual array languages. In: Brimkov, V.E., Barneva, R.P. (eds.) IWCIA 2017. LNCS, vol. 10256, pp. 156–169. Springer, Cham (2017). https://doi.org/10.1007/978-3-319-59108-7_13
25. Midya, A., Vaandrager, F., Thomas, D.G., Ghosh, C.: Simulating parallel internal column contextual array grammars using two-dimensional parallel restarting automata with multiple windows. In: Lukić, T., Barneva, R.P., Brimkov, V.E., Čomić, L., Sladoje, N. (eds.) IWCIA 2020. LNCS, vol. 12148, pp. 106–122. Springer, Cham (2020). https://doi.org/10.1007/978-3-030-51002-2_8
26. Mráz, F., Průša, D., Wehar, M.: Two-dimensional pattern matching against basic picture languages. In: Hospodár, M., Jirásková, G. (eds.) CIAA 2019. LNCS, vol. 11601, pp. 209–221. Springer, Cham (2019). https://doi.org/10.1007/978-3-030-23679-3_17
27. Narasimhan, R.: Labeling schemata and syntactic descriptions of pictures. Inf. Control **7**(2), 151–179 (1964)

28. Radhakrishnan, V., Chakravarthy, V., Krithivasan, K.: Some properties of matrix grammars- parallel image analysis. In: International Work shop on Parallel Image Processing and Analysis-Theory and Applications, pp. 213–225 (1999)
29. Radhakrishnan, V., Nagaraja, G.: Inference of even linear grammars and its application to picture description languages. Pattern Recogn. **21**(1), 55–62 (1988)
30. Rivest, R.L., Schapire, R.E.: Inference of finite automata using homing sequences. Inf. Comput. **103**(2), 299–347 (1993)
31. Rosenfeld, A.: Picture Languages: Formal Models for Picture Recognition. Academic Press, Cambridge (2014)
32. Rosenfeld, A., Siromoney, R.: Picture languages-a survey. Lang. Des. **1**(3), 229–245 (1993)
33. Salomaa, A.: On grammars with restricted use of productions. Suomalainen Tiedeakatemia (1969)
34. Sempere, J.M., García, P.: A characterization of even linear languages and its application to the learning problem. In: Carrasco, R.C., Oncina, J. (eds.) ICGI 1994. LNCS, vol. 862, pp. 38–44. Springer, Heidelberg (1994). https://doi.org/10.1007/3-540-58473-0_135
35. Stromoney, G., Siromoney, R., Krithivasan, K.: Abstract families of matrices and picture languages. Comput. Graph. Image Process. **1**(3), 284–307 (1972)
36. Takada, Y.: Grammatical inference for even linear languages based on control sets. Inf. Process. Lett. **28**(4), 193–199 (1988)
37. Takada, Y.: Algorithmic learning theory of formal languages and its applications. Ph.D. thesis, International Institute for Advanced Study of Social Information Science ... (1992)
38. Takada, Y.: A hierarchy of language families learnable by regular language learning. Inf. Comput. **123**(1), 138–145 (1995)
39. Takada, Y.: Learning formal languages based on control sets. In: Jantke, K.P., Lange, S. (eds.) Algorithmic Learning for Knowledge-Based Systems. LNCS, vol. 961, pp. 316–339. Springer, Heidelberg (1995). https://doi.org/10.1007/3-540-60217-8_15
40. Yokomori, T.: Learning non-deterministic finite automata from queries and counterexamples. In: Machine Intelligence, vol. 13, pp. 169–189 (1992)

Algebraic Properties of Parikh q-Matrices on Two-Dimensional Words

K. Janaki[1], R. Arulprakasam[1](✉), Meenakshi Paramasivan[2], and V. Rajkumar Dare[3]

[1] Department of Mathematics, Faculty of Engineering and Technology, SRM Institute of Science and Technology, Kattankulathur, Chennai 603203, India
{jk1063,arulprar}@srmist.edu.in
[2] FB IV - Informatikwissenschaften, Universität Trier, 54286 Trier, Germany
[3] Department of Mathematics, Madras Christian College, Chennai 600059, India

Abstract. Based on the idea of q−count of certain subwords of a word, the notion of Parikh q−matrix of a word over an ordered alphabet was introduced. On the other hand, with a two-dimensional picture array of symbols arranged in rows and columns, two kinds of upper triangular matrices, known as row and column Parikh q−matrices have also been introduced and investigated. Certain algebraic properties such as Parikh q−matrices commutators, alternate Parikh q−matrices and extending Parikh q−matrices have been investigated for one dimensional case, yet they do not suffice for two-dimensional words. In this paper, we introduce Parikh q−matrices commutators, alternate Parikh q−matrices and extending Parikh q−matrices for two-dimensional words and discuss their properties. We also derive the result used for transferring information with respect to subword occurrences derived from Parikh q−matrices to corresponding information derived from extending Parikh q−matrices.

Keywords: Subwords · Parikh matrix · M-ambiguity · Parikh q−matrix · Two-dimensional words

Mathematics Subject Classification: 68R15 · 68Q42 · 68R99 · 68Q45 · 68Q15 · 68Q25.

1 Introduction

The Parikh matrix mapping [16] was initially introduced as an extension of the Parikh vector [17]. Parikh matrices provide more structural information about words than Parikh vectors and are a useful tool in studying subword occurrences. A number of studies on various properties related to Parikh matrices have been extensively investigated in [1–3,14,15,18–21,23,25–28]. The alternate Parikh matrix defined and developed in [16] gives an interesting relation between the Parikh matrix and its inverse. It is shown that the inverse of the Parikh matrix is the alternate Parikh matrix of the mirror image of the word. Thus without computing the adjoint and determinant of the Parikh matrix one can

R. P. Barneva et al. (Eds.): IWCIA 2022, LNCS 13348, pp. 171–188, 2023.
https://doi.org/10.1007/978-3-031-23612-9_11

find its inverse. Parikh matrix of a word does not count scattered subwords of the word with repeated letters. To facilitate this, in [22], the concept of extending Parikh matrix mapping induced by a word instead of being defined with respect to an ordered alphabet.

The Parikh q–matrix which maps words to matrices with polynomial entries in q was introduced by Egecioglu et al. [12] as an extension of the Parikh matrix. Parikh q–matrix provides more information about words than the Parikh matrix. Two words having the same Parikh matrix can have different Parikh q–matrices. Since the Parikh q–matrix mapping is not injective, a number of studies on injectivity related to Parikh q–matrix have been investigated in [4,6,7,13]. The notion of the alternate Parikh q–matrix was introduced by Egecioglu and Ibarra in [12] and showed that the alternate Parikh q–matrix of the mirror image of the word is indeed the adjoint matrix of the Parikh q–matrix of the word. In [5], the notion of *extending Parikh* q–*matrix* concerning a word instead of an ordered alphabet was introduced and investigated some basic properties of this extending Parikh q–matrices.

On the other hand, a two-dimensional word or picture array is an arrangement of symbols from a finite alphabet in rows and columns. Literature has explored many combinatorial properties of arrays [9–11]. In [24], two types of Parikh matrices are defined, namely a row Parikh matrix and a column Parikh matrix for a picture array. These matrices extend the notion of a Parikh matrix to arrays. The notion of M–ambiguity of a picture array is introduced in [24] by considering two picture arrays to be M–equivalent if their row Parikh matrices and their column Parikh matrices are the same. More specifically, conditions that ensure M–ambiguity are established for binary and ternary words. The problem of reconstruction of two-dimensional binary images has been studied [29] based on Parikh matrices. Based on the notion of Parikh q–matrices on words and Parikh matrices of picture array, two types of Parikh q–matrices are introduced in [8]. Also the notions of q–row and q–column equivalences of two arrays and also several properties relating to q–ambiguity including conditions for q–ambiguity of row or column products for binary arrays are derived. However, Parikh q–matrix of an array does not q–count scattered subwords of the array with repeated letters. To facilitate this, in this paper, we introduce extending Parikh q–matrix of picture arrays induced by a word instead of being defined with respect to an ordered alphabet.

The remainder of this paper is structured as follows. Section 2 provides basics which are used in subsequent sections. In Sect. 3, we give a similar analogue of Parikh q–matrix commutators of words, called Parikh q–matrix commutators of array and discuss some basic properties. In Sect. 4, we introduce alternate Parikh q–matrix for picture arrays and investigate interrelation between the inverse of a Parikh q–matrix associated with an array and the Parikh q–matrix of the mirror image of an array. In Sect. 5, we introduce extending Parikh q–matrix for picture arrays and investigate the results used for transferring information with respect to subword occurrences derived from Parikh q–matrices to corresponding information derived from extending Parikh q−matrices. We end the paper with concluding remarks.

2 Preliminaries

In this section, we recollect fundamental definitions and notations of words, scattered subwords, Parikh matrix and Parikh q−matrix for two-dimensional words.

2.1 Subwords

Consider an alphabet $\Sigma = \{a_1, a_2, \cdots, a_k\}$ and the set of all words over Σ is Σ^*. For any word $x \in \Sigma^*$, the length of x is denoted by $|x|$. An ordered alphabet is an alphabet $\Sigma = \{a_1, a_2, \cdots, a_k\}$ with the total order relation $a_1 < a_2 < \cdots < a_k$ and it is denoted by Σ_k. The empty word is denoted by λ. A word $y \in \Sigma^*$ is called a *scattered subword* of x if there exist words $y_1, y_2, \cdots, y_n$ and $x_0, x_1, x_2, \cdots, x_n$ over Σ such that $y = y_1 y_2 \cdots y_n$ and $x = x_0 y_1 x_1 y_2 \cdots y_n x_n$. The number of occurrences of the word y as a scattered subword of the word x is denoted by $|x|_y$. For instance $|abbbaaab|_{aab} = 6$. Let a_{ij} be the word $a_i a_{i+1} \cdots a_j$ for $1 \leq i < j \leq k$ and if $i = j$ then $a_{ij} = a_i$. Let x, y be two letters in an alphabet Σ and $\delta_{a,b}$ be the Kronecker symbol regarding letters such that

$$\delta_{a,b} = \begin{cases} 1 & \text{if } a = b, \\ 0 & \text{if } a \neq b. \end{cases}$$

2.2 Parikh Matrix

Let $\mathcal{M}_k$ denote the set of all $k \times k$ upper triangular matrices with entries $\mathbb{N}$ and unit diagonal where $\mathbb{N}$ is the set of all non-negative integers.

Definition 1. *Let $\Sigma_k = \{a_1, a_2, \cdots, a_k\}$ be an ordered alphabet where $k \geq 1$. The Parikh matrix mapping denoted by ψ_k is the morphism $\psi_k : \Sigma_k^* \to \mathcal{M}_{k+1}$ defined as $\psi_k(a_l) = (m_{ij})_{1 \leq i, j \leq k+1}$ where*

- $m_{ii} = 1$ *for* $1 \leq i \leq k+1$
- $m_{l,(l+1)} = 1$

and all other entries are zero. For every word $x \in \Sigma_k^*$ such that $x = x_1 x_2 \cdots x_n$ with $x_i \in \Sigma_k$ then

$$\psi_k(x) = \psi_k(x_1)\psi_k(x_2) \cdots \psi_k(x_n).$$

Example 1. Let $x = acbc$ over Σ_3 then the Parikh matrix of x is

$$\psi_3(acbc) = \psi_3(a)\psi_3(c)\psi_3(b)\psi_3(c)$$
$$= \begin{bmatrix}1&1&0&0\\0&1&0&0\\0&0&1&0\\0&0&0&1\end{bmatrix}\begin{bmatrix}1&0&0&0\\0&1&0&0\\0&0&1&1\\0&0&0&1\end{bmatrix}\begin{bmatrix}1&0&0&0\\0&1&1&0\\0&0&1&0\\0&0&0&1\end{bmatrix}\begin{bmatrix}1&0&0&0\\0&1&0&0\\0&0&1&1\\0&0&0&1\end{bmatrix}$$
$$= \begin{bmatrix}1&1&1&1\\0&1&1&1\\0&0&1&2\\0&0&0&1\end{bmatrix}.$$

Two words $x, y \in \Sigma_k^*$ are said to be *M−equivalent* denoted by $x \sim_M y$ if and only if $\psi_k(x) = \psi_k(y)$. A word $z \in \Sigma_k^*$ is said to be *M−ambiguous* if there exists a word $w \neq z$ such that $z \sim_M w$. Otherwise z is called *M−unambiguous.*

2.3 Parikh q-Matrix

The notion of Parikh matrices is extended to a mapping called Parikh q−matrix mapping which takes its values in matrices with polynomial entries. The entries of the Parikh q−matrices are obtained by q−counting the number of occurrences of certain words as scattered subwords of a given word. The q-*counting of a scattered subword* a_{ij} of a word x represented by $S_{x,aij}$ is defined as follows:

Definition 2. *Let* $\Sigma_k = \{a_1, a_2, \cdots, a_k\}$ *be an ordered alphabet where* $k \geq 1$, $x \in \Sigma_k^*$ *and* a_{ij} *be a scattered subword of* x *for* $1 \leq i \leq j < k$. *Then*

$$S_{x,a_{ij}}(q) = \sum_{x=u_i a_i u_{i+1}\cdots u_j a_j u_{j+1}} q^{|u_i|_{a_i}+|u_{i+1}|_{a_{i+1}}+\cdots+|u_j|_{a_j}+|u_{j+1}|_{a_{j+1}}}.$$

Example 2. Let $x = baaabb$ be a word over Σ_2. Considering x as a word over Σ_3. Then

- For $i = 1$ and $j = 1$ we get $a_{ij} = a$ and $S_{x,a}(q) = q^{0+2} + q^{1+2} + q^{2+2} = q^2 + q^3 + q^4$
- For $i = 2$ and $j = 2$ we get $a_{ij} = b$ and $S_{x,b}(q) = q^{0+0} + q^{1+0} + q^{2+0} = 1 + q + q^2$
- For $i = 1$ and $j = 2$ we get $a_{ij} = ab$ and $S_{x,ab}(q) = q^{0+0+0} + q^{0+1+0} + q^{1+0+0} + q^{1+1+0} + q^{2+0+0} + q^{2+1+0} = 1 + 2q + 2q^2 + q^3$.

For any word $x \in \Sigma_k^*$, $S_{x,a_{ij}}(1) = |x|_{a_{ij}}$ for $1 \leq i \leq j \leq k$. Let $\mathcal{M}_k(q)$ denote the set of all $k \times k$ upper triangular matrices with entries $\mathbb{N}(q)$ and unit diagonal where $\mathbb{N}(q)$ is the set of all polynomials in the variable q with coefficients from $\mathbb{N}$.

Definition 3. *Let* $\Sigma_k = \{a_1, a_2, \cdots, a_k\}$ *be an ordered alphabet and* $x \in \Sigma_k^*$ *then the Parikh q-matrix mapping denoted by* ψ_q *is the morphism* $\psi_q : \Sigma_k^* \to \mathcal{M}_k(q)$ *defined as* $\psi_q(a_l) = (m_{ij})_{1 \leq i,\, j \leq k+1}$ *where*

- $m_{ll} = q$
- $m_{ii} = 1$ *for* $1 \leq i \leq k$, $i \neq l$

- $m_{l(l+1)} = 1$ *if* $l < k$

and all other entries are zero.

Definition 4. *Let* $\Sigma_k = \{a_1, a_2, \cdots, a_k\}$ *be an ordered alphabet and* $x \in \Sigma_k^*$ *then the principal diagonal entries of the matrix* $\psi_q(x)$ *is* $(q^{|x|_{a_1}}, q^{|x|_{a_2}}, \cdots, q^{|x|_{a_k}})$.

Note that the Parikh vector of x is given by the formal derivative of $(\mathsf{q}^{|x|_{a_1}}, \mathsf{q}^{|x|_{a_2}}, \cdots, \mathsf{q}^{|x|_{a_k}})$ with respect to q at $\mathsf{q}=1$. The entries of the q-matrices are obtained by q-counting the number of occurrences of certain words as scattered subwords of a given word.

Theorem 1. *[12] Let* $\Sigma_k = \{a_1, a_2, \cdots, a_k\}$ *be an ordered alphabet and* $x \in \Sigma_k^*$. *Then the Parikh* q*-matrix has the following properties*

- $m_{ij} = 0$ *for all* $1 \le j < i \le k$
- $m_{ii} = q^{|x|_{a_i}}$ *for* $1 \le i \le k$
- $m_{i(j+1)} = S_{x,a_{ij}}(q)$ *for all* $1 \le i \le j < k$.

Example 3. Let $x = acbc$ over Σ_3 then the Parikh q-matrix of x is

$$\begin{aligned}
\psi_{\mathsf{q}}(acbc) &= \psi_{\mathsf{q}}(a)\psi_{\mathsf{q}}(c)\psi_{\mathsf{q}}(b)\psi_{\mathsf{q}}(c) \\
&= \begin{bmatrix} \mathsf{q} & 1 & 0 \\ 0 & 1 & 0 \\ 0 & 0 & 1 \end{bmatrix} \begin{bmatrix} 1 & 0 & 0 \\ 0 & 1 & 0 \\ 0 & 0 & \mathsf{q} \end{bmatrix} \begin{bmatrix} 1 & 0 & 0 \\ 0 & \mathsf{q} & 1 \\ 0 & 0 & 1 \end{bmatrix} \begin{bmatrix} 1 & 0 & 0 \\ 0 & 1 & 0 \\ 0 & 0 & \mathsf{q} \end{bmatrix} \\
&= \begin{bmatrix} \mathsf{q} & \mathsf{q} & \mathsf{q} \\ 0 & \mathsf{q} & \mathsf{q} \\ 0 & 0 & \mathsf{q}^2 \end{bmatrix} \\
&= \begin{bmatrix} \mathsf{q}^{|x|_a} & \mathsf{S}_{x,a}(\mathsf{q}) & \mathsf{S}_{x,ab}(\mathsf{q}) \\ 0 & \mathsf{q}^{|x|_b} & \mathsf{S}_{x,b}(\mathsf{q}) \\ 0 & 0 & \mathsf{q}^{|x|_c} \end{bmatrix}.
\end{aligned}$$

The Parikh q-matrix of a word x over $\Sigma_k = \{a_1, a_2, \cdots, a_k\}$ coincides with the usual Parikh matrix, when the q-matrix is evaluated at $\mathsf{q}=1$ treating the word x as a word over $\Sigma_{k+1} = \{a_1, a_2, \cdots, a_{k+1}\}$. The Parikh matrix of the word $x = acbc$ over Σ_3 is a 4×4 upper triangular matrix given by

$$\psi_3(acbc) = \begin{bmatrix} 1 & 1 & 1 & 1 \\ 0 & 1 & 1 & 1 \\ 0 & 0 & 1 & 2 \\ 0 & 0 & 0 & 1 \end{bmatrix}.$$

By comparing Parikh matrix with Parikh q-matrix, add a new symbol d to Σ_3 to get $\Sigma_4 = \{a, b, c, d\}$ and compute the Parikh q-matrix of the word x treating it as a word over Σ_4. For example,

$$\begin{aligned}
\psi_{\mathsf{q}}(acbc) &= \psi_{\mathsf{q}}(a)\psi_{\mathsf{q}}(c)\psi_{\mathsf{q}}(b)\psi_{\mathsf{q}}(c) \\
&= \begin{bmatrix} \mathsf{q} & 1 & 0 & 0 \\ 0 & 1 & 0 & 0 \\ 0 & 0 & 1 & 0 \\ 0 & 0 & 0 & 1 \end{bmatrix} \begin{bmatrix} 1 & 0 & 0 & 0 \\ 0 & 1 & 0 & 0 \\ 0 & 0 & \mathsf{q} & 1 \\ 0 & 0 & 0 & 1 \end{bmatrix} \begin{bmatrix} 1 & 0 & 0 & 0 \\ 0 & \mathsf{q} & 1 & 0 \\ 0 & 0 & 1 & 0 \\ 0 & 0 & 0 & 1 \end{bmatrix} \begin{bmatrix} 1 & 0 & 0 & 0 \\ 0 & 1 & 0 & 0 \\ 0 & 0 & \mathsf{q} & 1 \\ 0 & 0 & 0 & 1 \end{bmatrix} \\
&= \begin{bmatrix} \mathsf{q} & \mathsf{q} & \mathsf{q} & 1 \\ 0 & \mathsf{q} & \mathsf{q} & 1 \\ 0 & 0 & \mathsf{q}^2 & \mathsf{q}+1 \\ 0 & 0 & 0 & 1 \end{bmatrix} \\
&= \begin{bmatrix} \mathsf{q}^{|x|_a} & \mathsf{S}_{x,a}(\mathsf{q}) & \mathsf{S}_{x,ab}(\mathsf{q}) & \mathsf{S}_{x,abc}(\mathsf{q}) \\ 0 & \mathsf{q}^{|x|_b} & \mathsf{S}_{x,b}(\mathsf{q}) & \mathsf{S}_{x,bc}(\mathsf{q}) \\ 0 & 0 & \mathsf{q}^{|x|_c} & \mathsf{S}_{x,c}(\mathsf{q}) \\ 0 & 0 & 0 & 1 \end{bmatrix}.
\end{aligned}$$

Two words $x, y \in \Sigma_k^*$ are said to be q-*equivalent* denoted by $x \sim_{\mathsf{q}} y$ if and only if $\psi_{\mathsf{q}}(x) = \psi_{\mathsf{q}}(y)$. A word $z \in \Sigma_k^*$ is said to be q-*ambiguous* if there exists a word $w \neq z$ such that $z \sim_{\mathsf{q}} w$. Otherwise z is called q-*unambiguous*. Note that if two words x, y are q−equivalent then they have same Parikh vector.

In [6], a notion called partial sum of two Parikh q−matrices was introduced, which was motivated by the notion in [15] considered for words.

Definition 5. *[6] Let x and y be the words over Σ_k. Let $\psi_q(x)$ and $\psi_q(y)$ be the corresponding Parikh q−matrices. Define $\psi_q(x) \oplus \psi_q(y) = M = (c_{ij})_{k\times k}$ where c_{ij} is the usual sum of the corresponding entries of the matrices $\psi_q(x)$ and $\psi_q(y)$ except for the elements in the main diagonal of M which are defined by $c_{ii} = \mathsf{q}^{|x|_{a_i} + |y|_{a_i}}$ for $1 \le i \le k$.*

Throughout the paper if there is a word x from Σ_k, we assume x to be a word from Σ_{k+1} and compute the Parikh q−matrix of x in $\mathcal{M}_{k+1}(\mathsf{q})$.

Definition 6. *Let $\Sigma_k = \{a_1, a_2, \cdots, a_k\}$ be an ordered alphabet and $x \in \Sigma_k^*$ then the alternate Parikh q-matrix mapping denoted by $\overline{\psi}_q$ is the morphism $\overline{\psi}_q : \Sigma_k^* \to \mathcal{M}_k(q)$ defined as $\overline{\psi}_q(a_l) = (m_{ij})_{1 \le i, j \le k}$ where*

- $m_{ll} = 1$
- $m_{ii} = q$ *for* $1 \le i \le k$, $i \neq l$
- $m_{l(l+1)} = -1$ *if* $l < k$

and all other entries are zero.

Example 4. Let $x = abba$ over Σ_3 then the alternate Parikh q−matrix of x is

$$\overline{\psi}_{\mathrm{q}}(abba) = \overline{\psi}_{\mathrm{q}}(a)\overline{\psi}_{\mathrm{q}}(b)\overline{\psi}_{\mathrm{q}}(b)\overline{\psi}_{\mathrm{q}}(a)$$
$$= \begin{bmatrix} 1 & -1 & 0 \\ 0 & \mathrm{q} & 0 \\ 0 & 0 & \mathrm{q} \end{bmatrix} \begin{bmatrix} \mathrm{q} & 0 & 0 \\ 0 & 1 & -1 \\ 0 & 0 & \mathrm{q} \end{bmatrix} \begin{bmatrix} \mathrm{q} & 0 & 0 \\ 0 & 1 & -1 \\ 0 & 0 & \mathrm{q} \end{bmatrix} \begin{bmatrix} 1 & -1 & 0 \\ 0 & \mathrm{q} & 0 \\ 0 & 0 & \mathrm{q} \end{bmatrix}$$
$$= \begin{bmatrix} \mathrm{q}^2 & -(\mathrm{q}+\mathrm{q}^2) & (\mathrm{q}+\mathrm{q}^2) \\ 0 & \mathrm{q}^2 & -(\mathrm{q}^2+\mathrm{q}^3) \\ 0 & 0 & \mathrm{q}^4 \end{bmatrix}.$$

Definition 7. *Let $\Sigma_k = \{a_1, a_2, \cdots, a_k\}$ be an ordered alphabet where $k \geq 1$, $x \in \Sigma_k^*$ and a_{ij} be a scattered subword of x for $1 \leq i \leq j < k$. Then the alternate q-counting of a scattered subword a_{ij} of a word x represented by $S_{x,aij}$ is defined as*

$$\overline{S}_{x,a_{ij}}(q) = (-1)^{i+j+1} \sum_{x=u_i a_i u_{i+1} \cdots u_j a_j u_{j+1}} q^{\sum_{t=i}^{j+1}(|u_t| - |u_t|_{a_t})}.$$

Example 5. Let $x = abba$ be a word over Σ_2. Considering x as a word over Σ_3. Then

- For $i = 1$ and $j = 1$ we get $a_{ij} = a$ and $\overline{\mathrm{S}}_{abba,a}(\mathrm{q}) = -(\mathrm{q}^{0+3-2} + \mathrm{q}^{3-1+0}) = -(\mathrm{q}^1 + \mathrm{q}^2)$
- For $i = 2$ and $j = 2$ we get $a_{ij} = b$ and $\overline{\mathrm{S}}_{abba,b}(\mathrm{q}) = -(\mathrm{q}^{1-0+2-0} + \mathrm{q}^{2-1+1}) = -(\mathrm{q}^3 + \mathrm{q}^2)$
- For $i = 1$ and $j = 2$ we get $a_{ij} = ab$ and $\overline{\mathrm{S}}_{abba,ab}(\mathrm{q}) = \mathrm{q}^{0+0+2-0} + \mathrm{q}^{0+1-1+1} = \mathrm{q} + \mathrm{q}^2$.

Definition 8. *Let $\Sigma_k = \{a_1, a_2, \cdots, a_k\}$ be an ordered alphabet and $x \in \Sigma_k^*$ and $u = u_1 u_2 \cdots u_t$ be a word of length t where $u_i \in \Sigma_k$ for all $1 \leq i \leq t$ then the extending Parikh q-matrix mapping induced by the word u denoted by ψ_q^u is the morphism $\psi_q^u : \Sigma_k^* \to \mathcal{M}_{t+1}(q)$ defined as follows. Assume that the letter $a \in \Sigma_k$ and $\psi_q^u(a) = (m_{ij})_{1 \leq i, j \leq t+1}$ where*

- $m_{ij} = 1$, *if* $i = j$
- *If* $\delta_{b_l,a} = 1$, $1 \leq l \leq t$ *then update the entries* $m_{ll} = q$
- $m_{l(l+1)} = 1$

and all other entries are zero.

Definition 9. *Let $\Sigma_k = \{a_1, a_2, \cdots, a_k\}$ be an ordered alphabet where $k \geq 1$, $x \in \Sigma_k^*$ and a_{ij} be a scattered subword of x for $1 \leq i \leq j < |u|$. Then the extending q-counting of a scattered subword a_{ij} of a word x with respect to the word u represented by $S_{x,aij}^u$ is defined as*

$$S_{x,a_{ij}}^u(q) = \sum_{x=u_i a_i u_{i+1} \cdots u_j a_j u_{j+1}} q^{\sum_{t=i}^{j+1} |u_t|_{a_t}}.$$

In [6], a notion called partial sum of two Parikh q−matrices was introduced, which was motivated by the notion in [15] considered for words.

Definition 10. *[6] Let x and y be the words over Σ_k. Let $\psi_q(x)$ and $\psi_q(y)$ be the corresponding Parikh q–matrices. Define $\psi_q(x) \oplus \psi_q(y) = M = (c_{ij})_{k \times k}$ where c_{ij} is the usual sum of the corresponding entries of the matrices $\psi_q(x)$ and $\psi_q(y)$ except for the elements in the main diagonal of M which are defined by $c_{ii} = q^{|x|_{a_i} + |y|_{a_i}}$ for $1 \le i \le k$.*

Throughout the paper if there is a word x from Σ_k, we assume x to be a word from Σ_{k+1} and compute the Parikh q–matrix of x in $\mathcal{M}_{k+1}(q)$.

2.4 Two Dimensional Words

Let $\Sigma_k = \{a_1 < a_2 < \cdots < a_k\}$ be an ordered alphabet and h, v be two positive integers. A two dimensional word (or picture array) X is a rectangular array of symbols over Σ_k in h rows and v columns which is in the form of

$$\begin{matrix} a_{11} & a_{12} & \cdots & a_{1v} \\ \vdots & \ddots & \ddots & \vdots \\ a_{h1} & a_{h2} & \cdots & a_{hv} \end{matrix}$$

where $a_{ij} \in \Sigma$ and $1 \le i \le h$, $1 \le j \le v$. The set of all picture arrays over Σ_k is denoted by $\wp$. Let $\circ$ and $\diamond$ be the symbol of column concatenation and row concatenation of picture arrays respectively in $\wp$. For $X, Y \in \wp$, $X \circ Y$ is defined if and only if X and Y have same number of rows and $X \diamond Y$ is defined if and only if X and Y have same number of columns.

2.5 Parikh q–Matrices of a Picture Array

Definition 11. *For $h, v \ge 1$, let $X \in \wp$ be a $h \times v$ array over Σ_k. Let x_i be the horizontal words in the h rows and y_j be the vertical words in the v columns. Let $\psi_q(x_i)$ and $\psi_q(y_j^t)$ be the Parikh q–matrix of x_i and y_j^t respectively. Then the row and column Parikh q–matrix $R_q(X)$ and $C_q(X)$ respectively are defined as*

$$\begin{aligned} R_q(X) &= \psi_q(x_1) \oplus \psi_q(x_2) \oplus \cdots \oplus \psi_q(x_h) \\ &= \bigoplus_{i=1}^{h} \psi_q(x_i) \\ C_q(X) &= \psi_q(y_1^t) \oplus \psi_q(y_2^t) \oplus \cdots \oplus \psi_q(y_v^t) \\ &= \bigoplus_{j=1}^{v} \psi_q(y_j^t). \end{aligned}$$

Definition 12. *For $h, v \ge 1$, let $X \in \wp$ be a $h \times v$ array over Σ_k. Let x_i be the horizontal words in the h rows and y_j be the vertical words in the v columns. Let $u_{i,j}$ be a scattered subword of X where $1 \le i \le j \le k$. Then the row and column q–counting scattered subword $u_{i,j}$ of an array X denoted by $R(S_{X,u_{i,j}}(q))$ and $C(S_{X,u_{i,j}}(q))$ respectively and defined as*
$R(S_{X,u_{i,j}}(q)) = \sum_{i=1}^{h} S_{x_i,u_{i,j}}(q)$
$C(S_{X,u_{i,j}}(q)) = \sum_{j=1}^{v} S_{y_j^t,u_{i,j}}(q)$.

3 Extending Parikh q-Matrix of Picture Arrays

In this section we define row and column extending Parikh q−matrix of a picture array. And also we derive the result used for transferring information with respect to subword occurrences derived from Parikh q−matrices to corresponding information derived from extending Parikh q−matrices.

Definition 13. *For $h, v \geq 1$, let $X \in \wp$ be a $h \times v$ array over Σ_k. Let x_i be the horizontal words in the h rows and y_j be the vertical words in the v columns. Let $u = u_1 u_2 ... u_t$ be a word of length t over Σ_k. Let $\psi_q^u(x_i)$ and $\psi_q^u(y_j^t)$ be the extending Parikh q−matrix of x_i and y_j^t respectively. Then the row and column extending Parikh q−matrix with respect to u are denoted as $R_q^u(X)$ and $C_q^u(X)$ respectively and defined as*

$$\begin{aligned} R_q^u(X) &= \psi_q^u(x_1) \oplus \psi_q^u(x_2) \oplus \cdots \oplus \psi_q^u(x_h) \\ &= \oplus_{i=1}^h \psi_q^u(x_i) \\ C_q^u(X) &= \psi_q^u(y_1^t) \oplus \psi_q^u(y_2^t) \oplus \cdots \oplus \psi_q^u(y_v^t) \\ &= \oplus_{j=1}^v \psi_q^u(y_j^t). \end{aligned}$$

Example 6. Let $X = \begin{matrix} a\ a\ b\ a\ b \\ b\ a\ a\ a\ b \\ b\ a\ b\ a\ b \\ a\ a\ b\ b\ b \end{matrix}$ be the array and $u = aba$ be the word over Σ_2.
Then the row extending Parikh q−matrix of X with respect to the word aba is

$$\begin{aligned} R_q^u(X) &= \bigoplus_{i=1}^{4} \psi_q^u(x_i) \\ &= \psi_q^u(x_1) \oplus \psi_q^u(x_2) \oplus \psi_q^u(x_3) \oplus \psi_q^u(x_4) \\ &= \begin{bmatrix} q^3 & 2q^3+q^2 & 3q^2+2q & q+1 \\ 0 & q^2 & 2q & 1 \\ 0 & 0 & q^3 & q^2+q+1 \\ 0 & 0 & 0 & 1 \end{bmatrix} \oplus \begin{bmatrix} q^3 & q^3+q^2+q & q^2+q+1 & 0 \\ 0 & q^2 & q^3+q & q^2+q+1 \\ 0 & 0 & q^3 & q^2+q+1 \\ 0 & 0 & 0 & 1 \end{bmatrix} \oplus \\ & \begin{bmatrix} q^2 & 2q^2 & 3q & 1 \\ 0 & q^3 & 3q^2 & 2q+1 \\ 0 & 0 & q^2 & q+1 \\ 0 & 0 & 0 & 1 \end{bmatrix} \oplus \begin{bmatrix} q^2 & q^4+q^3 & q^3+2q^2+2q+1 & 0 \\ 0 & q^3 & q^2+q+1 & 0 \\ 0 & 0 & q^2 & q+1 \\ 0 & 0 & 0 & 1 \end{bmatrix} \\ &= \begin{bmatrix} q^{10} & q^4+4q^3+4q^2+q & q^3+6q^2+8q+2 & q+2 \\ 0 & q^{10} & q^3+4q^2+4q+1 & q^2+3q+3 \\ 0 & 0 & q^{10} & 2q^2+4q+4 \\ 0 & 0 & 0 & 1 \end{bmatrix}. \end{aligned}$$

The column extending Parikh q−matrix of X with respect to the word aba is

$$C_q^u(X) = \bigoplus_{j=1}^{5} \psi_{\mathsf{q}}^u(y_j^t)$$
$$= \psi_{\mathsf{q}}^u(y_1^t) \oplus \psi_{\mathsf{q}}^u(y_2^t) \oplus \psi_{\mathsf{q}}^u(y_3^t) \oplus \psi_{\mathsf{q}}^u(y_4^t) \oplus \psi_{\mathsf{q}}^u(y_5^t)$$
$$= \begin{bmatrix} \mathsf{q}^2 & \mathsf{q}^2+\mathsf{q} & \mathsf{q}^2+\mathsf{q} & \mathsf{q}+1 \\ 0 & \mathsf{q}^2 & \mathsf{q}^2+\mathsf{q} & \mathsf{q}+1 \\ 0 & 0 & \mathsf{q}^2 & \mathsf{q}+1 \\ 0 & 0 & 0 & 1 \end{bmatrix} \oplus \begin{bmatrix} \mathsf{q}^4 & \mathsf{q}^3+\mathsf{q}^2+\mathsf{q}+1 & 0 & 0 \\ 0 & 1 & 0 & 0 \\ 0 & 0 & \mathsf{q}^4 & \mathsf{q}^3+\mathsf{q}^2+\mathsf{q}+1 \\ 0 & 0 & 0 & 1 \end{bmatrix} \oplus$$
$$\begin{bmatrix} \mathsf{q} & \mathsf{q}^2 & \mathsf{q}+1 & 0 \\ 0 & \mathsf{q}^3 & \mathsf{q}^2+2\mathsf{q} & 1 \\ 0 & 0 & \mathsf{q} & 1 \\ 0 & 0 & 0 & 1 \end{bmatrix} \oplus \begin{bmatrix} \mathsf{q}^3 & \mathsf{q}^3+\mathsf{q}^2+\mathsf{q} & \mathsf{q}^2+\mathsf{q}+1 & 0 \\ 0 & \mathsf{q} & 1 & 0 \\ 0 & 0 & \mathsf{q}^3 & \mathsf{q}^2+\mathsf{q}+1 \\ 0 & 0 & 0 & 1 \end{bmatrix} \oplus$$
$$\begin{bmatrix} 1 & 0 & 0 & 0 \\ 0 & \mathsf{q}^4 & \mathsf{q}^3+\mathsf{q}^2+\mathsf{q}+1 & 0 \\ 0 & 0 & 1 & 0 \\ 0 & 0 & 0 & 1 \end{bmatrix}$$
$$= \begin{bmatrix} \mathsf{q}^{10} & 2\mathsf{q}^3+4\mathsf{q}^2+3\mathsf{q}+1 & 2\mathsf{q}^2+3\mathsf{q}+2 & \mathsf{q}+1 \\ 0 & \mathsf{q}^{10} & \mathsf{q}^3+3\mathsf{q}^2+4\mathsf{q}+2 & \mathsf{q}+2 \\ 0 & 0 & \mathsf{q}^{10} & \mathsf{q}^3+2\mathsf{q}^2+3\mathsf{q}+4 \\ 0 & 0 & 0 & 1 \end{bmatrix}.$$

Definition 14. *For $h, v \geq 1$, let $X \in \wp$ be a $h \times v$ array over Σ_k. Let x_i be the horizontal words in the h rows and y_j be the vertical words in the v columns. Let $u = u_1u_2...u_t$ be a word of length t over Σ_k and $u_{i',j'}$ be a scattered subword of X where $1 \leq i' \leq j' \leq t$. Then the row and column q−counting scattered subword $u_{i',j'}$ of an array X with respect to a word u respectively and defined as*

$$R(\mathcal{S}^u_{X,u_{i',j'}}(\mathsf{q})) = \sum_{i=1}^{h} \mathcal{S}^u_{x_i,u_{i',j'}}(\mathsf{q}) \text{ and } C(\mathcal{S}^u_{X,u_{i',j'}}(\mathsf{q})) = \sum_{j=1}^{v} \mathcal{S}^u_{y_j^t,u_{i',j'}}(\mathsf{q}).$$

Example 7. Let $X = \begin{matrix} a\,a\,b\,a\,b \\ b\,a\,a\,a\,b \\ b\,a\,b\,a\,b \\ a\,a\,b\,b\,b \end{matrix}$ be the array and $u = bab$ be the word over Σ_2.

Then the row q−counting scattered subword ba of X with respect to the word bab is

$$R(\mathsf{S}^u_{X,ba}(\mathsf{q})) = \sum_{i=1}^{4} \mathsf{S}^u_{x_i,ba}(\mathsf{q})$$
$$= \mathsf{S}^u_{x1,ba}(\mathsf{q}) + \mathsf{S}^u_{x2,ba}(\mathsf{q}) + \mathsf{S}^u_{x3,ba}(\mathsf{q}) + \mathsf{S}^u_{x4,ba}(\mathsf{q})$$
$$= (\mathsf{q}^1) + (\mathsf{q}^1 + \mathsf{q}^2 + \mathsf{q}^3) + (3\mathsf{q}^2) + (0)$$
$$= 2\mathsf{q}^1 + 4\mathsf{q}^2 + \mathsf{q}^3.$$

The column q−counting scattered subword ba of X with respect to the word bab is

$$\begin{aligned} C(\mathrm{S}^u_{X,ba}(\mathsf{q})) &= \sum_{j=1}^{5} \mathrm{S}^u_{y^t_j,ba}(\mathsf{q}) \\ &= \mathrm{S}^u_{y^t_1,ba}(\mathsf{q}) + \mathrm{S}^u_{y^t_2,ba}(\mathsf{q}) + \mathrm{S}^u_{y^t_3,ba}(\mathsf{q}) + \mathrm{S}^u_{y^t_4,ba}(\mathsf{q}) + \mathrm{S}^u_{y^t_5,ba}(\mathsf{q}) \\ &= (1+\mathsf{q}^1) + (0) + (\mathsf{q}^2) + (0) + (0) \\ &= 1 + \mathsf{q}^1 + \mathsf{q}^2. \end{aligned}$$

Theorem 2. *Let X be an array and u be a word of length t over Σ_k. For every row and column extending Parikh q−matrix, $R^u_q(X)$ and $C^u_q(X)$ respectively, let $\Sigma_t = \{a_1 < a_2 < \cdots a_t\}$ be an ordered alphabet and let X' be an array over Σ_t such that*

$$R^u_q(X) = R_q(X')$$
$$C^u_q(X) = C_q(X')$$

can be effectively constructed.

Proof. Consider the word u of length t over Σ_k such that number the occurrences of the letters in u by $1, 2, \cdots, t$ which constitute the ordered aplhabet Σ_t ($\Sigma_t = \{1 < 2 < \cdots < t\}$). Consider a morphism δ of Σ^*_k into Σ^*_t is defined as follows. Let $a \in \Sigma$ which appear in u in positions $q_1, q_2, \cdots, q_s$ for $1 \le s \le t$ where $q_i < q_{i+1}$, $1 \le i \le s-1$. Then we define $\delta(a) = q_s q_{s-1} \cdots q_1$. Choose $X' = \delta(X)$ for $X \in \wp$ over Σ^*_k. For instance, let

$$X = \begin{matrix} a & b & b & a & b \\ b & a & a & b & a \\ a & b & a & a & b \\ a & b & a & b & b \\ b & b & b & a & a \end{matrix}$$

and $u = abab$ then $\delta(a) = 31$ and $\delta(b) = 42$. Hence for X we obtain

$$X' = \begin{matrix} 3\,1\,4\,2\,4\,2\,3\,1\,4\,2 \\ 4\,2\,3\,1\,3\,1\,4\,2\,3\,1 \\ 3\,1\,4\,2\,3\,1\,3\,1\,4\,2. \\ 3\,1\,4\,2\,3\,1\,4\,2\,4\,2 \\ 4\,2\,4\,2\,4\,2\,3\,1\,3\,1 \end{matrix}$$

For the ordered alphabet $\Sigma_4 = \{1, 2, 3, 4\}$, we have $R_q(X')$=

$$\begin{bmatrix} q^{12} & 4q^3+5q^2+2q+1 & 7q^2+8q & 2q^2+2q+1 & 2q+2 \\ 0 & q^{13} & q^4+2q^3+6q^2+4q & q^3+7q^2+7q+1 & 4q+4 \\ 0 & 0 & q^{12} & 4q^3+6q^2+q+1 & 4q^2+6q+3 \\ 0 & 0 & 0 & q^{13} & 3q^2+5q+5 \\ 0 & 0 & 0 & 0 & 1 \end{bmatrix}$$

$=R_q(X)$. Similarly we show that

$$C_q(X') = \begin{bmatrix} q^{12} & 3q^3+7q^2+2q & 3q^2+5q+2 & q^2+6q+1 & q+2 \\ 0 & q^{13} & 3q^3+6q^2+4q & 2q^3+4q^2+7q & 2q^2+4q+5 \\ 0 & 0 & q^{12} & 3q^3+7q^2+2q & 3q^2+7q+5 \\ 0 & 0 & 0 & q^{13} & q^3+2q^2+5q+5 \\ 0 & 0 & 0 & 0 & 1 \end{bmatrix}$$

$$= C_q(X).$$

□

4 Alternate Parikh q–Matrix of Picture Array

In this section we define row and column alternate Parikh q–matrix of a picture array. Also we provide a relation between Parikh q–matrix and its alternative Parikh q–matrix of a picture array.

Definition 15. *For $h, v \geq 1$, let $X \in \wp$ be a $h \times v$ array over Σ_k. Let x_i be the horizontal words in the h rows and y_j be the vertical words in the v columns. Let $\overline{\psi}_q(x_i)$ and $\overline{\psi}_q(y_j^t)$ be the alternate Parikh q–matrix of x_i and y_j^t respectively. Then the row and column alternate Parikh q–matrix $\overline{R}_q(X)$ and $\overline{C}_q(X)$ respectively are defined as*

$$\overline{R}_q(X) = \bigoplus_{i=1}^{h} \overline{\psi}_q(x_i) \text{ and } \overline{C}_q(X) = \bigoplus_{j=1}^{v} \overline{\psi}_q(y_j^t).$$

Example 8. Consider the array $X = \begin{matrix} a\ a\ b\ a\ b \\ b\ a\ a\ a\ b \\ b\ a\ b\ a\ b \\ a\ a\ b\ b\ b \end{matrix}$ over Σ_2. Then the row alternate Parikh q–matrix of X is

$$\overline{R}_q(X) = \bigoplus_{i=1}^{4} \overline{\psi}_q(x_i)$$

$$= \overline{\psi}_q(x_1) \oplus \overline{\psi}_q(x_2) \oplus \overline{\psi}_q(x_3) \oplus \overline{\psi}_q(x_4)$$

$$= \begin{bmatrix} q^2 & -2q-q^2 & q^3+2q^2+2q \\ 0 & q^3 & -q^3-q^4 \\ 0 & 0 & q^5 \end{bmatrix} \oplus \begin{bmatrix} q^2 & -q-q^2-q^3 & q^3+q^2+q \\ 0 & q^3 & -q^3-q^4 \\ 0 & 0 & q^5 \end{bmatrix} \oplus$$

$$\begin{bmatrix} q^3 & -2q^2 & 2q^2+q^3 \\ 0 & q^2 & -q^2-q^3-q^4 \\ 0 & 0 & q^5 \end{bmatrix} \oplus \begin{bmatrix} q^3 & -1-q & 1+2q+2q^2+q^3 \\ 0 & q^2 & -q^2-q^3-q^4 \\ 0 & 0 & q^5 \end{bmatrix}$$

$$= \begin{bmatrix} q^{10} & -q^3-4q^2-4q-1 & 4q^3+7q^2+5q+1 \\ 0 & q^{10} & -4q^4-4q^3-2q^2 \\ 0 & 0 & q^{20} \end{bmatrix}.$$

The column alternate Parikh q–matrix of X is

$$\overline{C}_q(X) = \bigoplus_{j=1}^{5} \overline{\psi}_{\mathsf{q}}(y_j^t)$$
$$= \overline{\psi}_{\mathsf{q}}(y_1^t) \oplus \overline{\psi}_{\mathsf{q}}(y_2^t) \oplus \overline{\psi}_{\mathsf{q}}(y_3^t) \oplus \overline{\psi}_{\mathsf{q}}(y_4^t) \oplus \overline{\psi}_{\mathsf{q}}(y_5^t)$$
$$= \begin{bmatrix} \mathsf{q}^2 & -\mathsf{q}^2-\mathsf{q} & \mathsf{q}+\mathsf{q}^2 \\ 0 & \mathsf{q}^2 & -\mathsf{q}^2-\mathsf{q}^3 \\ 0 & 0 & \mathsf{q}^4 \end{bmatrix} \oplus \begin{bmatrix} 1 & -1-\mathsf{q}-\mathsf{q}^2-\mathsf{q}^3 & 0 \\ 0 & \mathsf{q}^4 & 0 \\ 0 & 0 & \mathsf{q}^4 \end{bmatrix} \oplus$$
$$\begin{bmatrix} \mathsf{q}^3 & -\mathsf{q} & \mathsf{q}+\mathsf{q}^2 \\ 0 & \mathsf{q} & -\mathsf{q}-\mathsf{q}^2-\mathsf{q}^3 \\ 0 & 0 & \mathsf{q}^4 \end{bmatrix} \oplus \begin{bmatrix} \mathsf{q} & -1-\mathsf{q}-\mathsf{q}^2 & 1+\mathsf{q}+\mathsf{q}^2 \\ 0 & \mathsf{q}^3 & -\mathsf{q}^3 \\ 0 & 0 & \mathsf{q}^4 \end{bmatrix} \oplus$$
$$\begin{bmatrix} \mathsf{q}^4 & 0 & 0 \\ 0 & 1 & -1-\mathsf{q}-\mathsf{q}^2-\mathsf{q}^3 \\ 0 & 0 & \mathsf{q}^4 \end{bmatrix}$$
$$= \begin{bmatrix} \mathsf{q}^{10} & -\mathsf{q}^3-3\mathsf{q}^2-4\mathsf{q}-2 & 3\mathsf{q}^2+3\mathsf{q}+1 \\ 0 & \mathsf{q}^{10} & -\mathsf{q}^4-3\mathsf{q}^3-3\mathsf{q}^2-2\mathsf{q}-1 \\ 0 & 0 & \mathsf{q}^{20} \end{bmatrix}$$

Definition 16. *For $h, v \geq 1$, let $X \in \wp$ be a $h \times v$ array over Σ_k. Let x_i be the horizontal words in the h rows and y_j be the vertical words in the v columns. Let $u_{i,j}$ be a scattered subword of X where $1 \leq i \leq j \leq k$. Then the row and column alternate q–counting scattered subword $u_{i,j}$ of an array X denoted by $R(\overline{S}_{X,u_{i,j}}(q))$ and $C(\overline{S}_{X,u_{i,j}}(q))$ respectively and defined as*

$$R(\overline{S}_{X,u_{i,j}}(q)) = \sum_{i=1}^{h} S^u_{x_i,u_{i,j}}(q) \text{ and } C(\overline{S}_{X,u_{i,j}}(q)) = \sum_{j=1}^{v} S^u_{y_j^t,u_{i,j}}(q).$$

Theorem 3. *[6] Let x be a word over Σ_2 then $S_{x,a}(q)S_{x,b}(q) - q^{|x|_b}S_{x,ab}(q) = \overline{S}_{mi(x),ab}(q)$.*

We extend Theorem 3 for picture array as follows.

Theorem 4. *For $h, v \geq 1$, let $X \in \wp$ be an array over Σ_2. Let x_i be the horizontal words in the h rows and y_j be the vertical words in the v columns. Then*

(i) $\sum_{i=1}^{h} \left[S_{x_i,a}(q)S_{x_i,b}(q) - q^{|x_i|_b}S_{x_i,ab}(q)\right] = R\left(\overline{S}_{X_v,ab}(q)\right)$

(ii) $\sum_{j=1}^{v} \left[S_{y_j^t,a}(q)S_{y_j^t,b}(q) - q^{|y_j^t|_b}S_{y_j^t,ab}(q)\right] = C\left(\overline{S}_{X_v,ab}(q)\right)$.

Proof. Let X be the array in $\wp$ over Σ_2 with h number of rows and v number of columns such that $X = x_1 \diamond x_2 \diamond \cdots \diamond x_h$, $x_i \in \Sigma_2^*$, $1 \leq i \leq h$. Let x_i be a horizontal words over Σ_2 then by Theorem 3 we have $\mathsf{S}_{x_i,a}(\mathsf{q})\mathsf{S}_{x_i,b}(\mathsf{q}) - \mathsf{q}^{|x_i|_b}\mathsf{S}_{x_i,ab}(\mathsf{q}) = \left(\overline{\mathsf{S}}_{mi(x_i),ab}(\mathsf{q})\right)$. As there are h number of rows we have

$$\sum_{i=1}^{h} \left[\mathsf{S}_{x_i,a}(\mathsf{q})\mathsf{S}_{x_i,b}(\mathsf{q}) - \mathsf{q}^{|x_i|_b}\mathsf{S}_{x_i,ab}(\mathsf{q})\right] = R\left(\overline{\mathsf{S}}_{X_v,ab}(\mathsf{q})\right).$$

Similarly let $X = y_1^t \circ y_2^t \circ \cdots \circ y_v^t$, $y_j^t \in \Sigma_2^*$, $1 \leq j \leq v$. Let y_j^t be a vertical words over Σ_2 then we have $\mathrm{S}_{y_j^t,a}(\mathrm{q})\mathrm{S}_{y_j^t,b}(\mathrm{q}) - \mathrm{q}^{|y_j^t|_b}\mathrm{S}_{y_j^t,ab}(\mathrm{q}) = \left(\overline{\mathrm{S}}_{mi(y_j^t),ab}(\mathrm{q})\right)$. As there are v number of columns we have

$$\sum_{j=1}^{v}\left[\mathrm{S}_{y_j^t,a}(\mathrm{q})\mathrm{S}_{y_j^t,b}(\mathrm{q}) - \mathrm{q}^{|y_j^t|_b}\mathrm{S}_{y_j^t,ab}(\mathrm{q})\right] = C\left(\overline{\mathrm{S}}_{X_v,ab}(\mathrm{q})\right).$$

□

Theorem 5. *[5] For any word $x \in \Sigma_k^*$ treating x as a word over Σ_{k+1}, $\psi_q(x)\overline{\psi}_q(mi(x)) = q^{|x|}I_{k+1}$.*

The Theorem 5 does not hold for Picture array. For instance let $X = \begin{matrix} a & b & a & a & b \\ b & a & b & a & a \\ b & b & b & a & b \end{matrix}$ be the array over Σ_2 and X_v be the array obtained by reflecting X on its rightmost column *i.e.* $X_v = \begin{matrix} b & a & a & b & a \\ a & a & b & a & b \\ b & a & b & b & b \end{matrix}$. Then $R_\mathrm{q}(X)\overline{R}_\mathrm{q}(X_v)$=

$$\begin{bmatrix} \mathrm{q}^7 & \mathrm{q}^3 + 3\mathrm{q}^2 + 3\mathrm{q} & \mathrm{q}^2 + 2\mathrm{q} + 3 \\ 0 & \mathrm{q}^8 & \mathrm{q}^3 + \mathrm{q}^2 + 3\mathrm{q} + 3 \\ 0 & 0 & 1 \end{bmatrix}\begin{bmatrix} \mathrm{q}^8 & -\mathrm{q}^3 - 3\mathrm{q}^2 - 3\mathrm{q} & 3\mathrm{q}^3 + 4\mathrm{q}^2 + 3\mathrm{q} \\ 0 & \mathrm{q}^7 & -3\mathrm{q}^4 - 3\mathrm{q}^3 - \mathrm{q}^2 - \mathrm{q} \\ 0 & 0 & \mathrm{q}^{15} \end{bmatrix}$$

$\neq \mathrm{q}^{(3\times5)}I_3$.

Theorem 6. *For $h, v \geq 1$, let $X \in \wp$ be an array over Σ_2 treating X as an array over Σ_3. Let x_i be the horizontal words in the h rows and y_j be the vertical words in the v columns. Then*

(i) $\oplus_{i=1}^{h}\psi_q(x_i)\overline{\psi}_q(mi(x_i)) = q^{(h\times v)}I_3$
(ii) $\oplus_{j=1}^{v}\psi_q(y_j^t)\overline{\psi}_q(mi(y_j^t)) = q^{(h\times v)}I_3$.

Proof. Let X be the array in $\wp$ over Σ_2 with h number of rows and v number of columns such that $X = x_1 \diamond x_2 \diamond \cdots \diamond x_h$, $x_i \in \Sigma_2^*$, $1 \leq i \leq h$. Let x_i be a horizontal words over Σ_2 then by Theorem 5 we have $\psi_\mathrm{q}(x_i)\overline{\psi}_\mathrm{q}(mi(x_i)) = \mathrm{q}^{|x_i|}I_3$. As there are h number of rows we have

$$\bigoplus_{i=1}^{h}\psi_\mathrm{q}(x_i)\overline{\psi}_\mathrm{q}(mi(x_i)) = \mathrm{q}^{(h\times v)}I_3.$$

The proof of (ii) is similar. □

Theorem 7. *[4] Let x, y be two words over Σ_k then $x \sim_q y$ if and only if $mi(x) \sim_q mi(y)$*

We extend Theorem 7 for picture array as follows.

Theorem 8. *For $h, v \geq 1$, let X, Y be two arrays over Σ_2 such that $X \sim_q Y$ then $X_v \sim_q Y_v$.*

Proof. For $1 \leq i \leq h$, let x_i and y_i are the words in the ith rows of X and Y respectively such that $x_i \sim_q y_i$ implies $R_q(X) = R_q(Y)$ then by Theorem 7, we have $mi(x_i) \sim_q mi(y_i)$ which implies that $R_q(X_v) = R_q(Y_v)$. Similarly let x_j and y_j are the words in the jth columns of X and Y respectively such that $x_j \sim_q y_j$ implies $C_q(X) = C_q(Y)$ then by Theorem 7, we have $mi(x_j) \sim_q mi(y_j)$ which implies that $C_q(X_v) = C_q(Y_v)$. Therefore we have that $X_v \sim_q Y_v$. □

5 Parikh q-Matrix Commutator of Arrays

In this section we introduce Parikh q–matrix commutator of arrays and discuss some properties.

Definition 17. *For $h, v \geq 1$, let X and Y be two picture arrays over Σ_k are said to be*

1. *q–row equivalent if $R_q(X) = R_q(Y)$*
2. *q–column equivalent if $C_q(X) = C_q(Y)$.*

If the arrays X and Y are both having q–row equivalent and q–column equivalent then the arrays are q– equivalent (i.e. $X \sim_q Y$) also we can call that X as well as Y are q–ambiguous. Otherwise the arrays are said to be q–unambiguous.

Definition 18. *Let $X, Y \in \Sigma_2^{**}$. The set of all Parikh q–matrix commutators of a binary picture array X is defined as*

$$\phi_{com}^q(X) = \{Y | XY \sim_q YX\}$$

Theorem 9. *Let $X, Y \in \Sigma_2^{**}$. The following are equivalent.*

1. *$Y \in \phi_{com}^q(X)$*
2. *for all $i, j \geq 1$, the successive column concatenations $Y^i \in \phi_{com}^q(X^j)$ and the successive row concatenations $Y_i \in \phi_{com}^q(X_j)$*
3. *$Y_v \in \phi_{com}^q(X_v)$.*

Proof. $1 \Rightarrow 2$: $Y \in \phi_{com}^q(X) \Rightarrow XY \sim_q YX$. Therefore

$$\begin{aligned} Y^i X^j &\sim_q Y^{i-1} Y X X^{j-1} \\ &\sim_q Y^{i-1} X Y X^{j-1} \\ &\sim_q Y^{i-2} Y X Y X^{j-1} \\ &\sim_q Y^{i-2} X Y^2 X^{j-1}. \end{aligned}$$

Proceeding in this way we get $Y^i X^j \sim_q X^j Y^i$. Therefore $Y^i \in \phi_{com}^q(X^j)$. Similarly $Y_i \in \phi_{com}^q(X_j)$.
$2 \Rightarrow 3$: Taking $i = j = 1$, we get $YX \sim qXY$ and by Theorem 8, we have $(YX)_v \sim_q (XY)_v$ which implies that $Y_v X_v \sim_q X_v Y_v$. Therefore $Y_v \in \phi_{com}^q(X_v)$.
$3 \Rightarrow 1$: By Theorem 8, $Y \in \phi_{com}^q(X)$. □

Remark 1. Let $X, Y, Z \in \Sigma_2^{**}$. Then the following holds good

1. If $Z, Y \in \phi^{\mathtt{q}}_{com}(X)$ then the column concatenation $Z \cdot Y$ and $Y \cdot Z$ are in $\phi^{\mathtt{q}}_{com}(X)$
2. If $Z, Y \in \phi^{\mathtt{q}}_{com}(X)$ then the row concatenation $\begin{matrix} Z \\ \cdot \\ Y \end{matrix}$ and $\begin{matrix} Y \\ \cdot \\ Z \end{matrix}$ are in $\phi^{\mathtt{q}}_{com}(X)$
3. The successive column concatenation of X given by $X^2, X^3, X^4, \cdots\cdots \in \phi^{\mathtt{q}}_{com}(X)$
4. The successive row concatenation of X given by $X_2, X_3, X_4, \cdots\cdots \in \phi^{\mathtt{q}}_{com}(X)$
5. $X \in \phi^{\mathtt{q}}_{com}(X) \Rightarrow \phi^{\mathtt{q}}_{com}(X) \neq \emptyset$.

6 Conclusion

Several studies have been done on the combinatorial properties of picture arrays in relation to areas such as image processing, pattern recognition and computer vision. In this paper, we contribute to this field as well, by extending notions and concepts of Parikh q−matrices that have already been explored with one dimensional cases. It will be of interest to consider picture arrays to analyze the behaviour of these Parikh q-matrices under some array morphisms.

Acknowledgements. We would like to thank the unknown referees for their comments and suggestions on the manuscript in improving from an earlier version. The authors K. Janaki and R. Arulprakasam are very much thankful to the management, SRM Institute of Science and Technology for their continuous support and encouragement.

References

1. Atanasiu, A.: Binary amiable words. Int. J. Found. Comput. Sci. **18**(2), 387–400 (2007). https://doi.org/10.1142/S0129054107004735
2. Atanasiu, A., Atanasiu, R., Petre, I.: Parikh matrices and amiable words. Theor. Comput. Sci. **390**(1), 102–109 (2008). https://doi.org/10.1016/j.tcs.2007.10.022
3. Atanasiu, A., Martín-Vide, C., Mateescu, A.: On the injectivity of the parikh matrix mapping. Fundam. Informaticae **49**(4), 289–299 (2002). http://content.iospress.com/articles/fundamenta-informaticae/fi49-4-01
4. Bera, S., Ceterchi, R., Mahalingam, K., Subramanian, K.G.: Parikh q-matrices and q-ambiguous words. Int. J. Found. Comput. Sci. **31**(1), 23–36 (2020). https://doi.org/10.1142/S012905412040002X
5. Bera, S., Mahalingam, K.: Extending parikh q-matrices. Int. J. Comput. Appl. **975**, 8887 (2016)
6. Bera, S., Mahalingam, K.: Some algebraic aspects of parikh q-matrices. Int. J. Found. Comput. Sci. **27**(4), 479–500 (2016). https://doi.org/10.1142/S0129054116500118
7. Bera, S., Mahalingam, K.: On commuting parikh q-matrices. Fundam. Informaticae **172**(4), 327–341 (2020). https://doi.org/10.3233/FI-2020-1907

8. Bera, S., Mahalingam, K., Pan, L., Subramanian, K.: Two-dimensional picture arrays and parikh q-matrices. In: Journal of Physics: Conference Series, vol. 1132, p. 012006. IOP Publishing (2018)
9. Berthé, V., Tijdeman, R.: Balance properties of multi-dimensional words. Theor. Comput. Sci. **273**(1–2), 197–224 (2002). https://doi.org/10.1016/S0304-3975(00)00441-2
10. Carpi, A., de Luca, A.: Repetitions and boxes in words and pictures. In: Karhumäki, J., Maurer, H.A., Paun, G., Rozenberg, G. (eds.) Jewels are Forever, Contributions on Theoretical Computer Science in Honor of Arto Salomaa, pp. 295–306. Springer, Cham (1999). https://doi.org/10.1007/978-3-642-60207-8_26
11. Carpi, A., de Luca, A.: Repetitions, fullness, and uniformity in two-dimensional words. Int. J. Found. Comput. Sci. **15**(2), 355–383 (2004). https://doi.org/10.1142/S0129054104002479
12. Egecioglu, O., Ibarra, O.H.: A matrix Q-analogue of the Parikh map. In: Levy, J.-J., Mayr, E.W., Mitchell, J.C. (eds.) TCS 2004. IIFIP, vol. 155, pp. 125–138. Springer, Boston, MA (2004). https://doi.org/10.1007/1-4020-8141-3_12
13. Egecioglu, Ö., Ibarra, O.H.: A q-analogue of the Parikh matrix mapping. In: Mukund, M., Rangarajan, K., Subramanian, K.G. (eds.) Formal Models, Languages and Applications [this volume commemorates the 75th birthday of Prof. Rani Siromoney]. Series in Machine Perception and Artificial Intelligence, vol. 66, pp. 97–111. World Scientific (2007). https://doi.org/10.1142/9789812773036_0007
14. Janaki, K., Arulprakasam, R., Dare, V.: Generalized Parikh matrices of picture array. J. Math. Comput. Sci. **11**(2), 1955–1969 (2021)
15. Mateescu, A.: Algebraic aspects of Parikh matrices. In: Karhumäki, J., Maurer, H., Păun, G., Rozenberg, G. (eds.) Theory Is Forever. LNCS, vol. 3113, pp. 170–180. Springer, Heidelberg (2004). https://doi.org/10.1007/978-3-540-27812-2_16
16. Mateescu, A., Salomaa, A., Salomaa, K., Yu, S.: A sharpening of the Parikh mapping. RAIRO Theor. Inf. Appl. **35**(6), 551–564 (2001). https://doi.org/10.1051/ita:2001131
17. Parikh, R.: On context-free languages. J. ACM **13**(4), 570–581 (1966). https://doi.org/10.1145/321356.321364
18. Poovanandran, G., Teh, W.C.: On m-equivalence and strong m-equivalence for Parikh matrices. Int. J. Found. Comput. Sci. **29**(1), 123–138 (2018). https://doi.org/10.1142/S0129054118500065
19. Poovanandran, G., Teh, W.C.: Strong (2 · t) and strong (3 · t) transformations for strong m-equivalence. Int. J. Found. Comput. Sci. **30**(5), 719–733 (2019). https://doi.org/10.1142/S0129054119500187
20. Salomaa, A.: On the injectivity of parikh matrix mappings. Fundam. Informaticae **64**(1–4), 391–404 (2005). http://content.iospress.com/articles/fundamenta-informaticae/fi64-1-4-33
21. Salomaa, A.: Parikh matrices: subword indicators and degrees of ambiguity. In: Böckenhauer, H.-J., Komm, D., Unger, W. (eds.) Adventures Between Lower Bounds and Higher Altitudes. LNCS, vol. 11011, pp. 100–112. Springer, Cham (2018). https://doi.org/10.1007/978-3-319-98355-4_7
22. Serbanuta, T.: Extending Parikh matrices. Theor. Comput. Sci. **310**(1–3), 233–246 (2004). https://doi.org/10.1016/S0304-3975(03)00396-7
23. Subramanian, K.G., Huey, A.M., Nagar, A.K.: On Parikh matrices. Int. J. Found. Comput. Sci. **20**(2), 211–219 (2009). https://doi.org/10.1142/S0129054109006528
24. Subramanian, K.G., Mahalingam, K., Abdullah, R., Nagar, A.K.: Two-dimensional digitized picture arrays and Parikh matrices. Int. J. Found. Comput. Sci. **24**(3), 393–408 (2013). https://doi.org/10.1142/S012905411350010X

25. Teh, W.C.: On core words and the Parikh matrix mapping. Int. J. Found. Comput. Sci. **26**(1), 123–142 (2015). https://doi.org/10.1142/S0129054115500069
26. Teh, W.C.: Parikh matrices and strong M-equivalence. Int. J. Found. Comput. Sci. **27**(5), 545–556 (2016). https://doi.org/10.1142/S0129054116500155
27. Teh, W.C.: Parikh-friendly permutations and uniformly Parikh-friendly words. Australas. J. Comb. **76**, 208–219 (2020). http://ajc.maths.uq.edu.au/pdf/76/ajc_v76_p208.pdf
28. Teh, W.C., Kwa, K.H.: Core words and Parikh matrices. Theor. Comput. Sci. **582**, 60–69 (2015). https://doi.org/10.1016/j.tcs.2015.03.037
29. Masilamani, V., Krithivasan, K., Subramanian, K.G., Huey, A.M.: Efficient algorithms for reconstruction of 2D-arrays from extended Parikh images. In: Bebis, G., et al. (eds.) ISVC 2008. LNCS, vol. 5359, pp. 1137–1146. Springer, Heidelberg (2008). https://doi.org/10.1007/978-3-540-89646-3_113

Adjunct Partial Array Token Petri Net Structure

T. Kalyani[1(✉)], K. Sasikala[2], D. G. Thomas[3], and K. Bhuvaneswari[4]

[1] Department of Mathematics, St. Joseph's Institute of Technology, Chennai 600119, India
kalphd02@yahoo.com

[2] Department of Mathematics, St. Joseph's College of Engineering, Chennai 600119, India

[3] Department of Science and Humanities (Mathematics Division), Saveetha School of Engineering, SIMATS, Chennai 602105, India

[4] Department of Mathematics, Sathayabama Institute of Science and Technology, Chennai 600119, India

Abstract. Adjunct Array Token Petri Net Structure (AATPNS) and Adjunct Hexagonal Array Token Petri Net Structure (AHATPNS) are recently introduced in the literature. Partial Array Token Petri Net Structure (PATPNS) is introduced to generate partial array languages. In this paper, we extend this model using some control device called inhibitor arc and propose Adjunct Partial Array Token Petri Net Structure (APATPNS). It is compared with some partial array generating and recognizing models with respect to the generative power.

Keywords: Petri net · Partial arrays · Inhibitor arcs · Picture languages

1 Introduction

In formal languages, picture processing, picture generation, pattern matching, etc., play an important role. Many picture generative devices namely matrix grammars, automata, array grammars, kolam grammars and pasting systems have been proposed in the literature to generate two-dimensional picture languages [3,9]. Array grammars have been used effectively in the field of character recognition [2]. Cluster analysis can be accomplished on a set of patterns on the basis of a selected similarity measure, which can be used for pattern recognition [7]. Clustering methods is one of the methods effectively used in classification and production.

Motivated by the study of words and array languages, Berstel and Boasson [1] introduced partial words and many characterizations of partial words were derived. Later on partial array languages were introduced in [14] and its combinatorial properties were studied.

Recently another mechanism to generate array languages called Array Token Petri Net Structure was introduced in the literature [6,8,10]. Many extensions of

R. P. Barneva et al. (Eds.): IWCIA 2022, LNCS 13348, pp. 189–203, 2023.
https://doi.org/10.1007/978-3-031-23612-9_12

Petri nets have also been defined. These are called high-level Petri nets: coloured Petri nets, nested Petri nets and object nets. Recently, time has been included in Petri nets as a specification, which led to the evolution of Time Petri Nets (TPN) and Stochastic Petri Nets (STN). Adjunct Array Token Petri Net Structure is defined using adjunct rules to increase the generative capacity of the model [4,5]. As an application of array token Petri nets clustering analysis is used for character recognition [7].

In this paper, we propose Adjunct Partial Array Token Petri Net Structure (APATPNS) model to generate partial array languages. A control mechanism called inhibitor arc is also introduced and it is compared with other generating devices of partial array languages with respect to the generative capacity [11–13].

2 Preliminaries

In this section, we recall the definitions of partial word, partial array, Petri net, basic puzzle partial array grammar, partial array token Petri net structure with examples.

Definition 1. [11] *Let Σ be a finite alphabet of symbols. A partial word pw of length m over Σ, is a partial function $pw : N \to \Sigma$. For $1 \leq i \leq m$, if $pw(i)$ is defined, then $i \in D(pw)$; otherwise $i \in h(pw)$. $D(pw)$ is the domain of pw and $H(pw)$ is the hole set of pw. A word over Σ is a partial word over Σ with an empty set of holes. The 'do not know' symbol is denoted by '$\Diamond$' which does not belong to Σ. $H(pw)$ is the set of positions in which '$\Diamond$' appears in pw.*

Definition 2. [11] *Let Σ be a finite alphabet. A rectangular arrangement of elements over $\Sigma \cup \{\Diamond\}$ is defined as a partial array over $\Sigma \cup \{\Diamond\}$.*

Definition 3. [11] *If B is a partial array of size $m \times n$ over $\Sigma \cup \{\Diamond\}$, then the companion of B (denoted by $B_\Diamond$) is the total function $B_\Diamond : Z_+^2 \to \Sigma \cup \{\Diamond\}$ defined by* $B_\Diamond(i,j) = \begin{cases} B(i,j), & (i,j) \in D(B) \\ \Diamond, & \textit{otherwise, where } \Diamond \notin \Sigma \end{cases}$.

Example 1. Let $B_\Diamond = \begin{pmatrix} a & \Diamond & b \\ \Diamond & b & \Diamond \\ b & \Diamond & a \end{pmatrix}$. $B_\Diamond$ is the companion of a partial array B of size (3, 3).
$D(B) = \{(1,1),(1,3),(2,2),(3,1),(3,3)\}$
$H(B) = \{(1,2),(2,1),(2,3),(3,2)\}$.

The set of all partial arrays over $\Sigma \cup \{\Diamond\}$ is denoted by Σ_p^{**}, where $\Sigma_p = \Sigma \cup \{\Diamond\}$. $\Sigma_p^{++} = \Sigma_p^{**} - \{\Lambda\}$, where Λ is the empty array. $\Sigma_p^{(\ell,m)}$ is the set of all partial arrays over $\Sigma \cup \{\Diamond\}$ of size (ℓ, m).

Definition 4. [13] *The structure of a Basic Puzzle Partial Array Grammar (BPPAG) is $BPG_p = (A, B \cup \{\Diamond\}, P, S)$ where A is a finite non empty set of non terminal symbols and B is a finite non empty set of terminal symbols. '$\Diamond$' is a 'do not know' symbol, where $\Diamond \notin A \cup B$, $S \in A$ is the axiom pattern and P is a set of rules of the following forms:*

$(i)\quad X \rightarrow \textcircled{x}Y$	$(ii)\quad X \rightarrow \textcircled{\Diamond}Y$	$(iii)\quad X \rightarrow \textcircled{Y}x$
$(iv)\quad X \rightarrow \textcircled{Y}\Diamond$	$(v)\quad X \rightarrow Y\textcircled{x}$	$(vi)\quad X \rightarrow Y\textcircled{\Diamond}$
$(vii)\quad X \rightarrow x\textcircled{Y}$	$(viii)\quad X \rightarrow \Diamond\textcircled{Y}$	$(ix)\quad X \rightarrow \begin{array}{c}\textcircled{x}\\ Y\end{array}$
$(x)\quad X \rightarrow \begin{array}{c}\textcircled{\Diamond}\\ Y\end{array}$	$(xi)\quad X \rightarrow \begin{array}{c}\textcircled{Y}\\ x\end{array}$	$(xii)\quad X \rightarrow \begin{array}{c}\textcircled{Y}\\ \Diamond\end{array}$
$(xiii)\quad X \rightarrow \begin{array}{c}Y\\ \textcircled{x}\end{array}$	$(xiv)\quad X \rightarrow \begin{array}{c}Y\\ \textcircled{\Diamond}\end{array}$	$(xv)\quad X \rightarrow \begin{array}{c}x\\ \textcircled{Y}\end{array}$
$(xvi)\quad X \rightarrow \begin{array}{c}\Diamond\\ \textcircled{Y}\end{array}$	$(xvii)\quad X \rightarrow \textcircled{x}$	$(xviii)\quad X \rightarrow \textcircled{\Diamond}$

where $X, Y \in A$ *and* $x, y \in B$.

While processing the derivations in the production rule $X \rightarrow \textcircled{x}Y$, the nonterminal X is replaced by the right-hand member whose left-hand side is X.

The replacement is possible only if the noncircled symbol of the production rule consists of a blank symbol. The blank symbol is represented by the letter '#', which is an unoccupied place where any symbol can be occupied as per the derivation. The language generated by BPPAG is denoted by $\mathcal{L}(BPPAG)$.

Definition 5. [12] *If* $C = (P, T, I, O)$ *is a Petri Net structure with partial arrays over* $(\Sigma \cup \{\Diamond\})^{**}$ *as initial markings.* $\mu_0 : P \rightarrow (\Sigma \cup \{\Diamond\})^{**}$ *label of at least one transition being catenation rule and a finite set of final places* $F \subset P$*, then the Petri net structure* C *is defined as a Partial Array Token Petri Net Structure (PATPNS).*

Definition 6. [12] *If* C *is a PATPNS, then the Partial array language generated by the Petri Net* C *is defined as*

$$PL(C) = \{A_\Diamond \in (\Sigma \cup \{\Diamond\})^{**} \ / \ A_\Diamond \ is \ in \ p \ for \ some \ p \ in \ F\}$$

with partial arrays over $(\Sigma \cup \{\Diamond\})^{**}$ *in some places as initial marking when all possible sequences of transitions are fired. The set of all partial arrays collected in the final places* F *is called the partial array language generated by* C. *Let* $\mathcal{L}(PATPNS) = \{PL(C)/C \ is \ a \ PATPNS\}$.

Firing Rules in PATPNS [12]

We define three different types of enabled transition in PATPNS. The pre and post condition for firing the transition in all the three cases are given below:

1. When all the input places of t_1 (without label) have the same partial array as token.

  - Each input place should have at least the required number of partial arrays.
  - Firing t_1 removes partial array from all the input places and moves the partial array to all its output places.

The graph in Fig. 1 shows the position of the partial array before the transition fires and Fig. 2 shows the position of the partial array after transition t_1 fires.

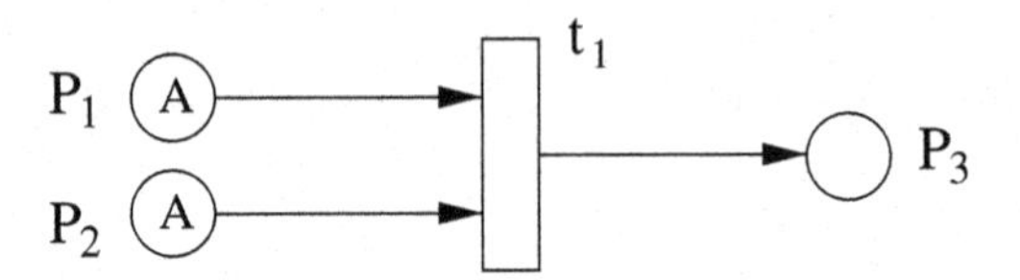

Fig. 1. Position of partial array before firing

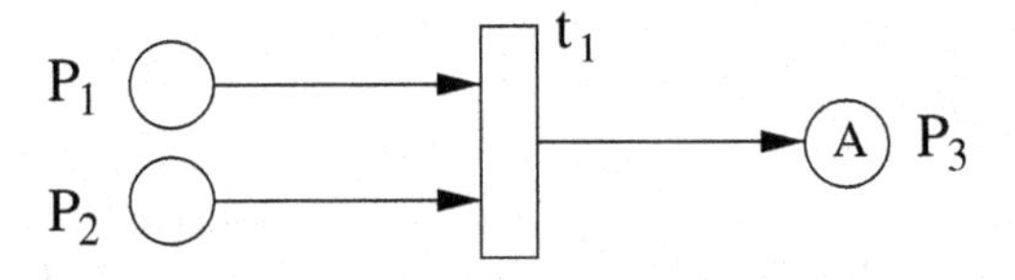

Fig. 2. Position of partial array after firing

2. When all the input places of t_1 have different partial arrays as token
   - The label of t_1 designates one of its input places.
   - The designated input place has sufficient number of partial arrays as tokens.
   - Firing t_1 removes partial array from all the input places and moves the partial array from the designated input place to all its output places.

The graph in Fig. 3 shows the position of the partial array before the transition fires and Fig. 4 shows the position of the partial array after transition t_1 fires. Since the designated place is P_1, the partial array in P_1 is moved to the output place.

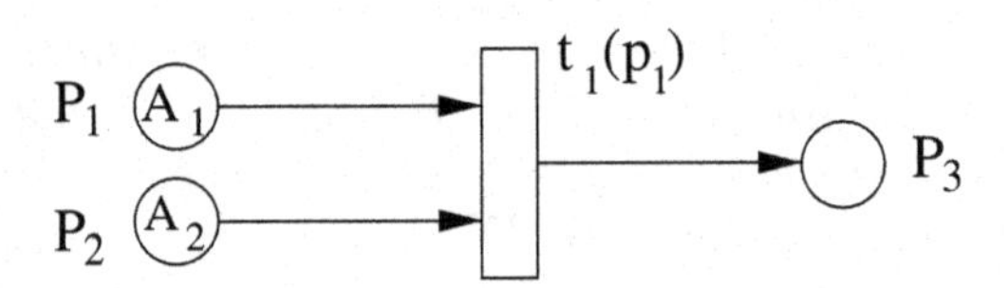

Fig. 3. Transition with label before firing

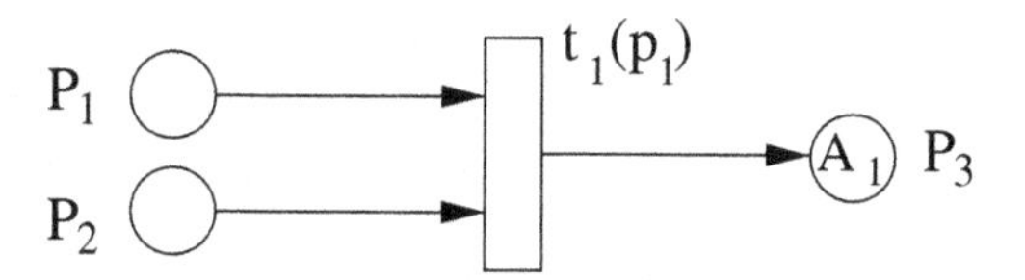

Fig. 4. Transition with label after firing

3. When all the input places of t_1 (with catenation rule as label) have the same partial array as token
   - Each input place should have at least the required number of partial arrays.
   - The condition for catenation should be satisfied.
   - The designated input place has sufficient number of partial arrays as tokens.
   - Firing t_1 removes partial array from all the input places P and the catenation is carried out in all its output places.

Catenation Rule as Label for Transitions: [12]
Column catenation rule is in the form $A \textcircled{|} B$. Here the partial array A denotes the $m \times n$ partial array in the input place of the transition. B is a partial array whose number of rows will depend on 'm', the number of rows of A. The number of columns of B is fixed. For example $A \textcircled{|} (x \;\; x)_m$ adds two columns of x after the last column of the partial array A which is in the input place. But $(x \;\; x)_m \textcircled{|} A$ would add two columns of x before the first column of A. 'm' always denote the number of rows of the input partial array A. Row catenation rule is in the form $A \ominus B$. Here again the partial array A denotes the $m \times n$ partial array in the input place of the transition. B is a partial array whose number of columns will depend on 'n', the number of columns of A. The number of rows of B is always fixed. For example $A \ominus \begin{bmatrix} x \\ x \end{bmatrix}^n$ adds two rows of x after the last row of the array A which is in the input place. But $\begin{bmatrix} x \\ x \end{bmatrix}^n \ominus A$ would add two rows of x before the first row of the partial array A. 'n' always denotes the number of columns of the input partial array A.

An example to explain row catenation rule is given below. The position of the partial array before the transition fires is shown in Fig. 5 and Fig. 6 shows the position of the partial array after transition t_1 fires. Since the catenation rule is associated with the transition, catenation takes place in P_3.

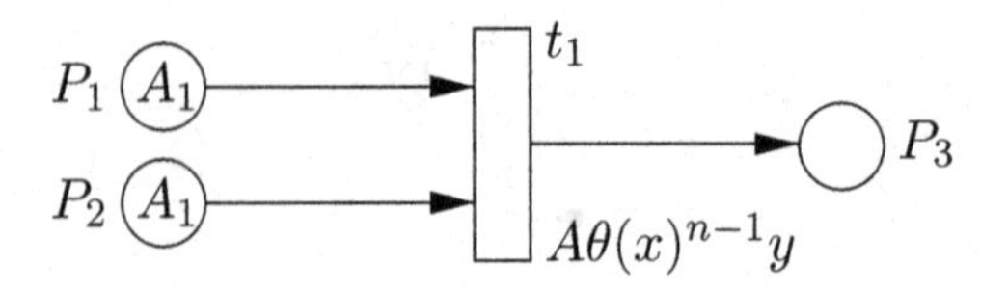

Fig. 5. Transition with catenation rule before firing

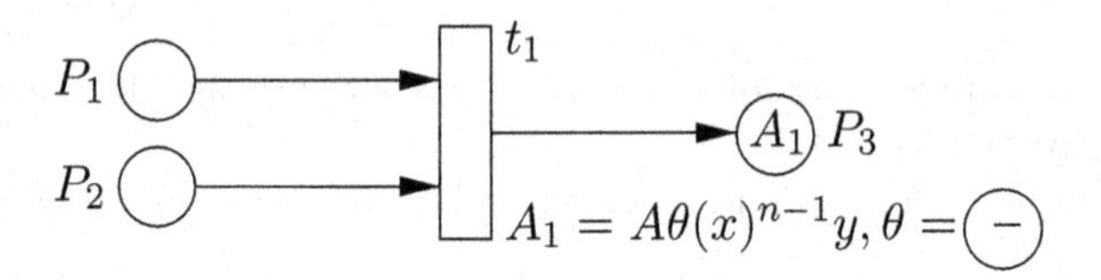

Fig. 6. Transition with catenation rule after firing

In $A_\diamondsuit = \begin{array}{ccc} a & a & a \\ a & \diamondsuit & a \\ a & a & a \end{array}$, the number of columns of A is 3, $n-1$ is 2, firing t_1 adds the row $x\ x\ y$ as the last row. Hence $A_{1\diamondsuit} = \begin{array}{ccc} a & a & a \\ a & \diamondsuit & a \\ a & a & a \\ x & x & y \end{array}$

Example 2. [12] Let $\Sigma = \{a\}$, $F = P_1$, where $S_\diamondsuit = \begin{array}{ccc} a & a & a \\ a & \diamondsuit & a \\ a & a & a \end{array}$, $Q_1 = (\diamondsuit)_m$, $Q_2 = (\diamondsuit)^n$ $Q_3 = (a)_m$, $Q_4 = (a)^n$

S is the initial partial array placed in P_1. The PATPNS is shown in Fig. 7. Derivations in PATPNS is given in the following tabular column.

Input place	Transition	Output place
S	$A \textcircled{\mid} Q_1$	$\begin{array}{cccc} a & a & a & \diamondsuit \\ a & \diamondsuit & a & \diamondsuit \\ a & a & a & \diamondsuit \end{array}$
$\begin{array}{cccc} a & a & a & \diamondsuit \\ a & \diamondsuit & a & \diamondsuit \\ a & a & a & \diamondsuit \end{array}$	$Q_1 \textcircled{\mid} A$	$\begin{array}{ccccc} \diamondsuit & a & a & a & \diamondsuit \\ \diamondsuit & a & \diamondsuit & a & \diamondsuit \\ \diamondsuit & a & a & a & \diamondsuit \end{array}$
$\begin{array}{ccccc} \diamondsuit & a & a & a & \diamondsuit \\ \diamondsuit & a & \diamondsuit & a & \diamondsuit \\ \diamondsuit & a & a & a & \diamondsuit \end{array}$	$A \textcircled{-} Q_2$	$\begin{array}{ccccc} \diamondsuit & a & a & a & \diamondsuit \\ \diamondsuit & a & \diamondsuit & a & \diamondsuit \\ \diamondsuit & a & a & a & \diamondsuit \\ \diamondsuit & \diamondsuit & \diamondsuit & \diamondsuit & \diamondsuit \end{array}$

<table>
<tr><td>◇ a a a ◇
◇ a ◇ a ◇
◇ a a a ◇
◇ ◇ ◇ ◇ ◇</td><td>$Q_2 \ominus A$</td><td>◇ ◇ ◇ ◇ ◇
◇ a a a ◇
◇ a ◇ a ◇
◇ a a a ◇
◇ ◇ ◇ ◇ ◇</td></tr>
<tr><td>◇ ◇ ◇ ◇ ◇
◇ a a a ◇
◇ a ◇ a ◇
◇ a a a ◇
◇ ◇ ◇ ◇ ◇</td><td>$A \obar Q_3$</td><td>◇ ◇ ◇ ◇ ◇ a
◇ a a a ◇ a
◇ a ◇ a ◇ a
◇ a a a ◇ a
◇ ◇ ◇ ◇ ◇ a</td></tr>
<tr><td>◇ ◇ ◇ ◇ ◇ a
◇ a a a ◇ a
◇ a ◇ a ◇ a
◇ a a a ◇ a
◇ ◇ ◇ ◇ ◇ a</td><td>$Q_3 \obar A$</td><td>a ◇ ◇ ◇ ◇ ◇ a
a ◇ a a a ◇ a
a ◇ a ◇ a ◇ a
a ◇ a a a ◇ a
a ◇ ◇ ◇ ◇ ◇ a</td></tr>
<tr><td>a ◇ ◇ ◇ ◇ ◇ a
a ◇ a a a ◇ a
a ◇ a ◇ a ◇ a
a ◇ a a a ◇ a
a ◇ ◇ ◇ ◇ ◇ a</td><td>$A \ominus Q_4$</td><td>a ◇ ◇ ◇ ◇ ◇ a
a ◇ a a a ◇ a
a ◇ a ◇ a ◇ a
a ◇ a a a ◇ a
a ◇ ◇ ◇ ◇ ◇ a
a a a a a a a</td></tr>
<tr><td>a ◇ ◇ ◇ ◇ ◇ a
a ◇ a a a ◇ a
a ◇ a ◇ a ◇ a
a ◇ a a a ◇ a
a ◇ ◇ ◇ ◇ ◇ a
a a a a a a a</td><td>$Q_4 \ominus A$</td><td>a a a a a a a
a ◇ ◇ ◇ ◇ ◇ a
a ◇ a a a ◇ a
a ◇ a ◇ a ◇ a
a ◇ a a a ◇ a
a ◇ ◇ ◇ ◇ ◇ a
a a a a a a a</td></tr>
</table>

The firing of sequence $(t_1t_2t_3t_4t_5t_6t_7t_8)^k$, $k \geq 0$ puts a square partial arrays of size $4k + 3$ in P_1, where the boundaries of the squares are alternatively ◇'s and a's. The partial array language generated by the PATPNS is a square partial array of size $4k+3$, $k \geq 0$ where the boundaries are alternatively ◇'s on the odd numbered boundaries and a's on the even numbered boundaries.

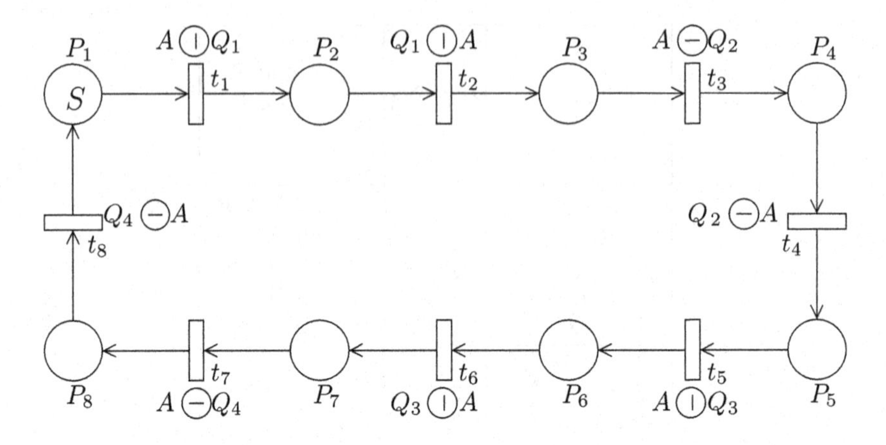

Fig. 7. PATPNS generating square partial arrays of size $4k+3$, $k \geq 0$

Example 3. [12] Consider a partial array language of square partial arrays of size $4k+3$, $k \geq 0$ whose boundaries are alternatively $\diamondsuit$'s on the odd numbered boundaries and a's on the even numbered boundaries.

$$BPG_{P_2} = (A, B \cup \{\diamondsuit\}, P, S)$$

where $A = \{X, Q_1, Q_2, Q_3, Q_4, Q_5, Q_6\}$, $B = \{a\}$, $S = X$ and P consists of the following rules:

(i) $X \rightarrow \textcircled{a}Q$ (ii) $Q \rightarrow \textcircled{a}Q$ (iii) $Q \rightarrow \begin{array}{c} Q_1 \\ \textcircled{a} \end{array}$

(iv) $Q_1 \rightarrow Q_2\textcircled{a}$ (v) $Q_2 \rightarrow Q_3\textcircled{\diamondsuit}$ (vi) $Q_3 \rightarrow \begin{array}{c} Q_4 \\ \textcircled{a} \end{array}$

(vii) $Q_4 \rightarrow \textcircled{a}Q_5$ $(viii)$ $Q_5 \rightarrow \textcircled{a}Q_5$ (ix) $Q_5 \rightarrow \textcircled{a}$

(x) $Q_3 \rightarrow \textcircled{\diamondsuit}Q_3$ (xi) $Q_5 \rightarrow \textcircled{\diamondsuit}Q_6$ (xii) $Q_6 \rightarrow \textcircled{a}Q_6$

$(xiii)$ $Q_6 \rightarrow \textcircled{\diamondsuit}Q$ (xiv) $Q_3 \rightarrow Q_2\textcircled{a}$ (xv) $Q_3 \rightarrow Q_3\textcircled{\diamondsuit}$

(xvi) $Q \rightarrow \textcircled{\diamondsuit}Q$

The first member of the language generated is shown below:

$$X \xrightarrow{(i)} \textcircled{a}\,Q \xrightarrow{(ii)} a\,\textcircled{a}\,Q \xrightarrow{(iii)} \begin{array}{ccc} & & Q_1 \\ a & a & \textcircled{a} \end{array} \xrightarrow{(iv)} \begin{array}{ccc} & Q_2 & \textcircled{a} \\ a & a & a \end{array}$$

$$\xrightarrow{(v)} \begin{array}{ccc} Q_3 & \textcircled{\diamondsuit} & a \\ a & a & a \end{array} \xrightarrow{(vi)} \begin{array}{ccc} Q_4 & & \\ a & \diamondsuit & a \\ a & a & a \end{array} \xrightarrow{(vii)} \begin{array}{ccc} \textcircled{a} & Q_5 & \\ a & \diamondsuit & a \\ a & a & a \end{array} \xrightarrow{(viii)} \begin{array}{ccc} a & \textcircled{a} & Q_5 \\ a & \diamondsuit & a \\ a & a & a \end{array} \xrightarrow{(ix)} \begin{array}{ccc} a & a & a \\ a & \diamondsuit & a \\ a & a & a \end{array}$$

3 Adjunct Partial Array Token Petri Net Structure

Here, the definition of adjunct array token Petri net structure is recalled and Adjunct Partial Array Token Petri Net Structure with and without inhibitor arcs is defined with examples.

If A and B are two partial arrays having same number of rows then $A \oplus B$ is the catenation of A and B columnwise. $A \ominus B$ is the row catenation of A and B, provided A and B are having the same number of columns.

Whereas, in row adjunction the partial array B can be joined to the partial array A after any row of A. In a similar manner, column adjunction can join partial array B after any column of A.

In partial array token Petri net structure, the partial arrays over the alphabet Σ_p are used as tokens in some input places.

Let $B \in \Sigma_p^{**}$ be a partial array of size $m \times n$ called host partial array. Let $D \in \Sigma_p^{**}$ be a partial array language consisting of partial arrays called adjunct partial arrays having fixed number of rows.

Definition 7. *An adjunct partial array D can be joined with a host partial array B by row adjunct rule (RAdR) in two ways as follows. The partial arrays B and D have the same number of columns.*

(i) By post rule denoted by (B, D, ar_j), the partial array D is adjoined to host partial array B after the jth row.
(ii) By pre rule (B, D, br_j), the adjunct partial array D is adjoined to the host partial array B before the jth row.

In a similar manner, the column adjunct rule (CAdR) can be defined.

(i) Post rule (B, D, ac_j) and Pre rule (B, D, bc_j) join D into B after the jth column of B and before the jth column of B respectively.

Definition 8. *An Adjunct partial Array Token Petri Net Structure (APATPNS) is a five tuple $S = (\Sigma, Q, M_0, \mu, F)$ where Σ is a given finite alphabet, $Q = (P, T, I, O)$ is a Petri net structure with partial arrays over $\Sigma \cup \{\lozenge\}$ as tokens, $M_0 : P \to \Sigma_p^{**}$, is the initial marking of the net, $\mu : T \to L$, a function from the set of transitions to the set of labels where catenation rules of the labels are either RAdR or CAdR and $F \subset P$ is a finite set of final places.*

The language generated by APATPNS S is
$L(S) = \{A \in (\Sigma \cup \{\lozenge\})^{**} / A$ *is the place q for some $q \in F$}.*

Example 4. Consider the Adjunct Partial Array Token Petri Net Structure $S_1 = (\Sigma, Q, M_0, \mu, F)$ where $\Sigma = \{a\}$, $Q = (P, T, I, O)$, $P = \{p_1, p_2\}$, $T = \{t_1, t_2\}$, $I(t_1) = \{p_1\}$, $I(t_2) = \{p_2\}$, $O(t_1) = \{p_2\}$, $O(t_2) = \{p_1\}$, M_0 is the initial marking; the partial array S is in p_1 and there is no partial array in p_2, $\mu(t_1) = (B, D_1, ac_1)$ and $\mu(t_2) = (B, D_2, br_1)$ and $F = \{p_1\}$. The Petri net diagram is shown in Fig. 8.

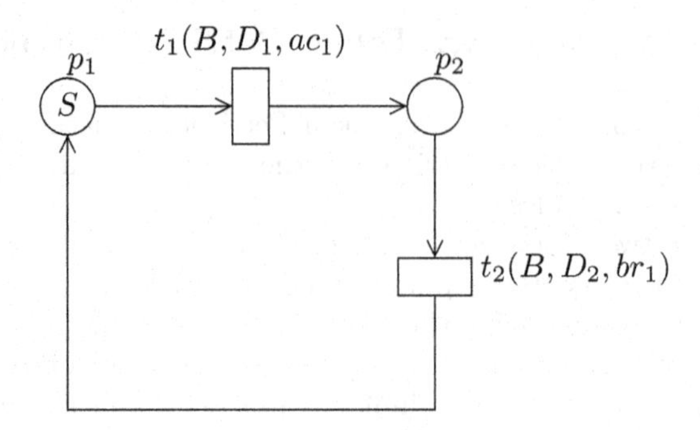

Fig. 8. APATPNS

The partial arrays used in the Petri net structure are defined as follows:

$$S = \begin{matrix} a & \lozenge \\ a & a \end{matrix}; D_1 = (a)_m \text{ and } D_2 = (\lozenge)^n.$$

Initially t_1 is the only enabled transition.

Firing of t_1 adjoins a column of a's after the first column of partial array S and puts the generated partial array in p_2, making t_2 enabled. Firing t_2 adjoins a row of $\lozenge$'s before the first row of the partial array in p_2 and puts the generated partial array in p_1. When the transitions t_1, t_2 fire, the partial array that reaches the output place p_1 is shown as below:

$$\begin{matrix} a & \lozenge \\ a & a \end{matrix} \overset{t_1}{\Longrightarrow} \begin{matrix} a & a & \lozenge \\ a & a & a \end{matrix} \overset{t_2}{\Longrightarrow} \begin{matrix} \lozenge & \lozenge & \lozenge \\ a & a & \lozenge \\ a & a & a \end{matrix}.$$

Firing the sequence $(t_1 t_2)^2$ generates the output partial array as

$$\begin{matrix} \lozenge & \lozenge & \lozenge & \lozenge \\ \lozenge & a & \lozenge & \lozenge \\ a & a & a & \lozenge \\ a & a & a & a \end{matrix}.$$

The partial array language L_1 generated by the Adjunct partial array token Petri net structure is the set of square partial arrays of size, 'n' consisting of only $\lozenge$'s above the main diagonal and only a's below the main diagonal except the first column and the main diagonal starts with a '$\lozenge$' followed by $n-1$ a's. The first column is of the form $\begin{pmatrix} (\lozenge)_{n-2} \\ a \\ a \end{pmatrix}$ where n is the size of the square partial array.

Example 5. Consider the Adjunct Partial Array Token Petri Net Structure $S_2 = (\Sigma, Q, M_0, \mu, F)$, where $\Sigma = \{a\}$, $Q = (P, T, I, O)$, $P = \{p_1, p_2, p_3, p_4, p_5, p_6, p_7\}$,

$T = \{t_1, t_2, t_3, t_4, t_5, t_6\}$, $I(t_1) = \{p_1, p_2\}$, $O(t_1) = \{p_3\}$, $I(t_2) = \{p_3\}$, $O(t_2) = \{p_4\}$, $I(t_3) = \{p_4\}$, $O(t_3) = \{p_2, p_5\}$, $I(t_4) = \{p_1, p_5\}$, $O(t_4) = \{p_6\}$, $I(t_5) = \{p_5, p_6\}$, $O(t_5) = \{p_1\}$, $I(t_6) = \{p_1, p_2\}$, $O(t_6) = \{p_7, p_2\}$. $\mu : T \rightarrow L$ is defined as $\mu(t_1) = p_2$, $\mu(t_2) = (A, B_1, ac_n)$, $\mu(t_3) = (A, B_2, ar_m)$, $\mu(t_4) = \mu(t_5) = \mu(t_6) = \lambda$. $F = \{p_7\}$. The Petri net graph is shown in Fig. 9.

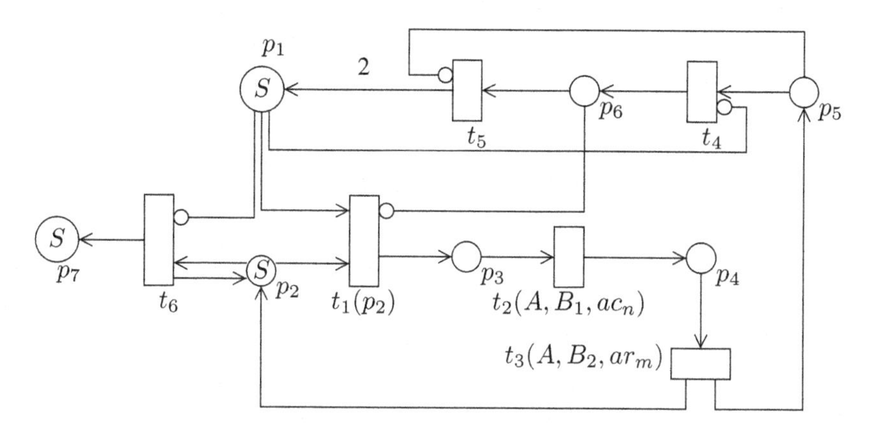

Fig. 9. APATPNS with inhibitor arcs

The partial arrays used in the Petri net structure are $S = \begin{matrix} a & \Diamond \\ a & \Diamond \end{matrix}$, $B_1 = \left(a\ \Diamond\right)_m$, $B_2 = \begin{pmatrix} \Diamond \\ a \end{pmatrix}^n$.

To being with only t_1 is enabled. After t_2 is enabled, the partial array B_1 is adjuncted with S after column 2 and the result is placed in p_4, now transition t_3 is enabled, the partial array B_2 is adjuncted with the resultant partial array after 2nd row. Thus firing of sequence of transitions $t_1t_2t_3$ results in a square partial array of $\{a, \Diamond\}$ of size 4×4 in p_2 and p_5.

$$\begin{matrix} a & \Diamond \\ a & \Diamond \end{matrix} \xrightarrow{t_1t_2t_3} \begin{matrix} a & \Diamond & a & \Diamond \\ a & \Diamond & a & \Diamond \\ \Diamond & \Diamond & \Diamond & \Diamond \\ a & a & a & a \end{matrix}.$$

Firing the transition t_6 puts the above partial array of size 4×4 in p_7. Firing t_4 pushes the partial array to p_6, emptying p_5. In this position t_5 is enabled. Firing t_5 puts two copies of the same partial array in p_1. Now, there are two copies in p_1, therefore to empty p_1, the sequence of transitions $t_1t_2t_3$ has to fire two times. The firing of sequence $t_4t_5(t_1t_2t_3)^2t_6$ puts a square partial array over $\{a, \Diamond\}$ of size 8×8 in p_7. The inhibitor input p_1 makes sure that a square partial array over $\{a, \Diamond\}$ of size 6×6 does not reach p_7. Thus this APATPNS generates a language of square partial arrays as in the following figure of size $(2^n, 2^n)$, $n \geq 1$ (Fig. 10).

$$\begin{array}{cccccccc}
a & \Diamond & a & \Diamond & a & \Diamond & a & \Diamond \\
a & \Diamond & a & \Diamond & a & \Diamond & a & \Diamond \\
\Diamond & \Diamond & \Diamond & \Diamond & a & \Diamond & a & \Diamond \\
a & a & a & a & a & \Diamond & a & \Diamond \\
\Diamond & \Diamond & \Diamond & \Diamond & \Diamond & \Diamond & a & \Diamond \\
a & a & a & a & a & a & a & \Diamond \\
\Diamond & \Diamond & \Diamond & \Diamond & \Diamond & \Diamond & \Diamond & \Diamond \\
a & a & a & a & a & a & a & a
\end{array}$$

Fig. 10. Square partial array of size $(2^3, 2^3)$

4 Comparative Results

In this section, we compare APATPNS with (i) Partial Array Token Petri Net Structure (PATPNS) and (ii) Basic Puzzle Partial Array Grammar System (BPPAG) for generative powers. The notation $\mathcal{L}(X)$ denotes the family of all partial array languages generated by the above systems X.

Theorem 1. *The family of languages generated by PATPNS is properly contained in the family of languages generated by APATPNS.*

Proof. The row catenation in PATPNS can be handled by the following row adjunct rules (i) By post rule (B, D, ar_m), the partial array D is adjoined to host partial array B after the mth row (ii) By pre rule (B, D, br_1), the partial array D is adjoined to the host partial array B before the 1st row.

The column catenation in PATPNS can be handled by the following column adjunct rules (i) By post rule (B, D, ac_n) and (ii) By pre rule (B, D, bc_1) join D into B, after the nth column of B and before the 1st column of B respectively.

Hence $\mathcal{L}(PATPNS)$ is a subset of $\mathcal{L}(APATPNS)$, this is also evident from the following example.

Consider a partial array language of square partial arrays of size $4k+3$, $k \geq 0$ whose boundaries are alternatively $\Diamond$'s on the odd numbered boundaries and a's on the even numbered boundaries (Example 2). This partial array language is generated by APATPNS.

Now, let us consider a APATPNS generating this partial array language $S_3 = (\Sigma, Q, M_0, \mu, F)$ where $\Sigma = \{a\}$, $Q = (P, T, I, O)$, $P = \{p_1, p_2, p_3, p_4, p_5, p_6, p_7, p_8\}$, $T = \{t_1, t_2, t_3, t_4, t_5, t_6, t_7, t_8\}$, $I(t_1) = \{p_1\}$, $I(t_2) = \{p_2\}$, $I(t_3) = \{p_3\}$, $I(t_4) = \{p_4\}$, $I(t_5) = \{p_5\}$, $I(t_6) = \{p_6\}$, $I(t_7) = \{p_7\}$, $I(t_8) = \{p_8\}$, $O(t_1) = \{p_2\}$, $O(t_2) = \{p_3\}$, $O(t_3) = \{p_4\}$, $O(t_4) = \{p_5\}$, $O(t_5) = \{p_6\}$, $O(t_6) = \{p_7\}$, $O(t_7) = \{p_8\}$, $O(t_8) = \{p_1\}$. M_0 is the initial marking, the partial array S is in p_1 and there is no partial array in the remaining places $\mu(t_1) = (B, D_1, ac_n)$, $\mu(t_2) = (B, D_1, bc_1)$, $\mu(t_3) = (B, D_2, ar_m)$, $\mu(t_4) = (B, D_2, br_1)$, $\mu(t_5) = (B, D_3, ac_n)$, $\mu(t_6) = (B, D_3, bc_1)$, $\mu(t_7) = (B, D_4, ar_m)$, $\mu(t_8) = (B, D_4, br_1)$ and $F = \{p_1\}$.

The Petri net graph is shown in Fig. 11.

The partial array language given in Example 5 generated by APATPNS cannot be generated by PATPNS, since in PATPNS, the axiom array can only be

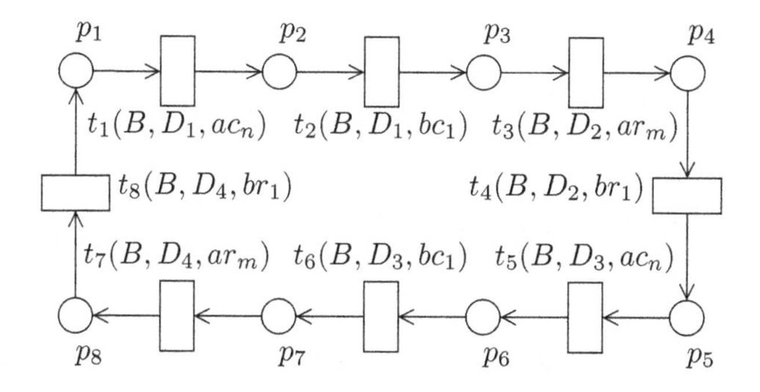

Fig. 11. APATPNS generating square partial arrays of size $4k + 3$, $k \geq 0$

concatenated either rowwise or columnwise but any row or column cannot be inserted which proves a proper containment.

Theorem 2. *The family of languages generated by BPPAG is properly contained in the family of languages generated by APATPNS.*

Proof. (a) The basic puzzle partial array grammar rules

(i) $X \rightarrow \begin{array}{c}\textcircled{x}\\ Y\end{array}$ (ii) $X \rightarrow \begin{array}{c}\textcircled{$\Diamond$}\\ Y\end{array}$ (iii) $X \rightarrow \begin{array}{c}Y\\ \textcircled{$\Diamond$}\end{array}$ (iv) $X \rightarrow \begin{array}{c}\textcircled{Y}\\ x\end{array}$

(v) $X \rightarrow \begin{array}{c}Y\\ \textcircled{x}\end{array}$ (vi) $X \rightarrow \begin{array}{c}Y\\ \textcircled{$\Diamond$}\end{array}$ (vii) $X \rightarrow \begin{array}{c}x\\ \textcircled{Y}\end{array}$ (viii) $X \rightarrow \begin{array}{c}\Diamond\\ \textcircled{Y}\end{array}$

(ix) $X \rightarrow \textcircled{x}$ (x) $X \rightarrow \textcircled{$\Diamond$}$

can be handled by the following row adjunct rules (i) (B, D, ar_m) and (B, D, br_1).

(b) The basic puzzle partial array grammar rules of the forms

(i) $X \rightarrow \textcircled{x}Y$ (ii) $X \rightarrow \textcircled{$\Diamond$}Y$ (iii) $X \rightarrow \textcircled{Y}\,x$ (iv) $X \rightarrow Y\,\textcircled{$\Diamond$}$
(v) $X \rightarrow Y\,\textcircled{x}$ (vi) $X \rightarrow Y\,\textcircled{$\Diamond$}$ (vii) $X \rightarrow \Diamond\,\textcircled{Y}$ (viii) $X \rightarrow x\,\textcircled{Y}$
(ix) $X \rightarrow \textcircled{x}$ (x) $X \rightarrow \textcircled{$\Diamond$}$

can be handled by the following column adjunct rules (B, D, ac_n) and (B, D, bc_1). Hence $\mathcal{L}(BPPAG)$ is a subset of $\mathcal{L}(APATPNS)$. This is also evident from the following example.

The partial array language of square partial arrays of size $4k+3$, $k \geq 0$ whose boundaries are alternatively $\Diamond$'s on the odd numbered boundaries and a's on the even numbered boundaries is generated by BPPAG (Example 3). This partial array language is also generated by APATPNS as shown in Theorem 1. Hence the two families intersect.

The language of partial arrays given in Example 5 cannot be generated by BPPAG, since in BPPAG, the partial array can be extended in right, left, up or down provided the place is empty for the symbol to be occupied. Any row or column cannot be inserted into a partial array. This proves proper containment.

5 Conclusion

In this paper we have considered Partial Array Token Petri Net structure with adjunction rules and also we have given example for APATPNS with inhibitor arcs and proved that inhibitor arcs is more powerful in generative power than APATPNS. APATPNS is compared with PATPNS and BPPAG and it is proved that APATPNS is more powerful in generative power when compared with these two models.

Adjunct Partial Array Token Petri net P system can be defined and its generative power can be tested with our earlier models. Also as an application of adjunct partial array token Petri structure various kolam patterns can be generated and character recognition can be done using clustering analysis. This is our future work.

It would be of interest in deciding whether a given image belongs to the language generated by a given APATPNS. This membership problem is equivalent to the Petri net reachability problem, which states that whether from the given initial configuration there exists a sequence of valid execution sets that reaches the given final configuration. The complexity of the petri net reachability problem is unsettled since the year 1960 and it is one of the most prominent open question in the theory of verification. We are exploring the membership problem for our future study.

References

1. Berstel, J., Boasson, L.: Partial words and a theorem of Fine and Wilf. Theoret. Comput. Sci. **218**(1), 135–141 (1999)
2. Fernau, H., Freund, R., Holzer, M.: Character recognition with k-head finite array automata. In: Amin, A., Dori, D., Pudil, P., Freeman, H. (eds.) SSPR /SPR 1998. LNCS, vol. 1451, pp. 282–291. Springer, Heidelberg (1998). https://doi.org/10.1007/BFb0033246
3. Giammarresi, D., Restivo, A.: Two - Dimensional Languages. Handbook of Formal Languages, pp. 215–267 (1997)
4. Kamaraj, T., Lalitha, D., Thomas, D.G.: A comparative study on adjunct array token Petri nets with some classes of array grammars. Appl. Math. Sci. **7**(135), 6705–6713 (2013)
5. Kamaraj, T., Lalitha, D., Thomas, D.G., Thambuarj, R., Atulya, K.: Nagar: adjunct hexagonal array token Petri nets and hexagonal picture languages. Math. Appl. **3**, 45–59 (2014)
6. D., L.: Rectangular array languages generated by a Petri net. In: Sethi, I.K. (ed.) Computational Vision and Robotics. AISC, vol. 332, pp. 17–27. Springer, New Delhi (2015). https://doi.org/10.1007/978-81-322-2196-8_3
7. Lalitha, D., Rangarajan, K.: An application of array token Petri nets to clustering analysis for syntactic patterns. Int. J. Comput. Appl. **42**(16), 21–25 (2012)
8. Lalitha, D., Rangarajan, K., Thomas, D.G.: Rectangular arrays and Petri nets. In: Barneva, R.P., Brimkov, V.E., Aggarwal, J.K. (eds.) IWCIA 2012. LNCS, vol. 7655, pp. 166–180. Springer, Heidelberg (2012). https://doi.org/10.1007/978-3-642-34732-0_13

9. Nivat, M., Saoudi, A., Subramanian, K.G., Siromoney, R., Dare, V.R.: Puzzle grammar and context - free array grammars. Int. J. Pattern Recogn. Artif. Intell. **05**(05), 663–676 (1991)
10. Peterson, J.L.: Petri Net Theory and Modeling of Systems. Prentice Hall Inc., Englewood Cliffs (1981)
11. Sasikala, K., Kalyani, T., Thomas, D.G.: Partial array grammars and partial array - rewriting P systems. Math. Eng. Sci. Aero Space **11**(1), 227–236 (2020)
12. Sasikala, K., Sweety, F., Kalyani, T., Thomas, D.G.: Partial array token Petri net and P system. In: Freund, R., Ishdorj, T.-O., Rozenberg, G., Salomaa, A., Zandron, C. (eds.) CMC 2020. LNCS, vol. 12687, pp. 135–152. Springer, Cham (2021). https://doi.org/10.1007/978-3-030-77102-7_8
13. Sweety, F., Sasikala, K., Kalyani, T., Thomas, D.G.: Partial array - rewriting P systems and basic puzzle partial array grammar. In: AIP Conference Proceedings, vol. 2277, p. 030003 (2020)
14. Vijaya Chitra, S., Sasikala, K.: Squares in partial arrays. In: AIP Conference Proceedings, vol. 2112, pp. 20–34 (2019)

2D Oxide Picture Languages and Their Properties

Helen Vijitha Ponraj[1](✉), Robinson Thamburaj[2](✉), and Meenakshi Paramasivan[3](✉)

[1] Department of Mathematics, Rajalakshmi Engineering College, Chennai 602105, India
helenvijitha.p@rajalakshmi.edu.in
[2] Department of Mathematics, Madras Christian College, Chennai 600059, India
robinson@mcc.edu.in
[3] FB IV - Informatikwissenschaften, Universität Trier, 54286 Trier, Germany
meena_maths@yahoo.com

Abstract. In the theory of formal languages, two-dimensional (picture) languages are a generalization of string languages to two dimensions. Pictures may be regarded as digitized finite arrays, occurring in studies concerning pattern recognition, image analysis, cellular automata, and parallel computing. Several studies have been done for generating and (or) recognizing rectangular, triangular, and hexagonal arrays using formal syntactic methods. Motivated by oxide molecular structures, the oxide pictures, a special class of two-dimensional pictures, are considered. Various generating and recognizing schemes, such as the Oxide Tiling System (OXTS), Oxide Wang System (OXWS), Oxide Tile Rewriting Grammar (OXTRG), and Oxide Sgraffito Automata (OXS), have been developed recently. It is found that the family of oxide picture languages recognizable by oxide tiling systems is closed under union, overlapping, half-turn, transpose, anti-transpose, and reflection (both along horizontal and vertical lines), but not closed under quarter-turn and anti-quarter-turn. This paper further discusses some language theoretic results as well.

Keywords: Two-dimensional languages · Oxide pictures · Oxide tiles

Mathematics Subject Classification: 68Q45 · 68R15

1 Introduction

Syntactic considerations of digital images have a tradition of about six decades. Two-dimensional picture languages generated by grammars or recognized by automata have been studied since the 1970s for their complications arising in the framework of pattern recognition and image analysis. Rani Siromoney and her co-authors in the early 1970s studied two-dimensional picture languages [7, 21, 26] where pictures are digitized finite arrays in a rectangular grid. These two-dimensional picture languages have connections with the generation of Kolam

R. P. Barneva et al. (Eds.): IWCIA 2022, LNCS 13348, pp. 204–225, 2023.
https://doi.org/10.1007/978-3-031-23612-9_13

patterns [28,31], which are traditional pieces of the south Indian style of painting. Dora Giammarresi and her co-authors studied two-dimensional picture languages and investigated its connection to Tiling systems [1,4] by defining local and recognizable picture languages.

Wang systems [16] are used as a formalism to recognize picture languages. It has been proved that the family of picture languages defined by Wang systems coincides with the family of picture languages recognized by tiling systems. In [18] triangular picture languages were studied by introducing triangular Wang automata based on triangular Wang tiles. It has been proved that triangular Wang automata with a specific scanning strategy recognize the class of triangular pictures recognized by triangular tiling systems. Two-dimensional hexagonal arrays on a triangular grid can be viewed or treated as its two-dimensional representations of three-dimensional rectangular parallelepipeds. Hexagonal arrays and hexagonal patterns are found in the literature on picture processing and scene analysis [2,20,25].

The study of picture languages has been carried out on appropriate grids: square grids, triangular grids, and hexagonal grids. To study oxide pictures, we define a hexo-triangular grid obtained from a triangular grid. We remove certain points in the triangular grid: starting from any arbitrarily chosen point remove the points of the grid alternatively in the directions of x, y, and z. Figure 4 is the punctured triangular grid obtained by removing the green grid points in Fig. 3. This punctured triangular grid can be viewed as a hexo-triangular grid with a hexagon whose neighbors are triangles. We name the hexo-triangular grid as oxide grid as it resembles the oxide network structure (see Fig. 2) proposed by Paul Manuel and Indira Rajasingh. Two-dimensional oxide pictures were considered and studied by introducing oxide tiling systems, oxide Wang systems, oxide tile rewriting grammars, and oxide Sgraffito automata (see [12–14]).

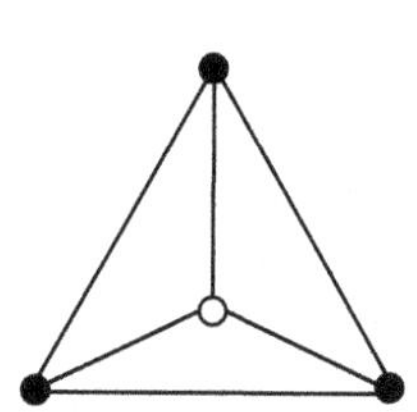

Fig. 1. SiO_4 - Silicon-oxygen Tetrahedron.

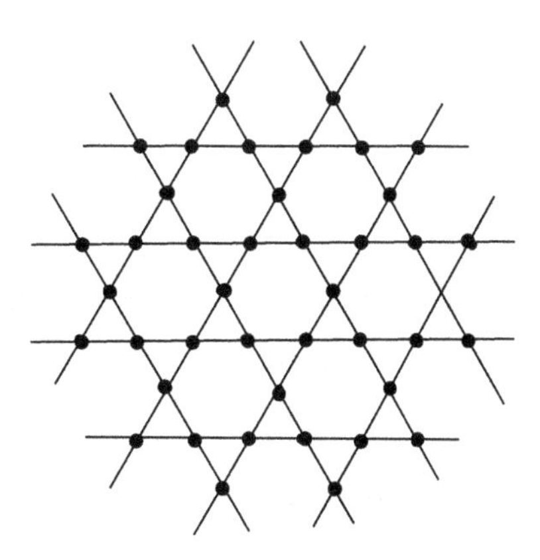

Fig. 2. An oxide network.

A preliminary version of this paper and some of its properties were presented by the authors in [15]. Now the paper is organized as follows: In Sect. 2 we recall some basics. In Sect. 3 we introduce two-dimensional oxide pictures and languages. A coordinate system is established for oxide pictures and types of oxide pictures with a special case namely, triangular oxide pictures. In Sect. 4 we prove some major results in connection with closure and non-closure under Boolean, unary, and binary operations. In Sect. 5 we conclude a few connections to further applications and implementations as directions to future research topics.

2 Preliminaries

In this section, we briefly recall the standard definitions and notations regarding two-dimensional oxide pictures and languages (see [12]). Let $\mathbb{N} := \{1, 2, 3, \ldots\}$ be the set of all natural numbers. For a finite alphabet Σ, a *string* or *word* (*over* Σ) is a finite sequence of symbols from Σ, and λ stands for the *empty string*. The notation Σ^+ denotes the set of all nonempty strings over Σ, and $\Sigma^* := \Sigma^+ \cup \{\lambda\}$. A *two-dimensional word* (also called a *picture*, a *matrix* or, an *array*) *over* Σ is a tuple

$$W := ((a_{1,1}, a_{1,2}, \ldots, a_{1,n}), (a_{2,1}, a_{2,2}, \ldots, a_{2,n}), \ldots, (a_{m,1}, a_{m,2}, \ldots, a_{m,n})),$$

where $m, n \in \mathbb{N}$ and, for every i, $1 \leq i \leq m$, and j, $1 \leq j \leq n$, $a_{i,j} \in \Sigma$ (see [3,11]).

An oxide array is a hexo-triangular array that is a hexagon surrounded by six triangles. The hexo-triangular arrangement of the oxide grid sensitized us to consider the pictures on this grid as oxide pictures and the languages of oxide pictures as oxide picture languages.

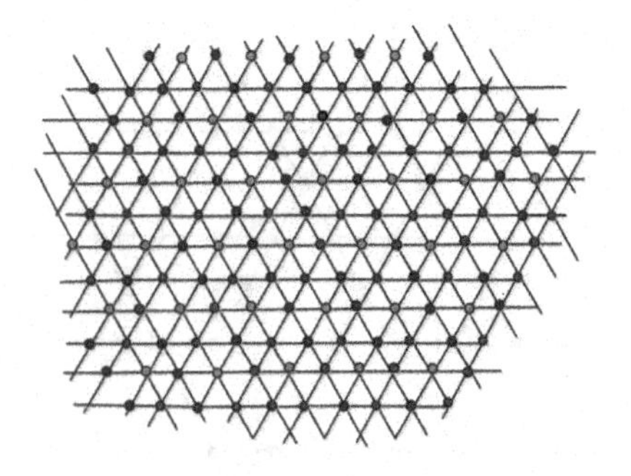

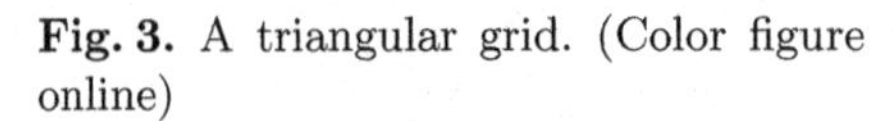

Fig. 3. A triangular grid. (Color figure online)

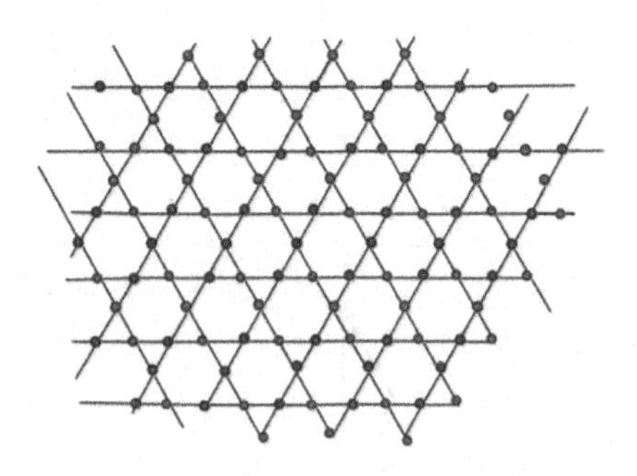

Fig. 4. An oxide (hexo-triangular) grid.

Definition 1. *An* oxide picture *over* Σ *is an oxide array of symbols. The set of all oxide pictures over* Σ *is denoted by* Σ^{**OX_p}. *An* oxide picture language *over* Σ *is a subset of* Σ^{**OX_p}.

Let $OX_p \in \Sigma^{**OX_p}$, we get the bordered version of OX_p denoted by $\hat{OX_p}$ when the special symbol $\# \notin \Sigma$ is added as the boundary to OX_p.

Example 1. An $OX_p \in \Sigma^{**OX_p}$ over $\Sigma = \{\mathtt{a}, \mathtt{b}, \mathtt{c}\}$ and its bordered version $\hat{OX_p}$ are given in Fig. 5.

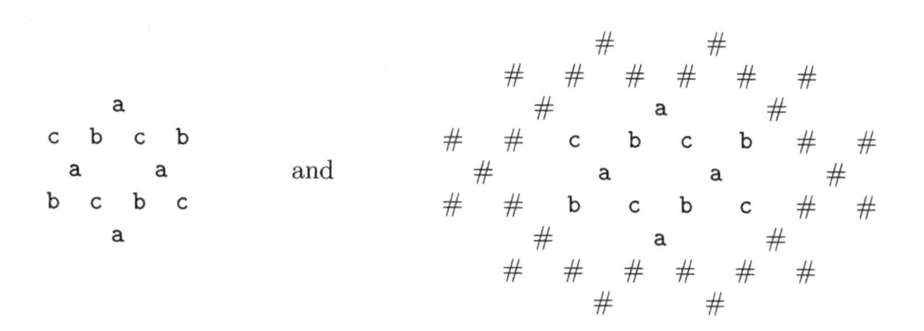

Fig. 5. An oxide picture over $\Sigma = \{\mathtt{a}, \mathtt{b}, \mathtt{c}\}$ and its bordered version.

Definition 2. *Let $\pi : \Gamma \rightarrow \Sigma$ be a mapping where Γ, Σ are finite set of alphabets. Let $OX_p \in \Gamma^{**OX_p}$ be an oxide picture of size (l, m, n). The* projection by mapping π of oxide picture OX_p *is the oxide picture $OX_{p'} \in \Sigma^{**OX_p}$ such that $OX_{p'}(x, y, z) = \pi(OX_p(x, y, z))$, for all $1 \leq x \leq l$, $1 \leq y \leq m$, $1 \leq z \leq n$.*

We will use $\pi(OX_p)$ instead of $\pi(OX_p(x, y, z))$ to indicate the projection of oxide picture OX_p by mapping π, when there is no ambiguity. It is natural to extend the definition of projection of an oxide picture to sets of oxide pictures.

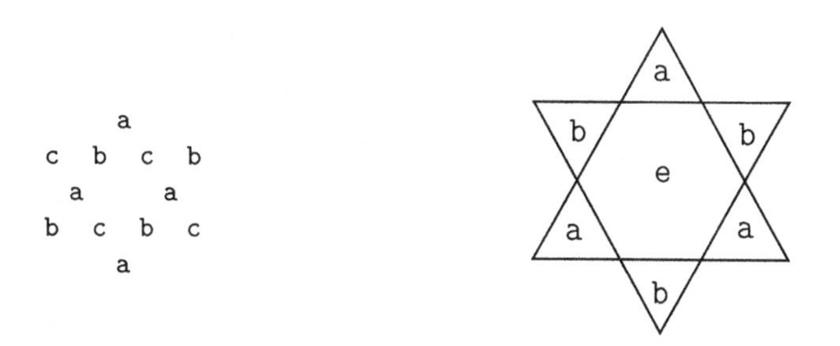

Fig. 6. An oxide picture and an oxide tile.

Definition 3. *Let $L \subseteq \Gamma^{**OX_p}$ be an oxide picture language. The* projection by mapping π of oxide picture language L *is the oxide picture language* $L' = \{OX_{p'} \mid OX_{p'} = \pi(OX_p)\ \forall OX_p \in L\} \subseteq \Sigma^{**OX_p}$.

By $\pi(L)$ we indicate the projection of oxide picture language L by mapping π, as in the case of oxide pictures. Given an oxide picture OX_p of size $(\ell_1 - 1, \ell_2 - 1, \ell_3 - 1)$ for $g \leq \ell_1 - 1, h \leq \ell_2 - 1$, and $k \leq \ell_3 - 1$, we denote by $B_{g,h,k}(OX_p)$ the set of all *oxide blocks* (or *oxide sub-pictures*) of OX_p of size (g, h, k). Each member of $B_{2,2,2}(OX_p)$ is called an *oxide tile* that resembles the star of David.

An oxide picture can be represented either by an oxide array or by an oxide tile as shown in Fig. 6. On the issue of labeling, an array is expressed with labeled vertices, a tile is labeled with its tiles (faces). The situation with oxide tiles is challenging as compared to rectangular and triangular arrays and tiles. In oxide tiles, we use two types of tiles, namely triangular tiles and hexagonal tiles.

Definition 4. *Let Σ be a finite alphabet. An oxide picture language $L \subseteq \Sigma^{**OX_p}$ is called* recognizable *if there exists an oxide local picture language L' (given by a set of oxide tiles) over an alphabet Γ and a mapping $\pi : \Gamma \rightarrow \Sigma$ such that $L = \pi(L')$.*

The family of all recognizable oxide picture languages is denoted by OXREC.

Definition 5 [12]**.** *An* oxide tiling system *(in short, OXTS) is a 4-tuple $\mathcal{OT} = (\Sigma, \Gamma, \pi, \Theta)$, where Σ and Γ are two finite alphabets, $\pi : \Gamma \rightarrow \Sigma$ is a projection and Θ is a finite set of oxide tiles over the alphabet $\Gamma \cup \{\#\}$.*

The oxide tiling system $\mathcal{OT}$ defines (recognizes) an oxide picture language L over the alphabet Σ as follows: $L = \pi(L')$ where $L' = L(\Theta)$ is the local oxide picture language over Γ corresponding to the set of oxide tiles Θ. We write $L = L(\mathcal{OT})$ and we say that L is the oxide picture language recognized by $\mathcal{OT}$. We will refer to the local oxide picture language $L' \subseteq \Gamma^{**OX_p}$ as the *underlying local oxide picture language* for L, while we call Γ the *local alphabet.* We say that an oxide picture language $L \subseteq \Sigma^{**OX_p}$ is *recognizable by oxide tiling systems* (or *oxide tiling recognizable*) if there exists an oxide tiling system $\mathcal{OT} = (\Sigma, \Gamma, \pi, \Theta)$ such that $L = L(\mathcal{OT})$.

We denote by $\mathcal{L}(OXTS)$ the family of all oxide picture languages recognizable by oxide tiling systems. In other words, $L \in \mathcal{L}(OXTS)$ if it is a projection of some local oxide picture language. Note that OXREC is exactly the family of oxide picture languages recognizable by oxide tiling systems $\mathcal{L}(OXTS)$.

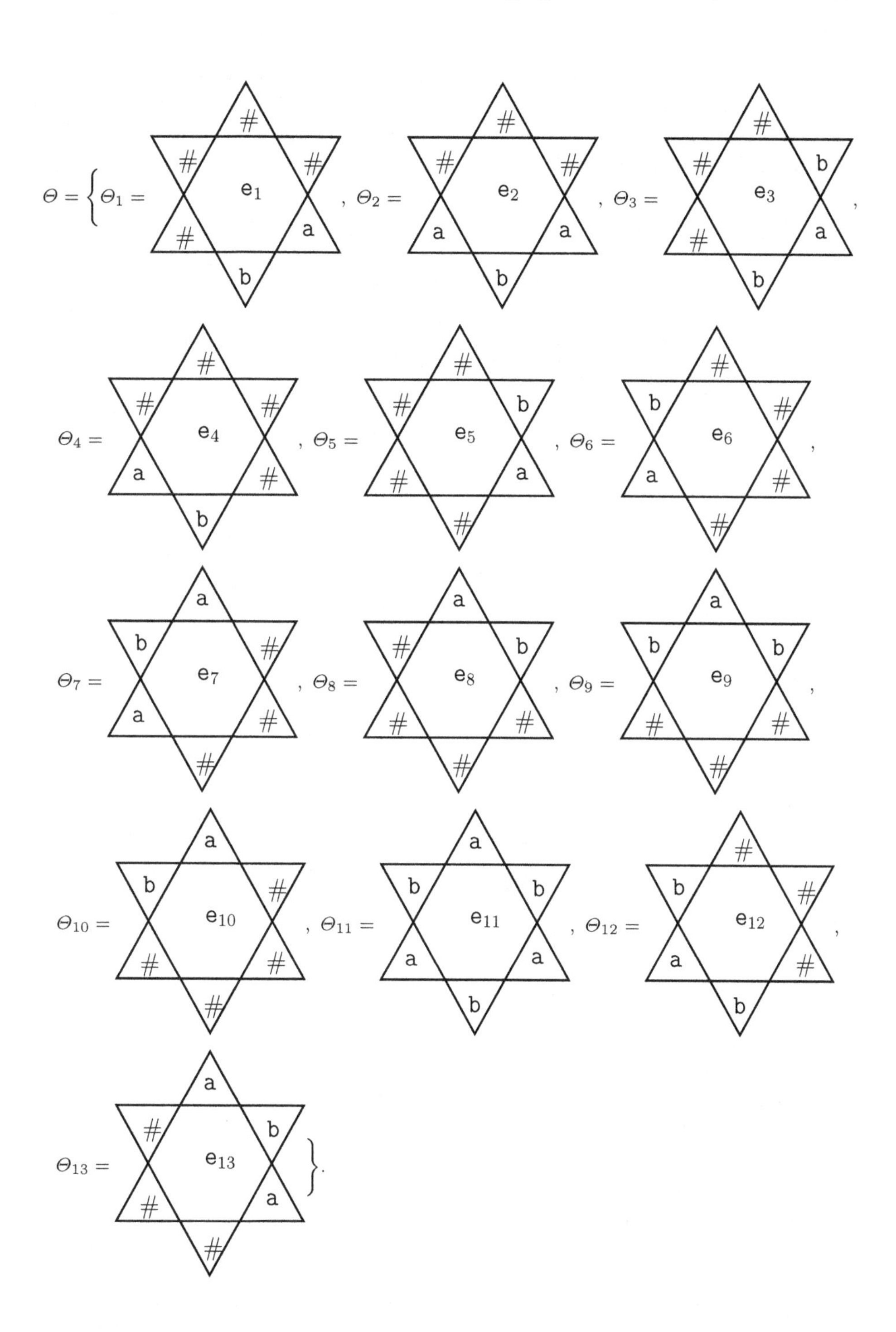

Θ = {Θ1 =
e1
, Θ2 =
e2
, Θ3 =
e3
Θ4 =
e4
, Θ5 =
e5
, Θ6 =
e6
Θ7 =
e7
, Θ8 =
e8
, Θ9 =
e9
Θ10 =
e10
, Θ11 =
e11
, Θ12 =
e12
Θ13 =
e13
}.

Example 2. Let Θ be a finite set of oxide tiles over $\Gamma = \{a, b\}$. Then the language $L \in \mathcal{L}(OXTS)$ has oxide pictures with symbols a and b in A and V triangles respectively, and e_i's in the hexagonal tile respectively. B stands for the border symbol #. Figure 7, 8, and 9 illustrates types of oxide pictures with borders viewed as oxide tiles (see Sect. 3.2).

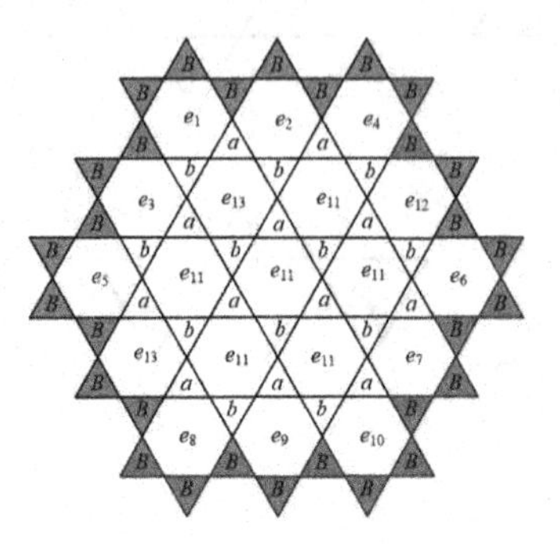

Fig. 7. Type I: $OX_p(4, 4, 4)$

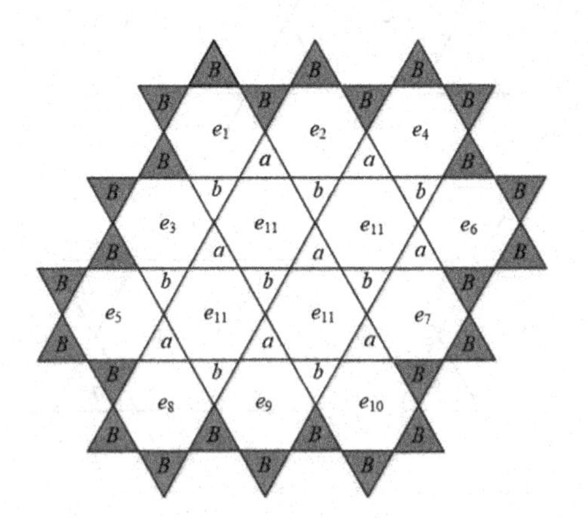

Fig. 8. Type II: $OX_p(4, 2, 4)$

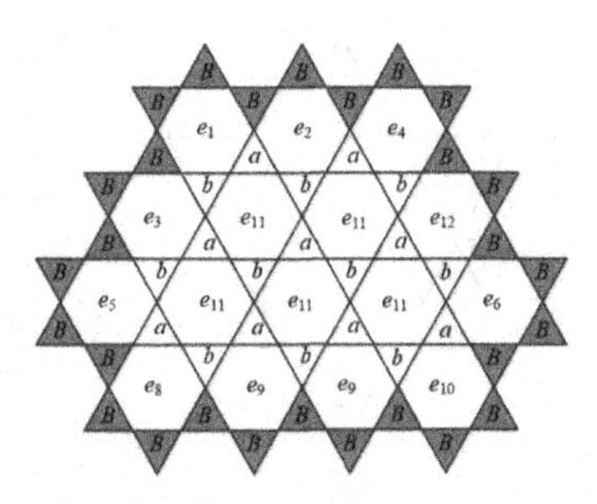

Fig. 9. Type III: $OX_p((4, 2), (2, 4), (4, 6))$

3 Two-Dimensional Oxide Pictures and Languages

Silicates are minerals that comprise metal oxides and sand. In chemistry, a silicate is a member of the family of anions: a silicon atom surrounded by four oxygen atoms, usually with the general formula $\left[\mathrm{SiO}_{4-x}^{(4-2x)-}\right]_n$, where $0 \leq x < 2$.

A silicon-oxygen tetrahedron (see Fig. 1) is the SiO_4 anionic group, or a silicon atom with four surrounding oxygen atoms arranged to define the corners of a tetrahedron. A silicate sheet is formed by the silicon ions, that are connected to get a silicate network. From this silicate network, a new network of oxygen network is obtained by removing the silicate ions alone. This new network of oxygen ions without silicon ions is named as oxide network (see Fig. 2). This network was identified by the graph theorists Paul Manuel and Indira Rajasingh [8] as a new network from the silicate network, when all the silicon nodes are deleted from a silicate network, we obtain an oxide network [19].

An oxide picture is viewed as a collection of symbols (an oxide array of elements) on an oxide grid. The boundary of the oxide pictures is filled with a special symbol # as explained in Fig. 5. Now, we propose a coordinate system for oxide pictures, the size of oxide pictures is defined in reference to the smallest hexagon that bounds the given oxide picture. We do classify the oxide pictures based on the sizes. We also realize a special case of oxide picture namely triangular oxide picture in this paper.

3.1 Coordinates and Size of Oxide Pictures

We fix the coordinates of each element of the oxide pictures similar to hexagonal pictures with respect to a triad $\swarrow x \quad y \searrow \overset{z}{\longrightarrow}$ of triangular axes x, y, and z. Given an oxide picture $OX_p \in \Sigma^{**OX_p}$, let ℓ_1 denote the number of elements in the hexagonal border of OX_p from the upper left vertex to the leftmost vertex in the x-direction, ℓ_2 denote the number of elements in the hexagonal border of OX_p from the upper right vertex to the rightmost vertex in the y-direction and ℓ_3 denote the number of elements in the hexagonal border of OX_p from the upper left vertex to the upper right vertex in the z-direction.

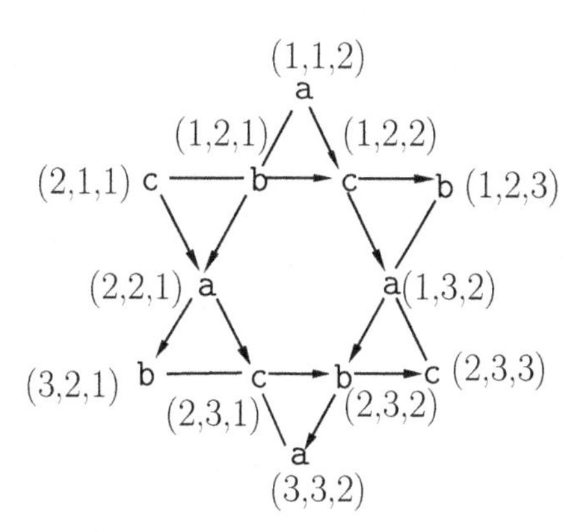

Fig. 10. Coordinates of elements of OX_p in Fig. 5.

The directions are fixed for some vertices only, as given in Fig. 10, whereas the coordinates are fixed with the origin of reference the outermost hexagon's upper left vertex having coordinates $(1, 1, 1)$. The triple $(\ell_1 - 1, \ell_2 - 1, \ell_3 - 1)$ is called the *size* of the oxide picture OX_p. Furthermore, if $1 \leq i \leq \ell_1$, $1 \leq j \leq \ell_2$, $1 \leq k \leq \ell_3$, where ℓ_1, ℓ_2, $\ell_3 \geq 3$ then let $OX_{p_{ijk}}$ denote the symbol in OX_p with coordinates (i, j, k). Here we can see that, for instance, the oxide picture in Fig. 5 of size $(2, 2, 2)$ has $OX_{p_{111}} = \lambda$, $OX_{p_{112}} = \mathsf{a}$, $OX_{p_{122}} = \mathsf{c}$ and $OX_{p_{123}} = \mathsf{b}$. The coordinates for the oxide picture in Example 5 are as in Fig. 10. In our study, there cannot be two rightmost vertices in the hexagon, which is well explained below with a few remarks.

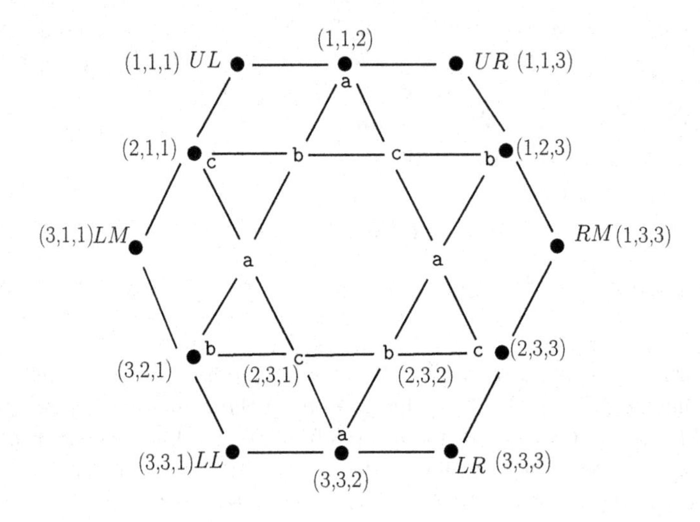

Fig. 11. The smallest bounding hexagon of OX_p in Example 5.

The coordinates are assigned to an oxide picture similar to hexagonal pictures [30]. Consider the oxider picture in Example 1. The smallest hexagon that bounds the given oxide picture is presented in Fig. 11.

Remark 1. In this section, we consider oxide pictures whose outer most hexagon has the same number of elements in both sides of x-direction, the same number of elements in both sides of y-direction and the same number of elements in both sides of z-direction. Also the center element of the hexagons in the oxide pictures are empty strings (that is, λ s), for instance $OX_{p_{131}} = \lambda$ in Example 1. To be more precise, in this section, we do consider the outer most hexagon which is of irregular shape as given in Fig. 12, also in Sect. 3.3 we will be encountering similar type of such hexagons (as the outermost hexagons).

Example 3. Let $\Sigma = \{0, 1, 2, 3\}$. Let $L_{(2,2,2k)}$ be the set of all oxide pictures of sizes $(2, 2, 2k)$, with $k \geq 1$. Some sample elements from $L_{(2,2,2k)}$ can be collected in L_{0123} as given as follows:

$$\begin{array}{ccccccccccccc}
 & & & 0 & & 0 & & 0 & & 0 & & & \\
 & & 1 & & 1 & & 1 & & 1 & & 1 & & \\
 & 2 & & 2 & & 2 & & 2 & & 2 & & 2 & \\
3 & & 3 & & 3 & & 3 & & 3 & & 3 & & 3 \\
 & 0 & & 0 & & 0 & & 0 & & 0 & & 0 &
\end{array}$$

Fig. 12. An irregular shape of hexagon.

$$L_{0123} = \left\{ \begin{array}{c} 0 \\ 0\ 1\ 1\ 0 \\ 2\quad\ 2 \\ 0\ 3\ 3\ 0 \\ 0 \end{array}, \begin{array}{c} 0\quad 0 \\ 0\ 1\ 1\ 1\ 1\ 0 \\ 2\quad\ 2\quad\ 2 \\ 0\ 3\ 3\ 3\ 3\ 0 \\ 0\quad 0 \end{array}, \begin{array}{c} 0\qquad 0\qquad 0 \\ 0\ \ 1\ \ 1\ \ 1\ \ 1\ \ 1\ \ 1\ \ 0 \\ 2\qquad 2\qquad 2\qquad 2 \\ 0\ \ 3\ \ 3\ \ 3\ \ 3\ \ 3\ \ 3\ \ 0 \\ 0\qquad 0\qquad 0 \end{array}, \cdots \right\}$$

Note that L_{0123} has the property that with z-direction elements at the top, one level below the top respectively are 0, 0 followed by 1s and ends with 0, at the middle are 2, and at the bottom, one level above the bottom respectively are 0, 0 followed by 3s and ends with 0. Here $L_{0123} \subsetneq L_{(2,2,2k)}$, for instance, all positions filled with 0s are not in L_{0123} where as it is an element in $L_{(2,2,2k)}$.

Now we define the concept of complete oxide pictures, which are the only permissible oxide pictures in our study.

Definition 6. *A* complete oxide picture *is an oxide picture, only if*

- *there exists a smallest bounding hexagon.*
- *there exists no single hole* (◊) *in any of the positions of the oxide arrays.*

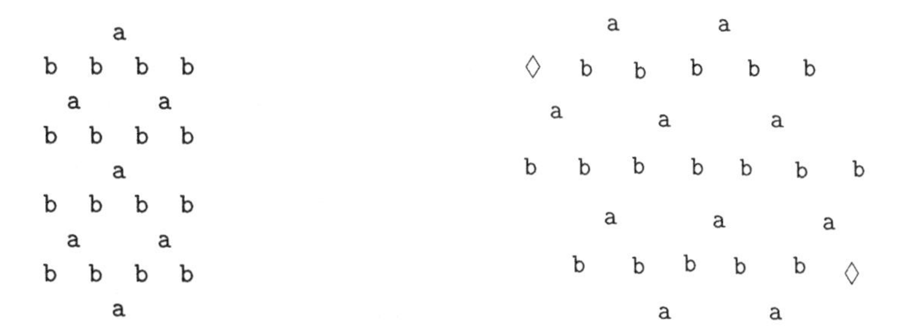

Fig. 13. No smallest bounding hexagon.

Fig. 14. An oxide picture with two holes.

For better readability, in Fig. 13, the oxide picture does not have the smallest hexagon. Figure 14 has two holes in the neighborhood position of the bounding hexagon, and a similar picture is depicted in Fig. 15 as well. We call such oxide pictures "incomplete oxide pictures" and we do not consider such oxide pictures. We use this symbol ◊ to denote holes in incomplete oxide pictures.

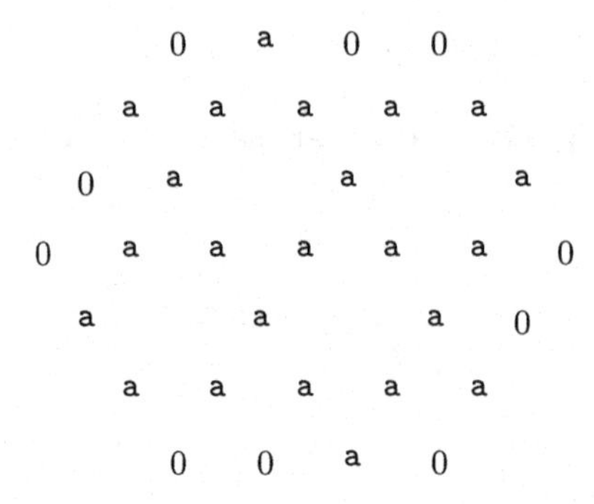

Fig. 15. Another incomplete oxide picture.

Remark 2. If $\ell_1 = \ell_2 = \ell_3$ where ℓ_1, ℓ_2, $\ell_3 \in \mathbb{N}$ are even numbers, then we observe that it sketches possible regular and (or) irregular hexagons, but violates to include complete oxide pictures. One can view this argument in Fig. 15, but it has a complete oxide picture of size $(2, 4, 2)$.

Note 1. No oxide pictures with size odd numbers exist, which is evident from Remark 2.

3.2 Types of Oxide Pictures

There are many types of oxide pictures depending on the size. For us, it is of interest to consider the oxide pictures that have the same size on the opposite sides of the x, y, and z directions.

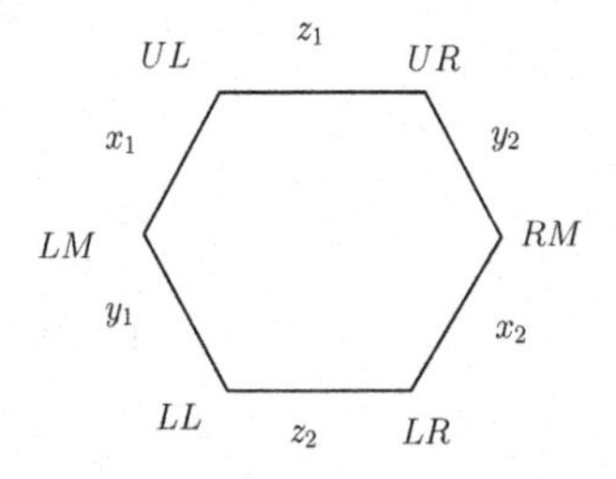

Let us now call the opposite sides of x-direction as x_1, x_2 similarly y_1, y_2 and z_1, z_2 for the y and z-directions. Among these pictures there are four types:

1. Type 1 (Regular Oxide Pictures I): $x_1 = x_2 = y_1 = y_2 = z_1 = z_2$.
2. Type 2 (Regular Oxide Pictures II): If $x_1 = x_2$, $y_1 = y_2$, $z_1 = z_2$ then
   - Case 1: $x_1 \neq y_1$, $y_1 \neq z_1$, $x_1 \neq z_1$
   - Case 2: $x_1 \neq y_1$, $y_1 \neq z_1$, $x_1 = z_1$
   - Case 3: $x_1 \neq y_1$, $y_1 = z_1$, $x_1 \neq z_1$
   - Case 4: $x_1 = y_1$, $y_1 \neq z_1$, $x_1 \neq z_1$
3. Type 3 (Regular Oxide Pictures III): If $x_1 \neq x_2$, $y_1 \neq y_2$, $z_1 \neq z_2$ then
   - Case 1a: $x_2 = y_1 = z_1 = 2$ and $x_1 = y_2 = z_2$.
   - Case 1b: $x_2 = y_1 = z_1 = 2$ and $x_1 = y_2$ and $x_1 \neq z_2$
   - Case 1c: $x_2 = y_1 = z_1 = 2$ and $x_1 \neq y_2$ and $x_1 \neq z_2$
     We will be focusing on a detailed study of this special case of oxide pictures in Sect. 4
   - Case 2: $x_1 = y_2$, $x_2 = y_1$, $z_1 \neq z_2$
     For the moment remaining cases are out of our focus.
4. Type 4 (Irregular Oxide Pictures): $x_1 \neq x_2 \neq y_1 \neq y_2 \neq z_1 \neq z_2$. We will be dealing with regular types of oxide pictures only.

If $\ell_1 = \ell_2 = \ell_3$ where ℓ_1, ℓ_2, $\ell_3 \in \mathbb{N}$ are odd numbers, then we say that an oxide picture OX_p is of size n, where $n = 2k$, $k \geq 0$, and it is denoted by $OX_p(n)$. The set of all oxide pictures Σ^{**OX_p} of size n is denoted by $\Sigma^{**OX_p(n)}$. Irregular oxide pictures also have the *size* $((l_1, l_2), (m_1, m_2), (n_1, n_2))$ similar to regular oxide pictures.

Example 4. An $OX_p(4)$ over $\Sigma = \{\mathtt{a}, \mathtt{b}\}$ and its bordered version are given in Fig. 16 which is type 1 of oxide pictures.

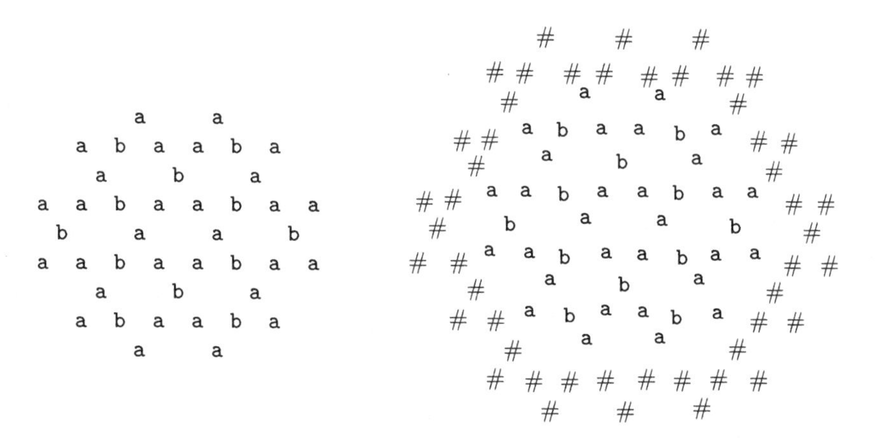

Fig. 16. An $OX_p(4)$ over $\Sigma = \{\mathtt{a}, \mathtt{b}\}$ and its bordered version.

Example 5. $OX_p(2,6,6)$ and $OX_p((6,2),(2,6),(6,10))$ over $\Sigma = \{\mathtt{a}\}$ are given in Fig. 17 and Fig. 18 which are of Type 2 (Case 3) and Type 3 (Case 2) oxide pictures.

Note 2. We would like to note that there are 2^{12} (i.e., 4096) oxide pictures of size 2 over $\Sigma = \{\mathtt{a}, \mathtt{b}\}$. There are 3^{12} (i.e., 531441) oxide pictures of size 2 over $\Sigma = \{\mathtt{a}, \mathtt{b}, \mathtt{c}\}$. One of them is illustrated in Example 1. Similarly, there are 2^{42} oxide pictures of size 4 over $\Sigma = \{\mathtt{a}, \mathtt{b}\}$ and one is illustrated in Fig. 16.

Remark 3. At time step $t \geq 1$ there are $|\Sigma|^{9t^2+3t}$ oxide pictures of size n, where $n = 2k$, $k \geq 1$ over Σ with $3t^2 - 3t + 1$ oxide tiles.

Remark 4. As a special case at time step $t = 0$, we get the empty oxide picture that is denoted by Λ_{OX_p} or $OX_p(0)$ that has $|\Sigma|^0$ possible pictures over Σ with 1 oxide tile which is an empty oxide tile. There are some sizes of oxide pictures, that can not be defined, for instance $(0,m,n)$, $(l,0,n)$, or $(l,m,0)$ where $l,m,n \in \mathbb{N}$.

Remark 5. We can generalize the number of oxide pictures and tiles with sizes as follows:

- At time step $t \geq 1$ there are $|\Sigma|^{O^{T_1}}$ oxide pictures of size (l,m,n) over Σ with O_{T_1} oxide tiles, where O^{T_1}, $O_{T_1} \in \mathbb{N}$.
- At time step $t \geq 1$ there are $|\Sigma|^{O^{T_{2,3}}}$ oxide pictures of size $((l_1,l_2),(m_1,m_2),(n_1,n_2))$ over Σ with $O_{T_{2,3}}$ oxide tiles, where $O^{T_{2,3}}$, $O_{T_{2,3}} \in \mathbb{N}$.

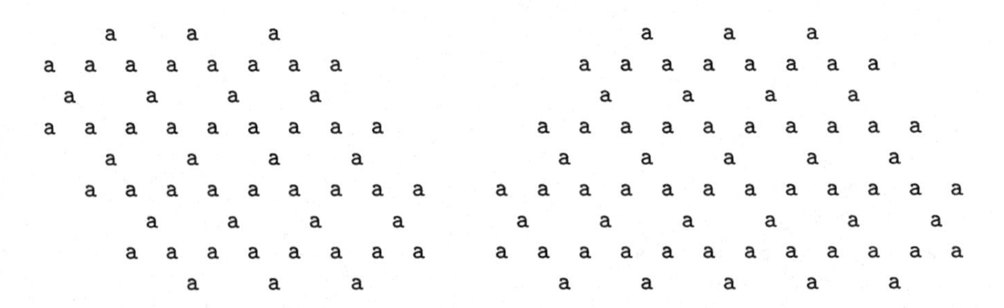

Fig. 17. An $OX_p(2,6,6)$ over $\Sigma = \{\mathtt{a}\}$.

Fig. 18. An $OX_p((6,2),(2,6),(6,10))$ over $\Sigma = \{\mathtt{a}\}$.

3.3 Triangular Oxide Pictures and Languages

Definition 7. *A* triangular oxide picture *over Σ is a triangular arrangement of oxide pictures. The set of all triangular oxide pictures over Σ is denoted by Σ^{**TOX_p}. A* triangular oxide picture language *over Σ is a subset of Σ^{**TOX_p}.*

Let $TOX_p \in \Sigma^{**TOX_p}$, we get the bordered version of TOX_p denoted by $\hat{TOX}_p$ when the special symbol $\# \notin \Sigma$ is added as the boundary to TOX_p. Given $TOX_p \in \Sigma^{**TOX_p}$ of size (l, m, n) we say that the TOX_p is of size n and it is denoted by $TOX_p(n)$ if $l = m = n$. The set of all triangular oxide pictures Σ^{**TOX_p} of size n is denoted by $\Sigma^{**TOX_p(n)}$.

Example 6. A $TOX_p \in \Sigma^{**TOX_p}$ over $\Sigma = \{\mathtt{a}, \mathtt{b}\}$ and its bordered version $\hat{TOX}_p$ are given in Fig. 19 and it is of size 6.

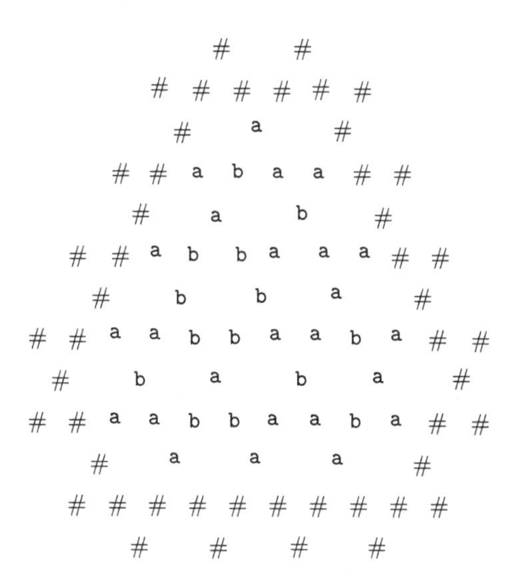

Fig. 19. $TOX_p((6,2),(2,6),(2,6))$ over $\Sigma = \{\mathtt{a}, \mathtt{b}\}$ with its bordered version.

Note 3. We note that there are 2^{12} (i.e., 4096) triangular oxide pictures of size 2 over $\Sigma = \{\mathtt{a}, \mathtt{b}\}$. There are 3^{12} (i.e., 531441) triangular oxide pictures of size 2 over $\Sigma = \{\mathtt{a}, \mathtt{b}, \mathtt{c}\}$. One of them is illustrated in Fig. 5. Similarly, there are 2^{39} triangular oxide pictures of size 6 over $\Sigma = \{\mathtt{a}, \mathtt{b}\}$ and one is illustrated in Example 6.

Remark 6. At time step $t \geq 1$ there are $|\Sigma|^{6t^2+9t-3}$ triangular oxide pictures of size n, where $n = 2k$, $k \geq 1$ over Σ with $2t^2 - t$ oxide tiles. Empty triangular oxide pictures can also be seen as a special case, also certain sizes of triangular oxide pictures are not defined.

Note 4. $OX_p(2, 6, 6)$ can become a $TOX_p((10, 2), (2, 10), (2, 10))$ over $\Sigma = \{\mathtt{a}\}$ if we add 13 nodes both in the y, z direction whereas $OX_p((6, 2), (2, 6), (6, 10))$ will become a $TOX_p((10, 2), (2, 10), (2, 10))$ over $\Sigma = \{\mathtt{a}\}$ if we add 13 nodes in the z-direction only.

Definition 8. *A triangular oxide tiling system, (TOXTS) is a 4-tuple* $\mathcal{TOT} = (\Sigma, \Gamma, \pi, \Theta)$*, where* Σ *and* Γ *are two finite alphabets,* $\pi : \Gamma \to \Sigma$ *is a projection and* Θ *is a finite set of triangular arrangement of oxide tiles over the alphabet* $\Gamma \cup \{\#\}$.

Example 7. Let $\Theta = \{\Theta_1, \Theta_2, \ldots, \Theta_{11}\}$ be a finite set of oxide tiles over $\Gamma = \{a, b\}$, where Θ_1 to Θ_{11} are as listed in Sect. 2 except for the last two tiles Θ_{12} and Θ_{13}. Then the language $L \in \mathcal{L}(OXTS)$ has only Type II oxide pictures ($x_1 = x_2$, $y_1 = y_2$, $z_1 = z_2$) with symbols a and b in A and V triangles respectively. We can observe that this tile set $\Theta = \{\Theta_1, \Theta_2, \ldots, \Theta_{11}\}$ does not recognize Type I and Type III oxide pictures.

4 Results

We do a comparison with the families $\mathcal{L}(OXTS)$ and $\mathcal{L}(TOXTS)$ and we also see the closure properties of these families on some operations.

Theorem 1. $\mathcal{L}(TOXTS) \subseteq \mathcal{L}(OXTS)$

Proof. Let L_1 be a oxide picture language over Σ, that is recognizable by an oxide tiling system $\mathcal{OT}_1 = (\Sigma, \Gamma, \pi, \Theta_1)$. Let L_2 be a triangular oxide picture language over Σ, that is recognizable by a triangular oxide tiling system $\mathcal{TOT}_2 = (\Sigma, \Gamma, \pi, \Theta_2)$. It is enough to show that $\Theta_2 \subseteq \Theta_1$. Let $\Theta_2 = \Theta_{2'} \cup \Theta_{\#2'}$ where

- $\Theta_{2'}$ is the set of oxide tiles without borders and
- $\Theta_{\#2'}$ is the set of bordered oxide tiles.

Let $\Theta_{\#2'} = \Theta_{\#21'} \cup \Theta_{\#22'}$ where

- $\Theta_{\#21'}$ is the set of bordered oxide tiles present along x_1, y_2, and z_2 and
- $\Theta_{\#22'}$ is the set of bordered oxide tiles present along with x_2, y_1, and z_1.

Similarly, let $\Theta_1 = \Theta_{1'} \cup \Theta_{\#1'}$ where

- $\Theta_{1'}$ is the set of oxide tiles without borders and
- $\Theta_{\#1'}$ is the set of bordered oxide tiles.

Similarly, let $\Theta_{\#1'} = \Theta_{\#11'} \cup \Theta_{\#12'}$ where

- $\Theta_{\#11'}$ is the set of bordered oxide tiles present along x_1, y_2, and z_2 and
- $\Theta_{\#12'}$ is the set of bordered oxide tiles present along with x_2, y_1, and z_1.

Since $\Theta_{\#22'}$ has $x_2 = y_1 = z_1 = 2$ where as $\Theta_{\#12'}$ has $x_2 = y_1 = z_1 \geq 2$ and for sufficiently large sizes $\Theta_{\#21'}$ has some tiles that are having same projection as of $\Theta_{\#11'}$. Without loss of generality, we assume that Θ_1' and Θ_2' are also having same projections. □

4.1 Boolean Operations

Theorem 2. *The family $\mathcal{L}(OXTS)$ is closed under union.*

Proof. Let $\mathcal{OT}_1 = (\Sigma, \Gamma_1, \pi_1, \Theta_1)$ and $\mathcal{OT}_2 = (\Sigma, \Gamma_2, \pi_2, \Theta_2)$ that recognize oxide picture languages L_1 and L_2 respectively. We construct an oxide tiling system $\mathcal{OT} = (\Sigma, \Gamma, \pi, \Theta)$ for $L = L_1 \cup L_2$. For this let $\Gamma = \Gamma_1 \cup \Gamma_2$, Γ_1 and Γ_2 are considered to be disjoint. Let $\Theta = \Theta_1 \cup \Theta_2$. Define a projection $\pi : \Gamma \rightarrow \Sigma$ in such a way that Γ_1 coincides with π_1 and Γ_2 coincides with π_2. □

Due to Theorem 1 we have the following:

Corollary 1. *The family $\mathcal{L}(TOXTS)$ is closed under union.*

The families $\mathcal{L}(OXTS)$, $\mathcal{L}(TOXTS)$ are not closed under complementation and are closed under intersection. As oxide pictures can also be seen as geometrical objects, several further unary and binary operations can be introduced [21].

4.2 Unary Operations

Definition 9. *Let $OX_p \in \Sigma^{**OX_p}$ then the operations for turns (rotations) are defined as follows:*

- *$H(OX_p)$ is the* half-turn,
- *$Q(OX_p)$ is the* quarter-turn,
- *$Q^{-1}(OX_p)$ is the* anti-quarter-turn.

It is natural to extend these turn (rotational) operations to the sets of oxide pictures and to lift these turn (rotational) operations to the families of oxide picture languages. Consider the oxide picture in Example 1, where

$$OX_p = \begin{array}{ccccc} & & \mathtt{a} & & \\ \mathtt{c} & \mathtt{b} & & \mathtt{c} & \mathtt{b} \\ & \mathtt{a} & & \mathtt{a} & \\ \mathtt{b} & \mathtt{c} & & \mathtt{b} & \mathtt{c} \\ & & \mathtt{a} & & \end{array} \quad \text{then } H(OX_p) = OX_p \text{ and}$$

$$Q(OX_p) = Q^{-1}(OX_p) = \begin{array}{ccccc} & \mathtt{b} & & \mathtt{c} & \\ & & \mathtt{a} & & \\ & \mathtt{c} & & \mathtt{b} & \\ \mathtt{a} & & & & \mathtt{a} \\ & \mathtt{b} & & \mathtt{c} & \\ & & \mathtt{a} & & \\ & \mathtt{c} & & \mathtt{b} & \end{array}$$

Theorem 3. *The family $\mathcal{L}(OXTS)$ is closed under half-turn.*

Proof. Let $L \in \mathcal{L}(OXTS)$ then there exists an $\mathcal{OT} = (\Sigma, \Gamma, \pi, \Theta)$ that recognizes L. Since $L \in \mathcal{L}(OXTS)$ there is an underlying local oxide picture language L' such that $L = \pi(L')$. Construct an oxide tiling system $\mathcal{OT}^H = (\Sigma, \Gamma_H, \pi_H, \Theta_H)$ that accepts L^H. We have to show $L^H \in \mathcal{L}(OXTS)$. This can be done by taking the local alphabet $\Gamma = \Gamma_H$. Let π_H be a projection from Γ_H to Σ. Since the family $\mathcal{L}(OXTS)$ is closed under projection, L' is also an underlying local language for L^H. □

Theorem 4. *The family $\mathcal{L}(OXTS)$ is not closed under quarter-turn and anti-quarter-turn.*

Proof. Let $L \in \mathcal{L}(OXTS)$ and $\mathcal{OT} = (\Sigma, \Gamma, \pi, \Theta)$ be an oxide tiling system that recognizes L. Then it is easy to show that no oxide tiling system recognizes quarter-turn and anti-quarter-turn of oxide pictures. □

Following Theorem 3 and Theorem 4 we have: The family $\mathcal{L}(TOXTS)$ is closed under half-turn. The family $\mathcal{L}(TOXTS)$ is not closed under quarter-turn and anti-quarter-turn.

Definition 10. *Let $OX_p \in \Sigma^{**OX_p}$ then the operations for reflections are defined as follows:*

- $R_x(OX_p)$, $R_y(OX_p)$, $R_z(OX_p)$ *are the* reflection along the diagonals *in x, y, z directions respectively,*
- $R_{x_1}(OX_p), R_{x_2}(OX_p)$, $R_{y_1}(OX_p), R_{y_2}(OX_p)$, $R_{z_1}(OX_p), R_{z_2}(OX_p)$ *are the* reflection along the sides *in both sides of x, y, z directions.*

It is natural to extend these reflection operations to the sets of oxide pictures and to lift these reflection operations to the families of oxide picture languages.

Theorem 5. *The families $\mathcal{L}(OXTS)$ and $\mathcal{L}(TOXTS)$ are closed under*

- *the diagonal reflections R_x, R_y, R_z*
- *the reflections along the sides R_{x_1}, R_{x_2}, R_{y_1}, R_{y_2}, R_{z_1}, R_{z_2}.*

Proof. The proof follows for the reflections R_z, R_{z_1}, and R_{z_2} as the families $\mathcal{L}(OXTS)$ and $\mathcal{L}(TOXTS)$ are closed under half-turn. Same arguments can be borrowed for other reflections as well R_x, R_y, R_{x_1}, R_{x_2}, R_{y_1} and R_{y_2}. □

4.3 Binary Operations

In this section, we will see some binary operations such as gluing and overlapping on various types of oxide pictures and languages. We define glueable A-arrays and glueable V-arrays as follows:

Definition 11. *Let A triangular arrays ($\triangle$) and V triangular arrays (∇) are of the form* $\begin{matrix} & \mathtt{x} & \\ \mathtt{x} & & \mathtt{x} \end{matrix}$ *and* $\begin{matrix} \mathtt{x} & & \mathtt{x} \\ & \mathtt{x} & \end{matrix}$ *respectively.*

- *Let* $A_1 = \begin{matrix} & \mathtt{a}_1 & \\ \mathtt{b}_1 & & \mathtt{c}_1 \end{matrix}$ *and* $A_2 = \begin{matrix} & \mathtt{a}_2 & \\ \mathtt{b}_2 & & \mathtt{c}_2 \end{matrix}$ *are A-triangular arrays. A_1 and A_2 are* glueable A-triangular arrays, *only if* $\mathtt{a}_1 = \mathtt{a}_2$, $\mathtt{b}_1 = \mathtt{b}_2$ *and* $\mathtt{c}_1 = \mathtt{c}_2$.
- *Let* $V_1 = \begin{matrix} \mathtt{x}_1 & & \mathtt{y}_1 \\ & \mathtt{z}_1 & \end{matrix}$ *and* $V_2 = \begin{matrix} \mathtt{x}_2 & & \mathtt{y}_2 \\ & \mathtt{z}_2 & \end{matrix}$ *are V-triangular arrays. V_1 and V_2 are* glueable V-triangular arrays, *only if* $\mathtt{x}_1 = \mathtt{x}_2$, $\mathtt{y}_1 = \mathtt{y}_2$ *and* $\mathtt{z}_1 = \mathtt{z}_2$.

– Let $T_1 = \begin{array}{ccccccc} & & \mathtt{a}_1 & & & & \\ \mathtt{b}_1 & \mathtt{c}_1 & & \mathtt{d}_1 & \mathtt{e}_1 & & \\ & \mathtt{f}_1 & & & & \mathtt{g}_1 & \\ \mathtt{h}_1 & \mathtt{i}_1 & & \mathtt{j}_1 & \mathtt{k}_1 & & \\ & & \mathtt{l}_1 & & & & \end{array}$ and $T_2 = \begin{array}{ccccccc} & & \mathtt{a}_2 & & & & \\ \mathtt{b}_2 & \mathtt{c}_2 & & \mathtt{d}_2 & \mathtt{e}_2 & & \\ & \mathtt{f}_2 & & & & \mathtt{g}_2 & \\ \mathtt{h}_2 & \mathtt{i}_2 & & \mathtt{j}_2 & \mathtt{k}_2 & & \\ & & \mathtt{l}_2 & & & & \end{array}$ are two oxide arrays. T_1 and T_2 are glueable oxide arrays, *only if* $\mathtt{a}_1 = \mathtt{a}_2$, $\mathtt{b}_1 = \mathtt{b}_2$ *and* $\mathtt{c}_1 = \mathtt{c}_2$, $\mathtt{d}_1 = \mathtt{d}_2$, $\mathtt{e}_1 = \mathtt{e}_2$ *and* $\mathtt{f}_1 = \mathtt{f}_2$, $\mathtt{g}_1 = \mathtt{g}_2$, $\mathtt{h}_1 = \mathtt{h}_2$ *and* $\mathtt{i}_1 = \mathtt{i}_2$, $\mathtt{j}_1 = \mathtt{j}_2$, $\mathtt{k}_1 = \mathtt{k}_2$ *and* $\mathtt{l}_1 = \mathtt{l}_2$.

We now define overlapping operations on oxide pictures using glueable A-triangular arrays, glueable V-triangular arrays, and glueable oxide arrays. Overlapping can happen between any two oxide pictures in the following ways:

Definition 12. *Let OX_{pA}, OX_{pB} are any two oxide pictures.*

- *if the rightmost V-triangular array and the rightmost A-triangular array of OX_{pA} are glueable with the leftmost V-triangular array and the leftmost A-triangular array of OX_{pB} respectively, then it is a* right overlapping of oxide pictures, *denoted by $OX_{pA} \textcircled{\scriptsize{I}}_R OX_{pB}$.*
- *if the leftmost V-triangular array and the leftmost A-triangular array of OX_{pA} are glueable with the rightmost V-triangular array and the rightmost A-triangular array of OX_{pB} respectively, then it is a* left overlapping of oxide pictures, *denoted by $OX_{pA} \textcircled{\scriptsize{I}}_L OX_{pB}$.*
- *if the rightmost (leftmost) oxide arrays of OX_{pA} are glueable with leftmost (rightmost) oxide arrays of OX_{pB} then it is an* overlapping of oxide pictures.

It is natural to extend these operations to the sets of oxide pictures and to lift these operations to the families of oxide picture languages.

Example 8. Let $OX_{pA} = \begin{array}{ccccc} & & \mathtt{b} & & \\ \mathtt{b} & \mathtt{b} & & \mathtt{a} & \mathtt{a} \\ & \mathtt{b} & & & \mathtt{a} \\ \mathtt{b} & \mathtt{b} & & \mathtt{a} & \mathtt{a} \\ & & \mathtt{b} & & \end{array}$ and $OX_{pB} = \begin{array}{ccccc} & & \mathtt{c} & & \\ \mathtt{a} & \mathtt{a} & & \mathtt{c} & \mathtt{c} \\ & \mathtt{a} & & & \mathtt{c} \\ \mathtt{a} & \mathtt{a} & & \mathtt{c} & \mathtt{c} \\ & & \mathtt{c} & & \end{array}$ be two oxide pictures of size $(2, 2, 2)$, the right overlapping of oxide pictures OX_{pA} and OX_{pB},

$$OX_{pA} \textcircled{\scriptsize{I}}_R OX_{pB} = \begin{array}{cccccc} & \mathtt{b} & & & \mathtt{c} & \\ \mathtt{b} & \mathtt{b} & \mathtt{a} & \mathtt{a} & \mathtt{c} & \mathtt{c} \\ \mathtt{b} & & \mathtt{a} & & \mathtt{c} & \\ \mathtt{b} & \mathtt{b} & \mathtt{a} & \mathtt{a} & \mathtt{c} & \mathtt{c} \\ & \mathtt{b} & & & \mathtt{c} & \end{array}.$$

Theorem 6. *The family $\mathcal{L}(OXTS)$ is closed under (right or (left)) overlapping operations.*

Proof. Consider two oxide picture languages L_1 and L_2 over Σ_1 and Σ_2. Let $\mathcal{OT}_1 = (\Sigma_1, \Gamma_1, \pi_1, \Theta_1)$ and $\mathcal{OT}_2 = (\Sigma_2, \Gamma_2, \pi_2, \Theta_2)$ be oxide tiling systems of L_1 and L_2 respectively. By definition of right overlapping, $OX_p \in L$ is composed of a pair of glueable oxide pictures $OX_{p_1} \in L_1$ and $OX_{p_2} \in L_2$ with the same number of rows of oxide tiles in the same direction such that the rightmost oxide tiles of OX_{p_1} are overlapped with the leftmost oxide tiles of OX_{p_2}. We define an oxide tiling system $\mathcal{OT} = (\Sigma, \Gamma, \pi, \Theta)$ of L as follows:

Let $\Sigma = \Sigma_1 \cup \Sigma_2$. Let us assume that Γ_1 and Γ_2 are disjoint and $\Gamma = \Gamma_1 \cup \Gamma_2$. Consider Θ, the set of oxide tiles over Γ that contains all the tiles from Θ_1 except those corresponding to the oxide tiles that are to the right borders and all the tiles from Θ_2 except those corresponding to the oxide tiles that are to the left borders. Some bordered oxide tiles are added in the upper and lower arrays when overlapping is done between the set of oxide pictures.

We define the three sets of tiles as follows: Θ_1' of Θ contains all the oxide tiles from Θ_1 except those corresponding to the oxide tiles that are to the right borders. Θ_2' of Θ contains all the oxide tiles from Θ_2 except those corresponding to the oxide tiles that are to the left borders. Θ_{12} contains all the bordered oxide tiles corresponding to the respective overlapping performed. Take $\Theta = \Theta_1' \cup \Theta_2' \cup \Theta_{12}$. Projection π is defined from $\Gamma \rightarrow \Sigma$ in such a way that elements of Γ_1 coincides with π_1 and of Γ_2 coincides with π_2. $\square$

Due to Theorem 6 we can also have the following:

Corollary 2. *The family $\mathcal{L}(TOXTS)$ is closed under (right or (left)) overlapping operations.*

5 Conclusions

Motivated by these types of silicate structures we have recently attempted to have a language theoretic investigation on the special class of oxide pictures, we further aim to compare and make a detailed study on Oxide Tile Rewriting Grammars and Oxide Wang Systems for other types of regular oxide picture languages also with Oxide Wang automata and Oxide Sgraffito Automata. On the other hand, the oxide grid is similar to one of the 8 semi-regular grids, mostly known as trihexagonal tiling/grid, some other ideas and results are in [6], also in [10] however picture languages are not studied in these cited papers, only some basic digital geometry, for instance, coordinate systems and distances.

In this paper, we set the stage with aspects such as size, category, closure properties, comparison with other generative models, etc. In the near future, we aim to implement through combinatorial algorithms with applications of oxide arrays through exploring several other language theoretic results through a comparative study on further properties. Our next focus is to study the state complexity and descriptional complexity for automata models and tiling systems that we have introduced in this paper via the Cut operation [5]. We further aim to achieve applications and implementations for 2D picture/tiling generation through a systematic approach listed below:

- Comparing automata, recognition schemes for 2D pictures, and 2D tiling patterns from [22–24].

- Studying the suitability of extending or applying schemes like pasting, a special kind of graph known as "map systems" that represents the cell structure of a plant, for oxide pictures and oxide tiles through connections from [9,27–29].
- Developing a modified Sgraffito automaton [17] for oxide pictures and oxide tiles (Fig. 20).

(a) Glass painting.

(b) Paper folding.

Fig. 20. The star of David in glass painting and paper folding.

Acknowledgements. The authors are grateful to the referees for their very useful comments, which helped to improve the presentation of the paper. The third author would like to express gratitude to the University of Trier, Germany, and Madras Christian College in Chennai, India, for visits in 2019 and 2020.

References

1. Anselmo, M., Giammarresi, D., Madonia, M.: A computational model for tiling recognizable two-dimensional languages. Theor. Comput. Sci. **410**(37), 3520–3529 (2009)
2. Dersanambika, K.S., Krithivasan, K., Martín-Vide, C., Subramanian, K.G.: Local and recognizable hexagonal picture languages. IJPRAI **19**(7), 853–871 (2005)
3. Fernau, H., Paramasivan, M., Schmid, M.L., Thomas, D.G.: Simple picture processing based on finite automata and regular grammars. J. Comput. Syst. Sci. **95**, 232–258 (2018)
4. Giammarresi, D., Restivo, A.: Two-dimensional languages. In: Rozenberg, G., Salomaa, A. (eds.) Handbook of Formal Languages, vol. III, pp. 215–267. Springer, Heidelberg (1997). https://doi.org/10.1007/978-3-642-59126-6_4
5. Holzer, M., Hospodár, M.: The range of state complexities of languages resulting from the cut operation. In: Martín-Vide, C., Okhotin, A., Shapira, D. (eds.) LATA 2019. LNCS, vol. 11417, pp. 190–202. Springer, Cham (2019). https://doi.org/10.1007/978-3-030-13435-8_14
6. Kovács, G., Nagy, B., Vizvári, B.: Weighted distances on the trihexagonal grid. In: Kropatsch, W.G., Artner, N.M., Janusch, I. (eds.) DGCI 2017. LNCS, vol. 10502, pp. 82–93. Springer, Cham (2017). https://doi.org/10.1007/978-3-319-66272-5_8

7. Krithivasan, K., Siromoney, R.: Array automata and operations on array languages. Int. J. Comput. Math. **4**(A), 3–40 (1974)
8. Manuel, P., Rajasingh, I.: Topological properties of silicate networks. In: 5th IEEE GCC Conference and Exhibition (2009)
9. Nagata, S., Thamburaj, R.: Digitalization of kolam patterns and tactile kolam tools. In: Mukund, M., Rangarajan, K., Subramanian, K.G. (eds.) Formal Models, Languages and Applications. Series in Machine Perception and Artificial Intelligence, vol. 66, pp. 354–363. World Scientific (2007)
10. Nagy, B.: Generalised triangular grids in digital geometry. Acta Math. Acad. Paedag. Nyiregyhaziensis **20**(1), 63–78 (2004)
11. Paramasivan, M.: Operations on graphs, arrays and automata. Ph.D. thesis, University of Trier, Germany (2018)
12. Ponraj, H.V., Thamburaj, R.: Oxide tiling system and oxide Wang system. Int. J. Curr. Trends Eng. Technol. **4**(2), 105–110 (2018)
13. Ponraj, H.V., Thamburaj, R.: Generative aspects of oxide pictures by oxide tile rewriting grammar. Int. J. Recent Technol. Eng. (IJRTE) **8**(3), 1537–1543 (2019)
14. Ponraj, H.V., Thamburaj, R.: Recognizability of oxide pictures by Sgraffito automata. J. Adv. Res. Dyn. Control Syst. (JARDCS) **11**(1), 285–293 (2019)
15. Ponraj, H.V., Thamburaj, R., Paramasivan, M.: Two-dimensional oxide picture languages. In: Proceedings of the 17th International Symposium on Artificial Intelligence and Mathematics 2022 (ISAIM 2022), Fort Lauderdale, Florida, USA, 3–5 January 2022 (2022)
16. de Prophetis, L., Varricchio, S.: Recognizability of rectangular pictures by Wang systems. J. Autom. Lang. Comb. **2**(4), 269–288 (1997)
17. Průša, D., Mráz, F., Otto, F.: Two-dimensional Sgraffito automata. RAIRO Theor. Inform. Appl. **48**(5), 505–539 (2014)
18. Rajaselvi, V.D., Kalyani, T., Dare, V.R., Thomas, D.G.: Recognizability of triangular picture languages by triangular Wang automata. In: Pant, M., Deep, K., Nagar, A., Bansal, J.C. (eds.) Proceedings of the Third International Conference on Soft Computing for Problem Solving. AISC, vol. 258, pp. 481–493. Springer, New Delhi (2014). https://doi.org/10.1007/978-81-322-1771-8_42
19. Simonraj, F., George, A.: Topological properties of few poly oxide, poly silicate, DOX and DSL networks. Int. J. Future Comput. Commun. **2**, 90–95 (2013)
20. Siromoney, G., Siromoney, R.: Hexagonal arrays and rectangular blocks. Comput. Graph. Image Process. **5**, 353–381 (1976)
21. Siromoney, G., Siromoney, R., Krithivasan, K.: Picture languages with array rewriting rules. Inf. Control (Now Inf. Comput.) **22**(5), 447–470 (1973)
22. Smith, T.J., Salomaa, K.: Recognition and complexity results for projection languages of two-dimensional automata. In: Jirásková, G., Pighizzini, G. (eds.) DCFS 2020. LNCS, vol. 12442, pp. 206–218. Springer, Cham (2020). https://doi.org/10.1007/978-3-030-62536-8_17
23. Smith, T.J., Salomaa, K.: Concatenation operations and restricted variants of two-dimensional automata. In: Bureš, T., et al. (eds.) SOFSEM 2021. LNCS, vol. 12607, pp. 147–158. Springer, Cham (2021). https://doi.org/10.1007/978-3-030-67731-2_11
24. Smith, T.J., Salomaa, K.: Decision problems and projection languages for restricted variants of two-dimensional automata. Theor. Comput. Sci. **870**, 153–164 (2021)
25. Subramanian, K.G.: Hexagonal array grammars. Comput. Graph. Image Process. **10**, 388–394 (1979)
26. Subramanian, K.G., Revathi, L., Siromoney, R.: Siromoney array grammars and applications. Int. J. Pattern Recognit. Artif. Intell. **3**, 333–351 (1989)

27. Thamburaj, R.: A study on circular languages, patterns and map systems. Ph.D. thesis, University of Madras, Chennai (2002)
28. Thamburaj, R.: Extended pasting scheme for kolam pattern generation. Forma **22**, 55–64 (2007)
29. Robinson, T., Jebasingh, S., Nagar, A.K., Subramanian, K.G.: Tile pasting systems for tessellation and tiling patterns. In: Barneva, R.P., Brimkov, V.E., Hauptman, H.A., Natal Jorge, R.M., Tavares, J.M.R.S. (eds.) CompIMAGE 2010. LNCS, vol. 6026, pp. 72–84. Springer, Heidelberg (2010). https://doi.org/10.1007/978-3-642-12712-0_7
30. Thomas, D.G., Begam, M.H., David, N.G., de la Higuera, C.: Hexagonal array acceptors and learning. In: Mukund, M., Rangarajan, K., Subramanian, K.G. (eds.) Formal Models, Languages and Applications. Series in Machine Perception and Artificial Intelligence, vol. 66, pp. 364–378. World Scientific (2007)
31. Yanagisawa, K., Nagata, S.: Fundamental study on design system of kolam pattern. Forma **22**, 31–46 (2007)

Lyndon Partial Words and Arrays with Applications

Meenakshi Paramasivan[1], R. Krishna Kumari[2(✉)], R. Arulprakasam[2(✉)], and V. Rajkumar Dare[3]

[1] FB IV - Informatikwissenschaften, Universität Trier, 54286 Trier, Germany
[2] Department of Mathematics, Faculty of Engineering and Technology, SRM Institute of Science and Technology, Kattankulathur, Chennai 603203, India
{kr1062,arulprar}@srmist.edu.in
[3] Department of Mathematics, Madras Christian College, Chennai 600059, India

Abstract. Lyndon words have been extensively studied in different contexts of free Lie algebra and combinatorics. We introduce Lyndon partial words, arrays and trees. We also study free monoid morphisms that preserve finite Lyndon partial words and check whether a morphism preserves or does not preserve the lexicographic order. We propose an algorithm to determine Lyndon partial words of given length over the binary alphabet. Image analysis in several way of scanning via automata and grammars has a significance in two-dimensional models, we connect 2D Lyndon partial words with few automata and grammar models.

Keywords: Lyndon partial words · 2D Lyndon words · 2D arrays

Mathematics Subject Classification: 68Q45 · 68Q70 · 68R15 · 68U10

1 Introduction

Lyndon words serve to be a useful tool for a variety of problems in combinatorics [2,19,21,23]. There are many applications of Lyndon words in semigroups, pattern matching, representation theory of certain algebras and combinatorics such as they are used to describe the generators of the free Lie algebras. All of these applications make use of the combinatorial properties of Lyndon words, in particular the factorisation theorem. Their role in factorising a string over an ordered alphabet was initially illustrated by Chen et al. [7]. Duval [9] presented a algorithm to derive a factorisation of strings over an ordered alphabet known as Lyndon factorisation. Lyndon trees [8] are associated with Lyndon words under the name of standard lexicographic sequences. The Lyndon arrays [4,13] of Lyndon words has recently become of interest since it could be used to efficiently compute all the maximal periodicities in a word. Lyndon trees have lately been shown to have solid connections with the structure in words and the Lyndon tree associated to a given word can be used as a basis for effective

R. P. Barneva et al. (Eds.): IWCIA 2022, LNCS 13348, pp. 226–244, 2023.
https://doi.org/10.1007/978-3-031-23612-9_14

computation of maximal repeats in that word [1,14]. A unique factorisation theorem for factoring a tree in terms of Lyndon trees is proved in [27]. The work in [24] characterised Lyndon morphisms and proved that they are order-preserving morphisms.

Partial words are nothing but words with holes over the alphabet and are considered in gene comparisons [12]. For instance, alignment of two DNA sequences which are genetic information carriers can be viewed as construction of two partial words that are compatible. In DNA computation, DNA strands are considered as finite words and are utilized for encoding information. While encoding, some part of information may be unseen or missing which are revealed by using partial words that denotes the positions of the missing symbols in a word. The study of partial words was initiated by Berstel and Boasson [3] and later the work was extended by Blanchet Sadri [5,6]. Both Lyndon and partial words have wide application in pattern matching. In this paper the concept of Lyndon partial words [17] is used and tree structure associated to Lyndon partial words is introduced. We also introduce free monoids morphisms that preserve finite Lyndon partial words and check whether a morphism preserves or does not preserve the lexicographic order. Image analysis is a rapidly growing technology. It is an important branch of science that investigates image descriptions that have relational structures that express relationships between image parts and describe their properties. Three steps are involved in image analysis: Importing the image via image acquisition tools; Analysis and manipulation of the image; Depending upon the image analysis, the output can be an altered image or report. In [10,11,22], the authors have derived an automaton model namely Boustrophedon finite automata (BFA) for picture processing, which is equivalent to Regular matrix grammars (RMGs). The paper has the following organisation. In Sect. 2 we recall some basics. In Sect. 3 we introduce Lyndon partial words and Lyndon partial arrays. A relation between the Lyndon partial words and trees is established. In Sect. 4 we characterise $\ell_{\Diamond}$-morphism and show that they are order-preserving morphism. In Sect. 5 we investigate few connections to 2D Lyndon words through 2D Lyndon partial words.

2 Basic Notations and Terminology

Let Σ termed as an *alphabet* represent a non-empty ($\neq \varnothing$) finite set of symbols (or letters). A *total word (or string)* is a sequence of letters over Σ. The *length (or size)* of a total word $x = x[1 \ldots n]$ is n. The length of a total word x is denoted by $|x|$. Σ^* denotes the set of all total words from Σ. $\Sigma^+ = \Sigma^* \setminus \{\lambda\}$ where λ denotes the empty word. A language L is a subset of Σ^*. The total word p is a *subword (or factor)* of q if there exists the total words x and y such that $q = xpy$. If $x, y \neq \lambda$ then p is a proper subword of q. If $x = \lambda$ then p is a prefix of q. If $y = \lambda$ then p is a suffix of q. An *ordered alphabet* is an alphabet with a total order so that comparisons of any two symbols from the alphabet can be computed in constant time. The *alphabetical order (lexicographical order)* $\prec$ on $(\Sigma^*, \prec)$ is defined by setting $u \prec v$ if at least one of the following conditions is satisfied:

1. u is a proper prefix of v,
2. there exists words x, y, z (possibly empty) and elements a and b of Σ such that $u = xay, v = xbz, a < b$.

The following Table 1 illustrates examples of alphabetical order of words over binary and tertiary alphabet.

Table 1. Alphabetical order of words

Alphabet	Order	Alphabetical order
a, b	$a < b$	$aa < aab < aba < b$
a, b, c	$a < b < c$	$ab < abb < abbc < abc < acbc$

A *Lyndon word* $l = l[1 \ldots n]$ is a primitive word which is non-empty $(\neq \varnothing)$ and less than all its rotations (conjugates) in the alphabetical order. Let Σ_1, Σ_2 be two alphabets. A morphism g is a mapping from Σ_1^* to Σ_2^* such that for all words x, y over Σ_1, $g(xy) = g(x)g(y)$ over Σ_2. The sequence of symbols that contains a number of "do not know symbols" or "holes" denoted as $\Diamond$ is termed as a *partial word.* A total word is a partial word with zero holes. Empty word is not a partial word. The symbol $\Diamond$ does not belong to the alphabet Σ but a standby symbol for the unknown letter. $\Diamond$ alone of any length cannot exist as a word. In other words, the hole of any length is neither a total word nor a partial word. A partial word $x = x[1 \ldots n]$ over Σ is a partial function $x : \{1, 2, \ldots, n\} \to \Sigma$. For $1 \leq i < n$ if $x(i)$ is defined, then we say $i \in D(x)$ (the domain of x), otherwise $i \in H(x)$ (the set of holes). The following definition is used in order to represent the positions of the holes of the partial words. The partial word of x denoted by $x_\Diamond$ is the total function $x_\Diamond : \{1, 2, \ldots, n\} \to \Sigma_\Diamond = \Sigma \cup \{\Diamond\}$ defined by

$$x_\Diamond(i) = \begin{cases} x(i) & \text{if } i \in D(x) \\ \Diamond & \text{if } i \in H(x). \end{cases}$$

The set of all partial words over $\Sigma_\Diamond$ is denoted as $\Sigma_\Diamond^*$. $\Sigma_\Diamond^+ = \Sigma_\Diamond^* \setminus \{\lambda\}$. A partial Language $L_\Diamond \subseteq \Sigma_\Diamond^*$ is a set of all partial words over $\Sigma_\Diamond$. A partial word $x = x[1 \ldots n]$ is *primitive (non-periodic)* if there exists no word y such that $x = y^i$ with $i \geq 2$. A partial word x is *unbordered* if no non-empty words p, q, y exist such that $x_\Diamond \subset py$ and $x_\Diamond \subset qp$. Unbordered partial words are primitive. Partial words that are not primitive are said to be *periodic* partial words.

3 Lyndon Partial Words

Here we introduce and study the generalisation of finite Lyndon partial words by using trees. A standard theorem for factoring a tree concerning trees associated to Lyndon partial words is proved. In [21], the authors have defined that a primitive partial word is a partial Lyndon word if and only if it is minimal in its conjugate class with respect to alphabetical order by assuming the order of $\Diamond$ as $\{a < b < \ldots < \Diamond\}$. The order of $\Diamond$ does not play a special role in the definition by studying properties of partial Lyndon words since the $\Diamond$ is considered as a letter with highest order which makes the definition similar to that of Lyndon words. In our definition of Lyndon partial word, the order of $\Diamond$ plays a special role in studying certain properties.

Definition 1. *A* Lyndon partial word $l_\Diamond = l_\Diamond[1\ldots n]$ *over the ordered alphabet* $\Sigma_{k,\Diamond} = \Sigma_k \bigcup\{\Diamond\} = \{a_1 < a_2 < \ldots < a_k\} \bigcup\{\Diamond\}, k > 1$ *is less than all its conjugates (rotations) with respect to the alphabetical order. Here the order of* $\Diamond$ *is considered as* $a_1 \leq \Diamond$, $\Diamond \leq a_k$ *and* $\Diamond$ *is compatible with all other elements of* Σ_k. *A* Lyndon partial language *over* Σ *is a subset of* $\Sigma^*_\Diamond$, *the set of all Lyndon partial words over* $\Sigma_\Diamond$.

Note 1. For readability we use $L_\Diamond$ notation for partial languages which shall not be confused with the $\ell_\Diamond$ notation for Lyndon partial languages.

Remark 1. Any Lyndon partial word is primitive but the converse may not be true. For instance $\Diamond abb$ is a primitive partial word but its conjugacy class $\{\Diamond abb, abb\Diamond, bb \Diamond a, b \Diamond ab\}$ does not contain a Lyndon partial word (see in Table 2 for length 4). This shows that the lexicographical order relation among Lyndon partial words is not always a total order relation but sometimes a partial order relation due to the presence of $\Diamond$. We remark $a \Diamond aabb$, $a \Diamond bbab$ that these two partial words are also not Lyndon partial words for our further references. Also note that $aa^* \Diamond + aa^* \Diamond bb^* + \Diamond(b + bbb^*)$ is the expression for the machine in Fig. 1. It is easy to think of another machine for $(a^*aa + a) \Diamond + aa^* \Diamond bb^* + \Diamond(b + bbb^*)$ instead of $aa^* \Diamond + aa^* \Diamond bb^* + \Diamond(b + bbb^*)$ in order to build the "basic blocks" of estimated elements in Table 2 also with elements like $aa^* \Diamond bb^*$ especially in odd lengths.

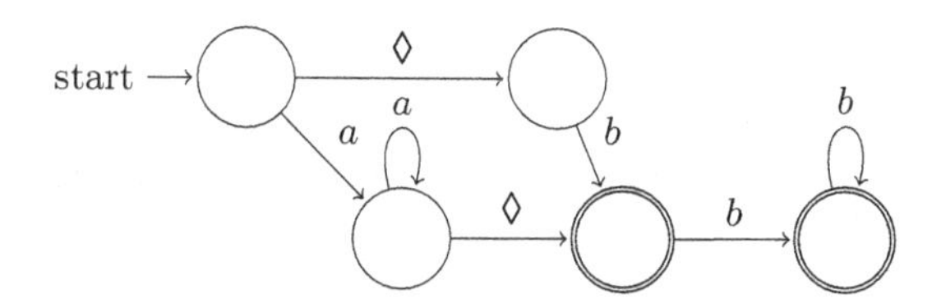

Fig. 1. Rough outline for some Lyndon partial words in Table 2

The following table shows the set of all Lyndon partial words with length at most five over the ordered alphabet $\Sigma_\Diamond = \{a < b\} \bigcup\{\Diamond\}$.

Remark 2. It is easy to observe that Lyndon partial words on binary alphabet takes the same integer sequence starting from 2, 3, 6, 9 by excluding the first three numbers namely 1, 2, 1 of that of Lyndon words as compared and evidenced in Table 2.

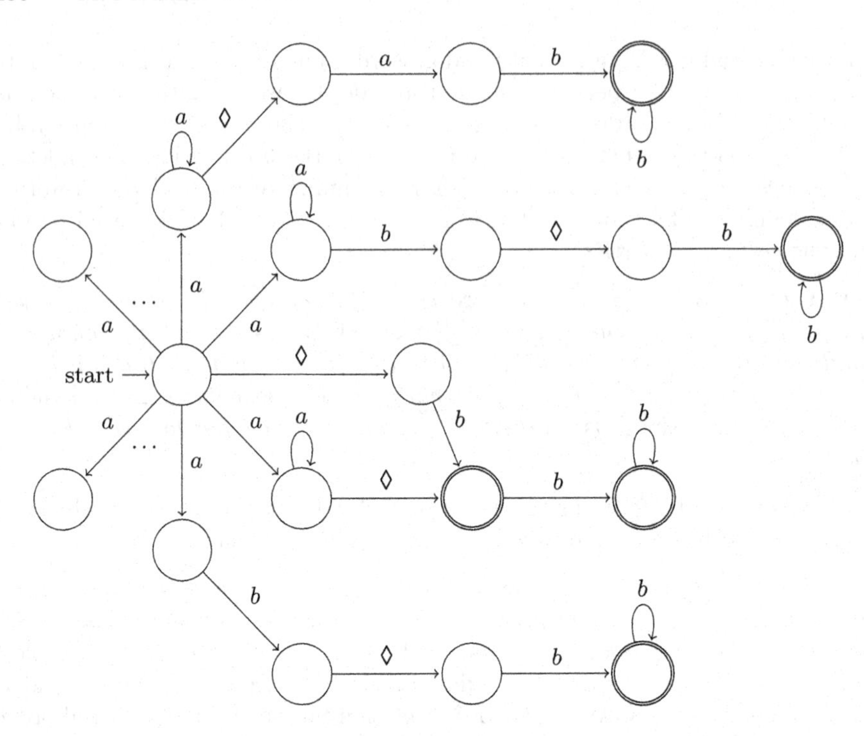

Fig. 2. Estimated automaton for some Lyndon partial words in Table 2

Table 2. Lyndon words along with Lyndon partial words

Length	Lyndon words	Lyndon partial words
0	λ	–
1	a, b	–
2	*ab*	$a\Diamond$, $\Diamond b$
3	*aab*, *abb*	$aa\Diamond$, $a\Diamond b$, $\Diamond bb$
4	*aaab*, *aabb*, *abbb*	$aaa\Diamond$, $aa\Diamond b$, $a\Diamond ab$, $a\Diamond bb$, $ab\Diamond b$, $\Diamond bbb$
5	*aaaab*, *aaabb*, *aabab*, *aabbb*, *ababb*, *abbbb*	$aaaa\Diamond$, $aaa\Diamond b$, $aa\Diamond ab$, $aa\Diamond bb$, $aab\Diamond b$, $a\Diamond abb$, $a\Diamond bbb$, $ab\Diamond bb$, $\Diamond bbbb$
⋮	…	⋱

Algorithm 1: To determine Lyndon partial words of given length over the binary alphabet

```
Input: Finite set of symbols/letters, Hole set, Length 'n'
Output: Collection of Lyndon partial words of given length
p.sort()
result = [-1]
k = len(p)
while result: do
    result[-1] + =1
    m = len(result)
if (m = =n) : then
    (''.join(s[i] for i in result))
while len(result)<n: do
    result.append(result[-m])
while result and result[-1] == k-1: do
    result.pop()
n = int(input("Enter the length of the partial word: "))
p = ['a','b'] ∪ ['◊']
lyndon partial words(p, n)
```

Theorem 1 [17]. *No proper subword exists as both prefix and a suffix of a Lyndon partial word.*

Theorem 2 [17]. *A partial word $l_\Diamond$ over $\Sigma_\Diamond^+$ belongs to $L_\Diamond$ if and only if $l_\Diamond < q_\Diamond$ for each proper suffix $q_\Diamond$ of $l_\Diamond$.*

Example 1. Consider a Lyndon partial word $l_\Diamond[1\ldots5] \in L_\Diamond$. $l_\Diamond = aa\Diamond bb < b = q_\Diamond$. Here $q_\Diamond$ is a proper suffix of $l_\Diamond$

Theorem 3 [17]. *Consider $p_\Diamond, q_\Diamond \in L_\Diamond$. Then $p_\Diamond q_\Diamond \in L_\Diamond$ if and only if $p_\Diamond < q_\Diamond$.*

Theorem 4. *Each Lyndon partial word $l_\Diamond$ over $\Sigma_\Diamond^+$ is unbordered but the converse is not true.*

Proof. Assume that $l_\Diamond$ has a non-overlapping border x. Then $l_\Diamond = xl_\Diamond x$. Let $i \geq 0$ be maximal such that $l_\Diamond = x^i l_\Diamond^1$. Then $l_\Diamond = x^{i+1} l_\Diamond^1 x$. Then $x^{i+2} l_\Diamond^1$ is lexicographically smaller than $x^{i+1} l_\Diamond^1$, a contradiction with $l_\Diamond = x^{i+1} l_\Diamond^1 x$ being Lyndon partial word. The following example illustrates that the converse is not true.

Example 2. Consider the unbordered partial word $l_\Diamond = bb\Diamond aa$ over $\Sigma_\Diamond$ which is also primitive. But $l_\Diamond$ is not a Lyndon partial word.

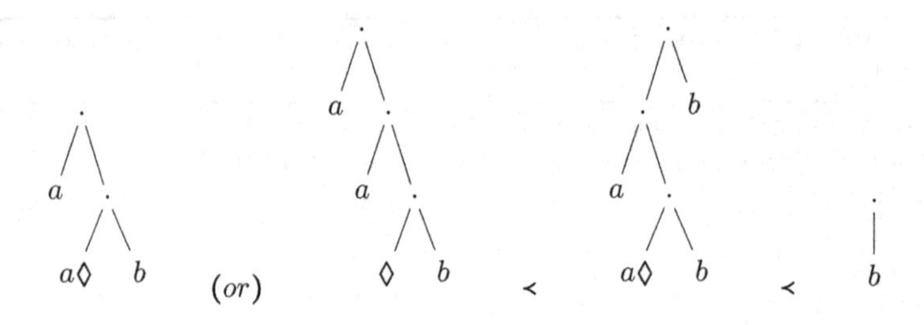

Fig. 3. An example for Theorem 3

Theorem 5 ***Factorisation Theorem*** [17]. *Any partial word $l_\Diamond$ over the alphabet $\Sigma_\Diamond^+$ can be uniquely written as $l_\Diamond = l_\Diamond^1 \ldots l_\Diamond^r$ with $l_\Diamond^1, \ldots, l_\Diamond^r \in L_\Diamond$ and $l_\Diamond^1 \geq \ldots \geq l_\Diamond^r$.*

Definition 2. *A Lyndon partial factor $l_\Diamond[i \ldots j]$ of a Lyndon partial word $l_\Diamond[i \ldots n]$ for any $j \leq n$ is* a maximal Lyndon partial factor *if it is Lyndon.*

Definition 3. *A* Lyndon partial array *(denoted as $l_\Diamond^A$) of $l_\Diamond[1 \ldots n]$ is an array of integers in the range $[1 \ldots n]$ such that, at each position $i = 1 \ldots n$ stores the length of the longest Lyndon partial factor of $l_\Diamond[1 \ldots n]$ starting at i.*

Example 3. Consider a Lyndon partial word $l_\Diamond[1 \ldots 7] = aabab\Diamond b$. The maximal Lyndon partial factor starting at position 1 is $aabab$, so $l_\Diamond^A[1] = 5$. The maximal Lyndon partial factor at position 2 is ab, so $l_\Diamond^A[2] = 3$. The maximal Lyndon partial factor starting at position 3 is b, so $l_\Diamond^A[3] = 3$. The maximal Lyndon partial factor starting at position 4 is ab, so $l_\Diamond^A[4] = 5$. The maximal Lyndon partial factor starting at position 5 is b, so $l_\Diamond^A[1] = 5$. The maximal Lyndon partial factor starting at position 6 is $\Diamond b$, so $l_\Diamond^A[6] = 7$. The maximal Lyndon partial factor starting at position 7 is b, so $l_\Diamond^A[7] = 7$. Therefore, $l_\Diamond^A = [5\ 3\ 3\ 5\ 5\ 7\ 7]$.

Position	1	2	3	4	5	6	7
Lyndon partial word	**a**	**a**	**b**	**a**	**b**	◊	**b**
Maximal Lyndon partial factor	**a**	**a**	**b**	**a**	**b**	◊	**b**
Lyndon partial array	5	3	3	5	5	7	7

Theorem 6. *If the positions i, j in $l_\Diamond[1 \ldots n]$ satisfy $1 \leq i < j \leq n$, then the two intervals $\langle i, l_\Diamond^A[i] \rangle$ and $\langle j, l_\Diamond^A[j] \rangle$ are not intersecting each other.*

Proof. Assume that the two intervals $\langle i, l_\Diamond^A[i]\rangle$ and $\langle j, l_\Diamond^A[j]\rangle$ are intersecting each other. Then $u_\Diamond = l_\Diamond[i \dots l_\Diamond^A[i]]$ and $v_\Diamond = l_\Diamond[j \dots l_\Diamond^A[j]]$ with longest length have a non-empty intersection. Then we can write $u_\Diamond = x'x, v_\Diamond = xx'$ for some empty x and x'. But then, we get $u_\Diamond \prec x \prec v_\Diamond \prec x'$ showing that $u_\Diamond x'$ is a Lyndon partial word. This contradicts the assumption that $u_\Diamond$ is of the longest length. Thus the two intervals $\langle i, l_\Diamond^A[i]\rangle$ and $\langle j, l_\Diamond^A[j]\rangle$ are not intersecting each other.

Theorem 7. *Consider a Lyndon partial word $l_\Diamond[1 \dots n]$ over the ordered alphabet $\Sigma_\Diamond$. Let $suf_{l_\Diamond}(i) = l_\Diamond[i \dots n]$ denote the suffix of $l_\Diamond$ beginning at position i. Then a Lyndon partial factor $l_\Diamond[i \dots j]$ is the maximal Lyndon partial factor of $l_\Diamond$ if and only if $suf_{l_\Diamond}(i) \prec suf_{l_\Diamond}(k)$ for any $1 < j \leq k$ and $suf_{l_\Diamond}(j+1) \prec suf_{l_\Diamond}(i)$.*

Proof. Assume that $l_\Diamond[1 \dots n]$ is Lyndon. Now to prove that $\mathrm{suf}_{l_\Diamond}(j+1) \prec \mathrm{suf}_{l_\Diamond}(i)$, consider for $j < n$, $\mathrm{suf}_{l_\Diamond}(j+1) \nprec \mathrm{suf}_{l_\Diamond}(i)$. Since $\mathrm{suf}_{l_\Diamond}(i)$ and $\mathrm{suf}_{l_\Diamond}(j+1)$ are distinct, it follows that $\mathrm{suf}_{l_\Diamond}(i) \prec \mathrm{suf}_{l_\Diamond}(j+1)$. Let P represent the longest common position of $(\mathrm{suf}_{l_\Diamond}(i), \mathrm{suf}_{l_\Diamond}(j+1)) + 1$. The following two cases arise:

1. If $P \leq j - i$.
   Here $i \leq i+P \leq j$. Thus $l_\Diamond[i \dots i+P1] = l_\Diamond[j+1 \dots j+P]$ and $l_\Diamond[i+P] \prec l_\Diamond[j+1+P]$, and so for $j \prec k \leq j+1+P, l_\Diamond[i \dots j+1+P] \prec l_\Diamond[k \dots j+1+P]$. Since $l_\Diamond[i \dots j]$ is a Lyndon partial word, $l_\Diamond[i \dots j] \prec l_\Diamond[k \dots j]$ and so $l_\Diamond[i \dots j+1+P] \prec l_\Diamond[k \dots j+1+P]$ for any $i \prec k \leq j$. Thus $l_\Diamond[i \dots j+1+P]$ is a Lyndon partial word, contradicting the assumption that $l_\Diamond[i \dots j]$ is the longest Lyndon partial factor starting at i.
2. If $j - i \leq P$. Let $P = r(j - i) + P_1$, where $P_1 \prec j - i$. Then $r \geq 1$ and $l_\Diamond[i + P] \prec l_\Diamond[j + 1 + P]$. This implies $l_\Diamond[i \dots j + 1 + P]$ is Lyndon, contradicting the assumption that $l_\Diamond[i \dots j]$ is the longest Lyndon partial factor starting at i. Thus $\mathrm{suf}_{l_\Diamond}(j+1) \prec \mathrm{suf}_{l_\Diamond}(i)$, as required.

Conversely, let $l_\Diamond[i \dots k]$ be a longest Lyndon partial factor of $l_\Diamond$ starting at position i. If $k < j$, then $\mathrm{suf}_{l_\Diamond}(k+1) \prec \mathrm{suf}_{l_\Diamond}(i)$, a contradiction since $k + 1 \leq j$. If $k > j$, then $\mathrm{suf}_{l_\Diamond}(i) \prec \mathrm{suf}_{l_\Diamond}(j+1)$ because $j + 1 \leq k$, which again gives us a contradiction. Thus $k = j$ and $l_\Diamond[i \dots j]$ is a longest Lyndon factor of x.

3.1 Tree Representation of a Lyndon Partial Word

Trees are non-linear data structure that are widely used for data organisation, sorting and pattern matching. The tree associated with a Lyndon partial word $l_\Diamond$ over $\Sigma = \{a, b\} \cup \{\Diamond\}$ with order $\{a \prec b\}$ denoted as ζ is represented in an hierarchical form. Since $l_\Diamond$ is a partial word, $|l_\Diamond| \geq 2$ such that the parent node (root) of $\zeta(l_\Diamond)$ has left child $\zeta(r_\Diamond)$ and right child $\zeta(s_\Diamond)$ where $(r_\Diamond, s_\Diamond)$ is an ordered pair of Lyndon partial words $r_\Diamond, s_\Diamond$ with $l_\Diamond = r_\Diamond s_\Diamond$ and $s_\Diamond$ is lexicographically smallest proper suffix of $l_\Diamond$. The tree ζ originates from the topmost node called root and the letters occur as node (vertices) connected by edges in each state. The leaf nodes are terminal nodes in the final state. They are also described as nodes with no child. The set of terminal nodes of ζ are denoted by $\delta(\zeta)$. ζ^+ represents the set of non-empty trees over Lyndon partial word. The

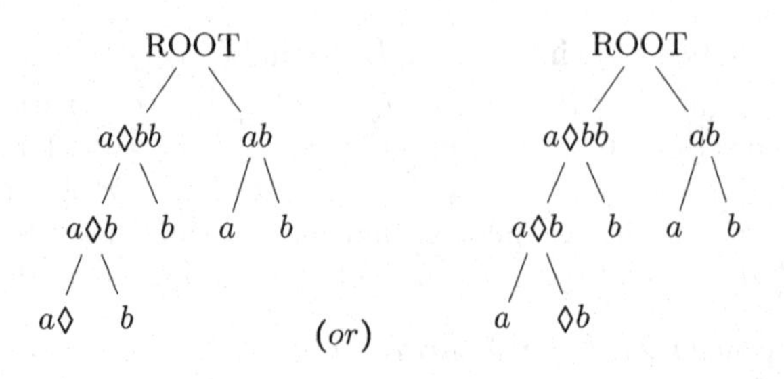

Fig. 4. Tree of $(a\Diamond bbab)$

tree of a Lyndon partial word is written as $\zeta = \zeta_m + v\zeta_n$ where $\zeta_m, \zeta_n \in \zeta^+$ and v denotes a node (vertex) such that $v \in \delta(\zeta_m)$. Figure 4 shows an illustration of a tree associated to the Lyndon partial word $l_\Diamond = a\Diamond bbab$.

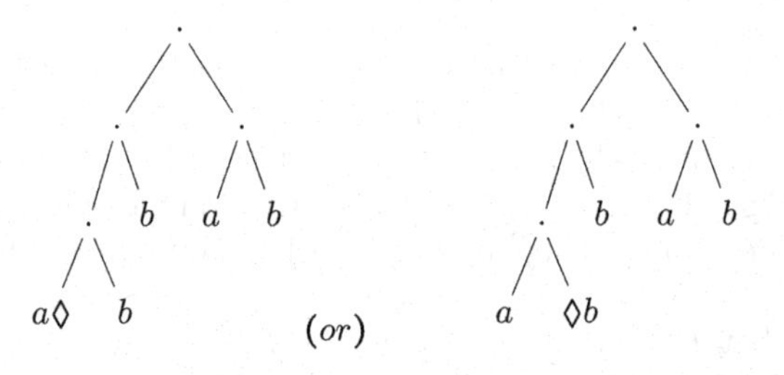

Fig. 5. Outline of tree of $(a\Diamond bbab)$

Precisely the tree $\zeta(l_\Diamond)$ in Fig. 4 is established as in Fig. 5

Definition 4. *A tree* ζ *associated with a Lyndon partial word is described with its minimal among all of its rotations.* $\Im$ *denotes set of such trees. A sub-tree of* ζ *is a tree with set of nodes as a subset of* ζ.

Theorem 8. *No proper subtree exists as both initial and terminal of the tree* ζ.

Proof. Let us consider ζ to be a tree of a Lyndon partial word over $\Sigma_\Diamond^+$. Let P be a subtree of ζ such that P is both initial and terminal of ζ. Then we have

$$\zeta = P + vQ, \ v \in \delta(P) \ and$$
$$\zeta = R + v'P, \ v' \in \delta(R)$$

where $v, v^{\prime}$ are nodes and $Q, R \in \zeta^{+}$. By Definition 4, we get

$$\zeta < Q + v^{\prime\prime}P,\ v^{\prime\prime} \in \delta(Q)\ and$$
$$\zeta < P + vR,\ v \in \delta(P)$$

Therefore we get $P + vQ \prec P + vR$ such that $Q \prec R$ and $R + v^{\prime}P \prec Q + v^{\prime\prime}P$ such that $R \prec Q$. This shows that $Q \prec R$ and $R \prec Q$ which is impossible. Therefore no proper subtree exists as both initial and terminal of the tree of a Lyndon partial word.

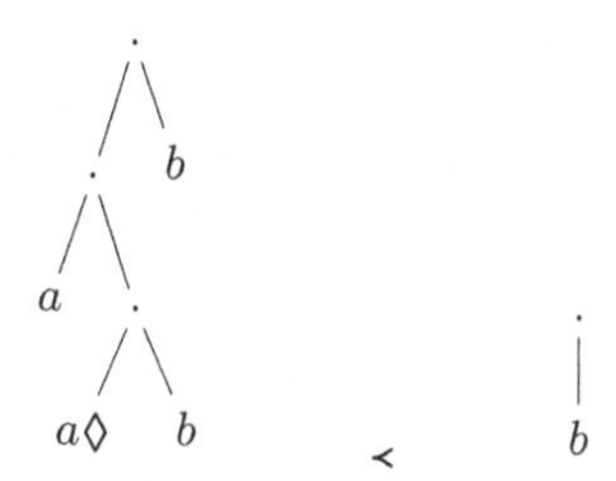

Fig. 6. Tree of $aa\Diamond bb \prec b$

Theorem 9. *ζ is a tree of a Lyndon partial word if and only if $\zeta = P + vQ, v \in \delta(P)$ where $\zeta, P, Q \in \Im$ and $P \prec Q$.*

Proof. Consider a tree ζ associated with a Lyndon partial word $u_{\Diamond}$. Let ζ have P and Q as left and right sub trees and let $p_{\Diamond}$ and $q_{\Diamond}$ be the segments of $u_{\Diamond}$ respectively floating from P and Q in $(u_{\Diamond}, \zeta)$. By Theorem 2 and Theorem 8, ζ is a tree of $u_{\Diamond}$ if and only if $\zeta \prec Q$. For instance, consider Example1 (see Fig. 6). Thus alike the proof of Theorem 3, $\zeta(u_{\Diamond})$ is a tree in Fig. 3.

Theorem 10. *Any tree ζ over the alphabet $\Sigma_{\Diamond}^{+}$ can be uniquely written as $\zeta = P_0 + v_1P_1 + v_2P_2 + ...v_kP_k, v_m \in \delta(v_nP_n)$ for some $n \geq m$ such that $P_0 \geq P_1 \geq P_2.... \geq P_k$.*

Proof. We have to show that a tree associated with any partial word factorises uniquely as a non-increasing product of trees. Now observe a factorisation of factors $\zeta = P_0 + v_1P_1 + v_2P_2 + ...v_kP_k$ with k minimal. If $v_iP_i \prec v_{(i+1)}P_{(i+1)}$ for some i then $\zeta = P_0 + v_1P_1 + v_2P_2 + ...v_iP_iv_{(i+1)}P_{(i+1)}....v_kP_k$ is a factorisation of factors in trees associated with Lyndon partial words since we have $v_iP_iv_{(i+1)}P_{(i+1)} \in \delta(v_nP_n)$ for some $n \geq i$. Now we have to prove the uniqueness. Let us assume that for any $v_iP_i, w_iQ_i \in \delta(v_nP_n)$ such that $P_0 + v_1P_1 + v_2P_2 + ...v_iP_iv_{(i+1)}P_{(i+1)}....\ v_kP_k = Q_0 + w_1Q_1 + w_2Q_2 + ...w_iQ_iw_{(i+1)}Q_{(i+1)}...w_kQ_k$, we have $P_0 \geq P_1 \geq P_2.... \geq P_k$ and $Q_0 \geq Q_1 \geq Q_2.... \geq Q_k$. Assume that v_1P_1 is longer than w_1Q_1. Then $v_1P_1 = w_1Q_1 + w_2Q_2 + ...w_iQ_ix$ with x a non empty prefix of $w_{(i+1)}Q_{(i+1)}$. Then $v_1P_1 \prec x \neq w_{(i+1)}Q_{(i+1)} \neq w_1Q_1v_1P_1$ which contradicts our assumption.

4 $\ell_\Diamond$ - Morphism

In this section we characterise $\ell_\Diamond$-morphism and show that they are order-preserving morphism. A non-empty morphism g over an ordered alphabet $\Sigma_{k,\Diamond}$ containing atleast two letters is an order-preserving morphism if for all partial words $r_\Diamond, s_\Diamond$ over $\Sigma_\Diamond$, $r_\Diamond < s_\Diamond \Rightarrow g(r_\Diamond) < g(s_\Diamond)$.

Definition 5. *Consider two ordered alphabets $U_\Diamond$ and $V_\Diamond$ each containing atleast two letters such that a morphism g from $U_\Diamond^*$ to $V_\Diamond^*$ is called a $\ell_\Diamond$- morphism if for any Lyndon partial words $l_\Diamond$ over $U_\Diamond$, $g(l_\Diamond)$ is a Lyndon partial word over $V_\Diamond$. In short a morphism that preserves the property of Lyndon partial words is defined as $\ell_\Diamond$ - morphism.*

Theorem 11. *A non-empty morphism g on $\Sigma_\Diamond^+$ containing atleast two letters is a $\ell_\Diamond$- morphism if and only if g is an order preserving morphism such that for each $u_\Diamond \in \Sigma_\Diamond$, $g(u_\Diamond)$ is a Lyndon partial word.*

Proof. Consider g to be a $\ell_\Diamond$- morphism. Then by Definition 5, for each $u_\Diamond \in \Sigma_\Diamond$, $g(u_\Diamond)$ is a Lyndon partial word. Let m, n be two integers such that $1 < m < n$ and given a least integer $p \geq 0$. Then, since g to be a $\ell_\Diamond$- morphism $g(u_m) < g(u_m u_n^p) < g(u_{m+1})$ such that $|g(u_m)| > |g(u_m u_n^p)| > |g(u_{m+1})|$. This implies that g is order preserving. Conversely, assume that g is an order preserving morphism such that for each $u_\Diamond \in \Sigma_\Diamond$, $g(u_\Diamond)$ is a Lyndon partial word. Now we have to show that for each m such that $1 < m < n$, $g(u_m u_n^p) < g(u_{m+1})$. Since g is a non-empty order preserving morphism, $g(u_m u_n^p u_{m+1})$ is a Lyndon partial word. Also $g(u_m) < g(u_m u_n^p u_{m+1}) < g(u_{m+1})$. Thus g is a $\ell_\Diamond$- morphism.

Corollary 1. *g is a $\ell_\Diamond$- morphism on $\Sigma_\Diamond = \{a, b\} \cup \{\Diamond\}$ if and only if $g(a)$ and $g(b)$ are Lyndon partial words with $g(a) < g(b)$.*

5 Two-Dimensional Lyndon Partial Words

The concept of Lyndon words are extended as two-dimensional Lyndon words in [20]. Those are useful to capture 2D horizontal periodicity of a matrix in which each row is highly periodic. It is also utilised to solve 2D horizontal suffix–prefix matching among a set of rectangular patterns efficiently. We introduce the following.

Definition 6. *A* two-dimensional row Lyndon partial word *is a horizontally primitive matrix which is least among its horizontal conjugates.*

Example 4. Consider a two-dimensional partial word $A = \begin{matrix} a & \Diamond & a \\ \Diamond & b & a \\ b & b & \Diamond \end{matrix}$ of size (3×3) over $\Sigma_\Diamond = \{a < b\} \cup \{\Diamond\}$. The horizontal conjugates of A are $\begin{matrix} a & \Diamond & a \\ \Diamond & b & a \\ b & b & \Diamond \end{matrix}$, $\begin{matrix} \Diamond & a & a \\ b & \Diamond & a \\ b & \Diamond & b \end{matrix}$, $\begin{matrix} a & a & \Diamond \\ a & \Diamond & b \\ \Diamond & b & b \end{matrix}$.

Here $\begin{matrix} a & a & \Diamond \\ a & \Diamond & b \\ \Diamond & b & b \end{matrix}$ is a two-dimensional row Lyndon partial word with each horizontal rows as minimal among its conjugates.

Definition 7. *A* regular two-dimensional Lyndon partial word *is a horizontally primitive matrix which is least of its horizontal conjugates by maintaining a regular order.*

In one-dimensional case Lyndon partial words of length 4 over binary alphabet are $aaa\Diamond$, $aa\Diamond b$, $a\Diamond ab$, $a\Diamond bb$, $ab\Diamond b$, $\Diamond bbb$. Now we can derive two-dimensional partial words as follows where there will be many 2D partial words, few sample of those 2D partial words are given in Example below which maintains a specific/regular ordering of Lyndon partial arrays.

Example 5. $\begin{matrix} a & a & a & \Diamond \\ a & a & a & \Diamond \end{matrix}, \begin{matrix} a & a & a & \Diamond \\ a & a & \Diamond & b \end{matrix}, \begin{matrix} a & a & a & \Diamond \\ a & \Diamond & a & b \end{matrix}, \begin{matrix} a & a & a & \Diamond \\ a & \Diamond & b & b \end{matrix}, \begin{matrix} a & a & a & \Diamond \\ \Diamond & b & b & b \end{matrix}, \begin{matrix} a & a & \Diamond & b \\ a & a & \Diamond & b \end{matrix}, \begin{matrix} a & a & \Diamond & b \\ a & \Diamond & a & b \end{matrix}, \begin{matrix} a & a & \Diamond & b \\ a & \Diamond & b & b \end{matrix},$
$\begin{matrix} a & a & \Diamond & b \\ \Diamond & b & b & b \end{matrix}, \begin{matrix} a & \Diamond & a & b \\ a & \Diamond & a & b \end{matrix}, \begin{matrix} a & \Diamond & a & b \\ a & \Diamond & b & b \end{matrix}, \begin{matrix} a & \Diamond & a & b \\ \Diamond & b & b & b \end{matrix}, \begin{matrix} a & \Diamond & b & b \\ a & \Diamond & b & b \end{matrix}, \begin{matrix} a & \Diamond & b & b \\ \Diamond & b & b & b \end{matrix}, \begin{matrix} \Diamond & b & b & b \\ \Diamond & b & b & b \end{matrix}$

$$\begin{matrix} a & a & a & \Diamond \\ a & a & a & \Diamond \\ a & a & a & \Diamond \end{matrix}, \cdots, \begin{matrix} \Diamond & b & b & b \\ \Diamond & b & b & b \\ \Diamond & b & b & b \end{matrix}, \begin{matrix} a & a & a & \Diamond \\ a & a & a & \Diamond \\ a & a & a & \Diamond \\ a & a & a & \Diamond \end{matrix}, \cdots, \begin{matrix} \Diamond & b & b & b \\ \Diamond & b & b & b \\ \Diamond & b & b & b \\ \Diamond & b & b & b \end{matrix}, \begin{matrix} a & a & a & \Diamond \\ a & a & a & \Diamond \\ a & a & a & \Diamond \\ a & a & a & \Diamond \\ a & a & a & \Diamond \end{matrix}, \cdots, \begin{matrix} a & a & a & \Diamond \\ a & a & \Diamond & b \\ a & \Diamond & a & b \\ a & \Diamond & b & b \\ \Diamond & b & b & b \end{matrix}, \cdots, \begin{matrix} \Diamond & b & b & b \\ \Diamond & b & b & b \\ \Diamond & b & b & b \\ \Diamond & b & b & b \\ \Diamond & b & b & b \end{matrix}.$$

One can observe that these 6 elements that follows are similar to many other elements which are NOT present in above collection: $\begin{matrix} \Diamond & a & a & a \\ b & a & a & \Diamond \\ b & a & \Diamond & a \\ b & a & \Diamond & b \\ b & \Diamond & b & b \end{matrix}, \begin{matrix} a & \Diamond & a & a \\ \Diamond & b & a & a \\ a & b & a & \Diamond \\ b & b & a & \Diamond \\ b & b & \Diamond & b \end{matrix}, \begin{matrix} a & a & \Diamond & b \\ a & \Diamond & a & b \\ a & \Diamond & b & b \\ \Diamond & b & b & b \\ a & a & a & \Diamond \end{matrix},$
$\begin{matrix} a & a & a & \Diamond \\ a & \Diamond & a & b \\ a & \Diamond & b & b \\ \Diamond & b & b & b \\ a & a & \Diamond & b \end{matrix}, \begin{matrix} a & a & a & \Diamond \\ a & a & \Diamond & b \\ a & \Diamond & b & b \\ \Diamond & b & b & b \\ a & \Diamond & a & b \end{matrix}, \begin{matrix} a & a & a & \Diamond \\ a & a & \Diamond & b \\ a & \Diamond & a & b \\ \Diamond & b & b & b \\ a & \Diamond & b & b \end{matrix}$ as these do not satisfy the property of being a regular 2D Lyndon partial word. Now it is of interest to identify specific/regular 2D Lyndon partial words among those elements.

Due to Remark 2, we see the following pattern collected as a 2D partial language which shall be named as L_{DD} for further references in our work based on this paper.

$$L_{DD} = \left\{ \begin{array}{cc} a & \Diamond \\ \Diamond & b \end{array}, \begin{array}{ccc} a & a & \Diamond \\ a & \Diamond & b \\ \Diamond & b & b \end{array}, \begin{array}{cccc} a & a & a & \Diamond \\ a & a & \Diamond & b \\ a & b & \Diamond & b \\ a & \Diamond & a & b \\ a & \Diamond & b & b \\ \Diamond & b & b & b \end{array}, \begin{array}{ccccc} a & a & a & a & \Diamond \\ a & a & a & \Diamond & b \\ a & a & b & \Diamond & b \\ a & a & \Diamond & a & b \\ a & a & \Diamond & b & b \\ a & b & \Diamond & b & b \\ a & \Diamond & a & b & b \\ a & \Diamond & b & b & b \\ \Diamond & b & b & b & b \end{array}, \begin{array}{cccccc} a & a & a & a & a & \Diamond \\ a & a & a & a & \Diamond & b \\ a & a & a & b & \Diamond & b \\ a & a & b & b & \Diamond & b \\ a & a & a & \Diamond & a & b \\ a & a & a & \Diamond & b & b \\ a & a & b & \Diamond & a & b \\ a & a & b & \Diamond & b & b \\ a & a & \Diamond & a & a & b \\ a & a & \Diamond & a & b & b \\ a & a & \Diamond & b & a & b \\ a & a & \Diamond & b & b & b \\ a & \Diamond & a & a & a & b \\ a & \Diamond & a & b & b & b \\ a & \Diamond & b & b & b & b \\ \Diamond & b & b & b & b & b \end{array}, \ldots, \right\}.$$

Fig. 7. Kambi Kolam (5 × 5)

Connections to Image Analysis

Kolam is one of several types of street/home art practices performed as ephemeral designs on thresholds in India. In Tamil Nadu, kolam designs that grace the thresholds of houses each day attract attention from different angles. In addition to their importance within Tamil culture, there are an unusual example of mathematical ideas expressed in a cultural context. In recent years, kolam figures have also attracted the attention of computer scientists interested in analysing and describing images with picture languages [26]. The kolam figures grouped into families attracted theoretical computer scientists who are involved in the analysis and description of images using picture languages [29]. These languages use basic units and formal rules for combining them. Studies of picture languages are similar to formal language theory, which dates back to the study of natural languages some 45 years ago. The analysis and specification of programming languages has been supported by formal language theory in subsequent decades.

Fig. 8. Kolams for 2D row Lyndon partial word

Figure 7 is a kambi kolam of (5×5). In Fig. 8 we embed the kolam in Fig. 7 with any two-dimensional row Lyndon partial word of size 5×5 over $\Sigma_\Diamond = \{a \prec b\} \cup \{\Diamond\}$. Among the 9 Lyndon partial words of length 5 mentioned in Table 2, if $aaa\Diamond b$,

Fig. 9. Kolam patterns to show sensitivity of $aaaa\Diamond$

$aa\Diamond ab$, $aa\Diamond bb$, $aab\Diamond b$, $a\Diamond abb$, $a\Diamond bbb$, $ab\Diamond bb$ occurs as ANY of the rows of a two-dimensional word, then the obtained two-dimensional word will become a two-dimensional row Lyndon partial word. Also no change in the fixed kolam pattern is allowed to occur while the $\Diamond$ in each row is replaced by the letters a or b of Σ

Fig. 10. Kolam patterns to show sensitivity of $\Diamond bbbb$

due to the sensitive cases that is explained in the below lines fixes the property of Definition 6.

The sensitive cases are the remaining two Lyndon partial words $aaaa\Diamond$ and $\Diamond bbbb$ (out of those 9 Lyndon partial words of length 5). Both of these Lyndon partial words gets converted to periodic total words, if the $\diamond$ in the former replaced by the letter a and if the $\diamond$ in the latter replaced by the letter b by violating the property of Lyndon word.

The Figs. 9 and 10 shows the changes in the kolam patterns of these sensitive cases of the two-dimensional row Lyndon partial words. The dots in each rows of the kolam patterns denotes each letter corresponding to the Lyndon partial words, but as a whole the kolam patterns are not two-dimensional row Lyndon partial words due to the sensitivity of the $\diamond$s.

6 Discussions

In [17], we introduced Lyndon partial words and proved that the language of all Lyndon partial words over the binary alphabet is not context free. Now in this paper, we introduced Lyndon partial arrays and trees associated with Lyndon partial words. Kolams, Celtic knots (indo-germanic patterns) were studied for decades by researchers worldwide since 1970s. We learned to find out very basic connections to two-dimensional formal languages through Lyndon partial words [17], arrays with few automata and grammar models in [22].

In near future, we will do a detailed study on several variants of 2D Lyndon partial words and their properties in an upgraded version of this paper. Motivated by the questions on the study of Lyndon partial words one can think of an automaton model as we tried to carve in Figs. 1, 2, as we remarked in this paper to find a sequence of Lyndon partial words (for binary alphabet) is not achieved so far as similar as the Lyndon word integer sequence namely (sequence A001037 in the OEIS). We aim to solve this problem by developing the borders with further connections to that of Descriptional Complexity of Formal Systems through studies on State Complexity [15,18].

Also time and space complexity of Lyndon partial words versus 2D Lyndon partial words shall be investigated similar to [20]. Closure properties on several operations shall be investigated along with the comparison of the family of languages with other several family of languages for both Lyndon partial words versus 2D Lyndon partial words with different variants shall be thoroughly investigated through more study with finite array automata and array grammars [16,25]. We also aim to connect with Computer Vision [28] to find out the lost or hidden information which is main application/implementation areas to connect with the partiality and sensitivity of holes in our study.

Acknowledgements. We would like to thank the unknown referees for their comments and suggestions on the manuscript in improving from an earlier version. The corresponding authors R. Krishna Kumari and R. Arulprakasam are very much thankful to the management, SRM Institute of Science and Technology for their continuous support and encouragement. Meenakshi Paramasivan would like to thank the financial support provided by CIRT (Center for Informatics Research and Technology) - University of Trier, Germany for the Celtic Studies in 2017–2018. The authors owe a big thanks to Prof. Rani Siromoney and her co-authors.

References

1. Bannai, H., Tomohiro, I., Inenaga, S., Nakashima, Y., Takeda, M., Tsuruta, K.: The "runs" theorem. SIAM J. Comput. **46**(5), 1501–1514 (2017)
2. Barcelo, H.: On the action of the symmetric group on the free lie algebra and the partition lattice. J. Comb. Theory Ser. A **55**(1), 93–129 (1990)
3. Berstel, J., Boasson, L.: Partial words and a theorem of Fine and Wilf. Theor. Comput. Sci. **218**(1), 135–141 (1999)
4. Bille, P., et al.: Space efficient construction of Lyndon arrays in linear time. In: Czumaj, A., Dawar, A., Merelli, E. (eds.) 47th International Colloquium on Automata, Languages, and Programming, ICALP 2020, 8–11 July 2020, Saarbrücken, Germany (Virtual Conference). LIPIcs, vol. 168, pp. 14:1–14:18. Schloss Dagstuhl - Leibniz-Zentrum für Informatik (2020)
5. Blanchet-Sadri, F.: Primitive partial words. Discret. Appl. Math. **148**(3), 195–213 (2005)
6. Blanchet-Sadri, F., Goldner, K., Shackleton, A.: Minimal partial languages and automata. RAIRO Theor. Inform. Appl. **51**(2), 99–119 (2017)
7. Chen, K.T., Fox, R.H., Lyndon, R.C.: Free differential calculus, IV. The quotient groups of the lower central series. Ann. Math. 81–95 (1958)
8. Crochemore, M., Russo, L.M.: Cartesian and Lyndon trees. Theoret. Comput. Sci. **806**, 1–9 (2020)
9. Duval, J.: Factorizing words over an ordered alphabet. J. Algorithms **4**(4), 363–381 (1983)
10. Fernau, H., Paramasivan, M., Schmid, M.L., Thomas, D.G.: Scanning pictures the boustrophedon way. In: Barneva, R.P., Bhattacharya, B.B., Brimkov, V.E. (eds.) IWCIA 2015. LNCS, vol. 9448, pp. 202–216. Springer, Cham (2015). https://doi.org/10.1007/978-3-319-26145-4_15
11. Fernau, H., Paramasivan, M., Schmid, M.L., Thomas, D.G.: Simple picture processing based on finite automata and regular grammars. J. Comput. Syst. Sci. **95**, 232–258 (2018)
12. Fischer, M.J., Paterson, M.S.: String-matching and other products. In: Karp, R.M. (ed.) Complexity of Computation, SIAM-AMS Proceedings, vol. 7, pp. 113–125 (1974)
13. Franek, F., Islam, A.S.M.S., Rahman, M.S., Smyth, W.F.: Algorithms to compute the Lyndon array. In: Holub, J., Zdárek, J. (eds.) Proceedings of the Prague Stringology Conference 2016, Prague, Czech Republic, 29–31 August 2016, pp. 172–184. Department of Theoretical Computer Science, Faculty of Information Technology, Czech Technical University in Prague (2016)
14. Hohlweg, C., Reutenauer, C.: Lyndon words, permutations and trees. Theor. Comput. Sci. **307**(1), 173–178 (2003)
15. Holzer, M., Kutrib, M.: Descriptional complexity - an introductory survey. In: Martín-Vide, C. (ed.) Scientific Applications of Language Methods, Mathematics, Computing, Language, and Life: Frontiers in Mathematical Linguistics and Language Theory, vol. 2, pp. 1–58. World Scientific/Imperial College Press (2010)
16. Krithivasan, K., Siromoney, R.: Array automata and operations on array languages. Int. J. Comput. Math. **4**, 3–30 (1974)
17. Kumari, R.K., Arulprakasam, R., Dare, V.: Language of Lyndon partial words. Comput. Sci. **15**(4), 1173–1177 (2020)
18. Kutrib, M., Wendlandt, M.: State complexity of partial word finite automata. In: Han, Y., Ko, S. (eds.) DCFS 2021. LNCS, vol. 13037, pp. 113–124. Springer, Cham (2021). https://doi.org/10.1007/978-3-030-93489-7_10

19. Lothaire, M.: Combinatorics on Words. Cambridge Mathematical Library, 2nd edn. Cambridge University Press, Cambridge (1997)
20. Marcus, S., Sokol, D.: 2D Lyndon words and applications. Algorithmica **77**(1), 116–133 (2017)
21. Nayak, A.C., Kapoor, K.: On the language of primitive partial words. In: Dediu, A.-H., Formenti, E., Martín-Vide, C., Truthe, B. (eds.) LATA 2015. LNCS, vol. 8977, pp. 436–445. Springer, Cham (2015). https://doi.org/10.1007/978-3-319-15579-1_34
22. Paramasivan, M.: Operations on graphs, arrays and automata. Ph.D. thesis, University of Trier, Germany (2018)
23. Reutenauer, C.: Free Lie Algebras. London Mathematical Society Monographs New Series (1993). Cohn, P.M., Dales, H.G. (eds.)
24. Richomme, G.: Lyndon morphisms. Bull. Belgian Math. Soc.-Simon Stevin **10**(5), 761–785 (2003)
25. Siromoney, G., Siromoney, R., Krithivasan, K.: Picture languages with array rewriting rules. Inf. Control **22**(5), 447–470 (1973)
26. Siromoney, R., Subramanian, K.G., Dare, V.R., Thomas, D.G.: Some results on picture languages. Pattern Recognit. **32**(2), 295–304 (1999)
27. Subramanian, K.G., Siromoney, R., Mathew, L.: Lyndon trees. Theor. Comput. Sci. **106**(2), 373–383 (1992)
28. Szeliski, R.: Computer Vision - Algorithms and Applications. Texts in Computer Science, 2nd edn. Springer, Cham (2022). https://doi.org/10.1007/978-3-030-34372-9
29. Waring, T.M.: Sequential encoding of Tamil kolam patterns. Forma **27**, 83–92 (2012)

Theory and Applications

Tomography Reconstruction Based on Null Space Search

Tibor Lukić(✉) and Tamara Kopanja

Faculty of Technical Sciences, University of Novi Sad, Novi Sad, Serbia
{tibor,kopanja.v11.2019}@uns.ac.rs

Abstract. The paper introduces a new tomography reconstruction approach for gray and binary image reconstruction. The proposed method intends to find a solution by searching for the best linear combination of the basis vectors of the null space of the projection matrix. One of the advantages of the proposed approach is that the projection error remains always extremely low, practically equal to zero, during the reconstruction process. The method applies a gradient based optimization algorithm. A short experimental evaluation, including three relevant and well-know algorithms for comparison, is presented.

Keywords: Tomography reconstruction · Null space · Energy minimization · Gradient based optimization · Regularization

1 Introduction

Tomography is a field of image processing which deals with reconstruction of unknown images from available projection data [7]. Mathematically, the image can be modeled by a function whose codomain or range may be continuous or discrete set. In the *Computerized Tomography* the image function has a continuous range. The *Discrete Tomography* (DT) [8,9] is a sub-field of tomography, where the range of the image function is a finite and discrete set. If this range contains just a few predefined intensity levels, then we talk about *Multi-Level Tomography* [12]. In particular, in *Binary Tomography* (BT), the unknown image contains only two different intensity values, usually zero and one.

Application spectrum for tomography methods is very wide. Tomography image reconstruction techniques are widely used in different industrial investigation problems, often in the form of nondestructive material testing [4]. Another vast field of its application is connected to the human radiology diagnostic procedures, like CT scanning. Great field of tomography application belongs to security screening techniques, for example, many airports use X-rays computed tomography for screening baggage [11]. BT methods are extremely useful in cases when we investigate the presence or absence of a specific material inside a homogeneous structure, for example in the determination of the presence of atoms in crystalline structures [10].

There are several successful methods, proposed in literature, for solving the tomography reconstruction problem. We just mention a few well-known iterative algorithms: *Algebraic Reconstruction Technique* (ART) [6], *Simultaneous*

R. P. Barneva et al. (Eds.): IWCIA 2022, LNCS 13348, pp. 247–259, 2023.
https://doi.org/10.1007/978-3-031-23612-9_15

Iterative Reconstruction Technique (SIRT) [5], *Discrete Algebraic Reconstruction Technique* (DART) [1,2], *Spectral Projected Gradient* (SPG) [15], *Difference of Convex functions* (DC) [21]. The common feature of these methods is that they change the pixel intensities of the current solution in attempt to reduce the projection error, in other words, the projection error is minimized by the convergence process of these methods. As a result of that, the projection error may be minimized with smaller or higher success. This issue motivated us to develop a reconstruction method where the projection error does not change during the iterative process.

In the reconstruction process the basic criteria is the matching of the reconstructed image with the measured projection data. The acquired data is often the most reliable information about the unknown image. Therefore, it is important that the projection error of the reconstructed image remains as small as possible. In this paper we propose a new reconstruction method which seeks to answer to this issue. In the proposed reconstruction process the projection error always remains unchanged and theoretically equals to zero - in practical applications close to zero.

The paper has the following structure. Section 2 gives the description of the basic tomography reconstruction problem. In Sect. 3 the new reconstruction method is presented. In Sect. 4 a short experimental evaluation of the proposed new reconstruction method is given. Finally, the conclusion is given in the Sect. 5.

2 Tomography Reconstruction Problem

In this paper we consider transmission tomography reconstruction model. The main characteristic of transmission tomography is that the both, source and detector are placed out of the considered object. Mathematically the problem of tomography reconstruction may be formulated by the following system of linear equations

$$Au = b, \quad A \in \mathbb{R}^{m\times n}, \quad u \in \mathbb{R}^n, \quad b \in \mathbb{R}^m, \tag{1}$$

where u represents the unknown image which should be reconstructed. In the case of binary tomography, components of vector u have only two different values, usually 0 and 1. The matrix A is so-called *projection matrix* and its rows hold information about a length of the projection ray passing through the pixel. The assumption is that each pixel is represented as a square with unit side length, see Fig. 1. Further, it represents calculation projection value for a given image from one projection direction denoted by angle θ. The another projection direction is obtained by rotation source-detector system around of the center of the circle. Each projection direction contributes a new parallel set of the projection rays. Detected projection values are placed in the projection vector b. It is not hard to see that the projection matrix A is spare, i.e. the majority of elements $a_{i,j}$ are equal to zero.

In the reconstruction problem, both the matrix A and the projection vector b are given, as a calculated or measured data. The task is to determine the unknown image u. The system (1) is often underdetermined ($m << n$), and consequently, in general case, we can count on an infinite number of possible solutions. The applied parallel beam projection geometry, see Fig. 1, allows us to assume that the matrix A is full row rank, that is $\text{rank}(A) = m$. Therefore, the general solution has $n - m$ degrees of freedom.

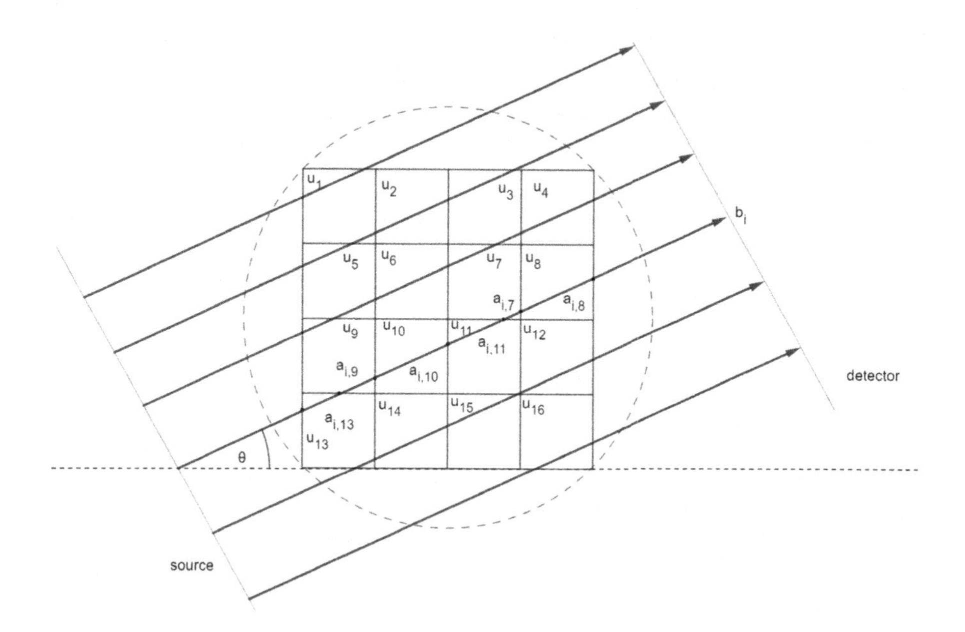

Fig. 1. The transmission parallel beam projection geometry. The i-th projection value is obtained as $b_i = a_{i,13}u_{13} + a_{i,9}u_9 + a_{i,10}u_{10} + a_{i,11}u_{11} + a_{i,7}u_7 + a_{i,8}u_8$.

3 Tomography Reconstruction Method Based on Null Space Search

In this section we introduce a new tomography reconstruction method for both gray-scale and binary image reconstructions. The main idea uses the fact that any solution of the projection linear system of Eqs. (1)

$$A\,u = b,$$

can be represented as a sum of one particular solution $u_p (A\,u_p = b)$ and an appropriate vector belongs to the null space of the matrix A, defined by

$$\mathcal{N}(A) = \{x \in \mathbb{R}^n \mid A\,x = 0\}.$$

Hence the set of all solutions can be represented as

$$u_p + z, \quad \text{where} \quad z \in \mathcal{N}(A).$$

Let us denote a basis of the vector space $\mathcal{N}(A)$ by the set of vectors $\{\mathbf{b}_1, \mathbf{b}_2, \ldots \mathbf{b}_k\}$, where $k = n - m$. Each solution of the system (1) may be represented in the following form

$$w(\alpha) = u_p + \alpha_1 \mathbf{b}_1 + \alpha_2 \mathbf{b}_2 + \ldots + \alpha_k \mathbf{b}_k, \qquad \text{where } \alpha = (\alpha_1, \alpha_2, \ldots, \alpha_k) \in \mathbb{R}^k \text{ and } \mathbf{b}_i \in \mathbb{R}^n \text{ for all } i = 1, \ldots, k. \tag{2}$$

Now we will choose coefficients $\alpha_1, \alpha_2, \ldots \alpha_k$ in α in a such way that the obtained solution $w(\alpha) \in \mathbb{R}^n$ lies in a predefined area, for example in the hyper cube $[0,1]^n$. To achieve this, let us look at the following unconstrained optimization problem

$$\arg\min_{\alpha} \sum_{i=1}^{n} W(w(\alpha)_i), \tag{3}$$

where W is a specially designed potential function. It is defined by

$$W(x) = \begin{cases} x^2, & x \leq 0 \\ (x-1)^2, & x \geq 1 \\ 0, & 0 < x < 1 \end{cases} . \tag{4}$$

W is a continuously differentiable function consisting of two square parabolas and a horizontal line. Its important property is that it is always greater than zero except for values between 0 and 1, when it is zero, see Fig. 2. Therefore, the solution α^* of the problem (3) gives a combination of coefficients of basis vectors in (2) for which the corresponding image solution $w(\alpha^*)$ belongs to the set $[0,1]^n$.

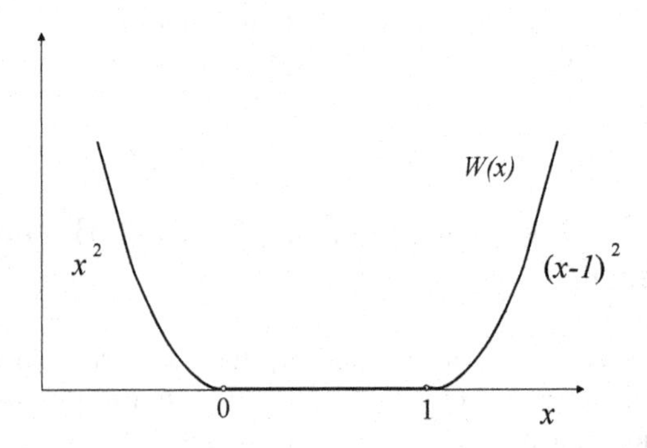

Fig. 2. Potential function W, designed for reconstruction of gray solution that belongs to the hyper cube $[0,1]^n$.

The gradient of the objective function $G(\alpha) := \sum_{i=1}^{n} W(w(\alpha)_i)$ in the optimization problem (3) can be determined by elementary calculus and in analytical way. It has the following form

$$\operatorname{grad} G(\alpha) = N^T \cdot [W'(w(\alpha)_1), W'(w(\alpha)_2), ..., W'(w(\alpha)_n)]^T,$$

where the matrix $N \in \mathbb{R}^{n \times k}$ represents the null space $\mathcal{N}(A)$ in a way that it contains its basis vectors in the following way

$$N = \begin{bmatrix} b_{11} & b_{12} & \dots & b_{1k} \\ b_{21} & b_{22} & \dots & b_{2k} \\ \vdots & \vdots & \vdots & \vdots \\ b_{n1} & b_{n2} & \dots & b_{nk} \end{bmatrix} = [\mathbf{b}_1 \ \mathbf{b}_2 \ \dots \ \mathbf{b}_k].$$

In practical applications, this matrix can be obtained by elementary calculus, or just by applying the fast Matlab command $N = null(A)$. For the minimization of the problem (3) different gradient type deterministic algorithms can be used. We suggest the Spectral Conjugate Gradient algorithm [3], which shows best performance in our experiments.

It is necessary to determine one solution (a particular solution) u_p of the system (1), since u_p is needed in the formula $w(\alpha)$ (2). We suggest the least norm solution u_{LN}, for this purpose. The algorithm of the Conjugate Gradient [19] provides a very fast and very accurate calculation of u_{LN}. Accordingly, the calculation of particular solution in (2) does not reduce the speed and accuracy of the whole reconstruction procedure.

The design of the proposed model (3) is such that the projection error of the solution ($\|A\,w(\alpha^*) - b\|$) is always extremely low, practically equal to zero. This is achieved by the manner of searching for the solution: the coefficients of basis vectors of the null space (2) are changing during the process, however, this change in the values has no effect on the projection error - this error always remains practically zero. We emphasize that this fact is one of the main advantages of the proposed reconstruction method. The projection data is the most accurate information about the solution in the tomography image reconstruction, hence their accordance with the reconstruction is extremely important. The proposed new method, which we will call *Null Space Search based Tomography* (NSST), has just this feature.

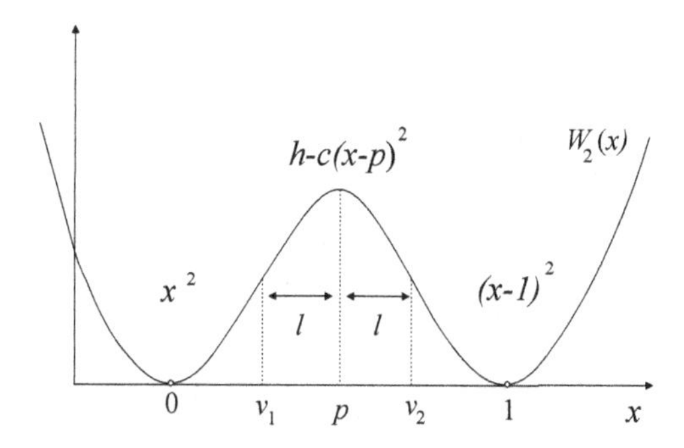

Fig. 3. Potential function W_2 (5), designed for reconstruction of binary solutions.

In a case when our goal is to find the binary solution of the tomography reconstruction problem (1), this can be achieved if we replace the potential function (4) in the proposed NSST method with the two well potential function defined by

$$W_2(x) = \begin{cases} x^2, & x \le v_1 \\ (x-1)^2, & x \ge v_2 \\ h - c(x-p)^2, & v_1 < x < v_2 \end{cases}, \tag{5}$$

where for the given parameter l we set

$$p = \frac{1}{2},\ v_1 = p - l,\ v_2 = p + l,\ c = \frac{1}{2l} - 1 \text{ and}$$

$$h = \frac{1}{4}(1 - 2l)$$

Function W_2 has 2 minima in the points 0 and 1, such that $W_2(0) = W_2(1) = 0$, see Fig. 3. It is a piece-wise quadratic function, where the constants h and c are determined in such a way that the function W_2 is continuously-differentiable for each x. Therefore, the following optimizing problem

$$\arg\min_{\alpha} \sum_{i=1}^{n} W_2(w(\alpha)_i), \tag{6}$$

can be solved in the same deterministic manner as it is suggested for the model (3) - by a gradient based optimization algorithm.

The solution α^* of (6) determines a binary image solution $w(\alpha^*)$ of the tomography problem (1). This means that the proposed reconstruction model (6) defines a variant of the NSST method which belongs to the binary tomography reconstruction methods. The pseudo-algorithm of this method is presented in Algorithm 1. The proposed approach envisages reconstruction by the sequence of optimization steps, where the parameter l, starting from its largest possible value $\frac{1}{2}$ (for which W_2 reduces to W), gradually decreased in each step. This process slowly enforces the "binarization" of the current solution. The whole process is terminated when the (almost) binary solution is achieved, which is controlled by the exit condition of the *while* loop.

In many proposed tomography reconstruction models, different types of regularization approaches are applied in terms to enhance the quality of reconstructions, see [13,14,18]. The quadratic total variation type regularization, which in continuous case has a form

$$\iint_{\Omega} \|\nabla u(x,y)\|^2\, dxdy, \tag{8}$$

has an isotropic diffusion type effect on an applied image $u(x,y)$, and it is often used in tomography reconstruction [16,17,22]. These good experiences motivated us to adapt and add this regularization to the our model as well. Accordingly, we propose the following regularized reconstruction model

$$\arg\min_{\alpha} \sum_{i=1}^{n} W_2(w(\alpha)_i) + \mu \sum_{i=1}^{n} (w(\alpha)_i - w(\alpha)_r)^2 + (w(\alpha)_i - w(\alpha)_b)^2, \tag{9}$$

Algorithm 1: NSST algorithm for binary reconstruction

Parameters: $\epsilon_{out} = 0.1$, $l = \frac{1}{2}$, $l_\Delta = 0.001$, $\alpha^{init} = (0, 0, \ldots 0)$.
while $| w(\alpha^{init})^T \cdot ((1, 1, \ldots, 1) - w(\alpha^{init})) | > \epsilon_{out}$
do

```
/* Solve by SCG algorithm: */
```

$$\alpha^{new} = \arg\min_{\alpha} \sum_{i=1}^{n} W_2(w(\alpha)_i) \tag{7}$$

$\alpha^{init} = \alpha^{new}$

$l \leftarrow l - l_\Delta$

end

where indices r and b point to neighbour pixels right and below from $w(\alpha)_i$, respectively. The parameter $\mu > 0$ balances between intensity of influence of two different terms in the proposed energy function (9), its value is set to 0.01 in our experiments. Let us denote the second term by $H(\alpha) = \sum_{i=1}^{n}(w(\alpha)_i - w(\alpha)_r)^2 + (w(\alpha)_i - w(\alpha)_b)^2$. This function is an adapted discrete version of the operator (8). The analytical expression of its gradient is given by $\operatorname{grad} H(\alpha) = \left[\frac{\partial H(\alpha)}{\partial \alpha_1}, \frac{\partial H(\alpha)}{\partial \alpha_2}, \ldots, \frac{\partial H(\alpha)}{\partial \alpha_k}\right]^T$, where

$$\frac{\partial H(\alpha)}{\partial \alpha_i} = 2\sum_{l=1}^{l=k}(w(\alpha)_l - w(\alpha)_r)(b_{li} - b_{ri}) + (w(\alpha)_l - w(\alpha)_b)(b_{li} - b_{bi}).$$

Therefore, the gradient of the energy function in (9) is determined analytically and it can be easily used in gradient based minimization algorithms. To minimize the model (9), we apply the same approach as required by the Algorithm 1, but in this case the model (7) is replaced by the regularized model (9).

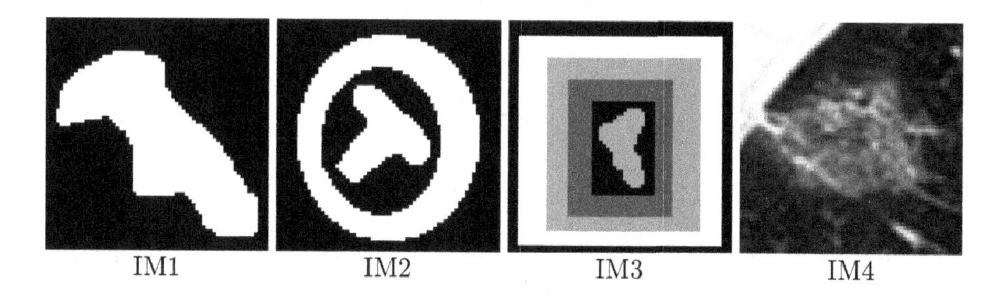

Fig. 4. Original test images used in experiments.

4 Experimental Evaluation

In this Section a short experimental evaluation of the proposed NSST method is presented. Four test images are used in the experimental work, shown in Fig. 4. Images denoted by IM1 and IM2 are binary phantoms, while IM3 is a phantom with gray pixel intensities. Test image IM4 shows a fragment of a CT image of a human lung with stain caused by COVID-19 disease. All test images (64×64) have the same pixel intensity range of $[0, 1]$. These images are used as originals in the reconstruction experiments.

Table 1. Experimental results for IM1 and IM2 images, using three different reconstruction methods. The abbreviation d indicates the number of taken projection directions.

d		IM1				IM2			
		2	3	4	6	2	3	4	6
NSST	E_P	2.22e−12	8.61e−08	6.44e−08	8.26e−08	1.73e−12	7.24e−08	8.63e−08	8.52e−08
	E_R	313.51	18.5770	5.30e−07	1.19e−06	1296	573.89	6.16	8.51e−07
	rE_R	23.93%	1.42%	≈0%	≈0%	78.83%	34.91%	0.37%	≈0%
SPG	E_P	2.82	0	0	0	4.89	5.39	0	0
	E_R	21	0	0	0	1184	595	0	0
	rE_R	1.60%	0%	0%	0%	72.02%	36.19%	0%	0%
DC	E_P	23.49	6.91	6.69	7.87	23.49	17.96	16.50	10.65
	E_R	1325	27	12	10	1325	1007	236	30
	rE_R	1.01.15%	2.06%	0.92%	0.76%	80.60%	61.25%	14.36%	1.82%

The performance of the proposed NSST reconstruction method is compared with performances of three well-known reconstruction procedures: with the SPG [14,15] and the DC [21,22] algorithms for BT case, and with the SIRT [5,20] algorithm for gray image reconstructions. All considered algorithms are implemented in Matlab environment.

The quality of the obtained reconstructions is expressed by the following three error measure functions

$$E_P(u^r) = \|Au^r - b\|,$$

$$E_R(u^r) = \sum_{i=1}^{n} |u_i^r - u_i^*|,$$

$$rE_R(u^r) = \frac{E_R(u^r)}{n_O} \cdot 100\,\%,$$

where u^r is the reconstructed image, while u^* denotes the original image and n_O is the number of object pixels in u^*. Function E_P is called *projection error* and its measures the accordance of the reconstruction with the given projection data. The *reconstruction error* E_R expresses the distance of u^r from the original image u^*. In the case of binary reconstructions, the function E_R express the

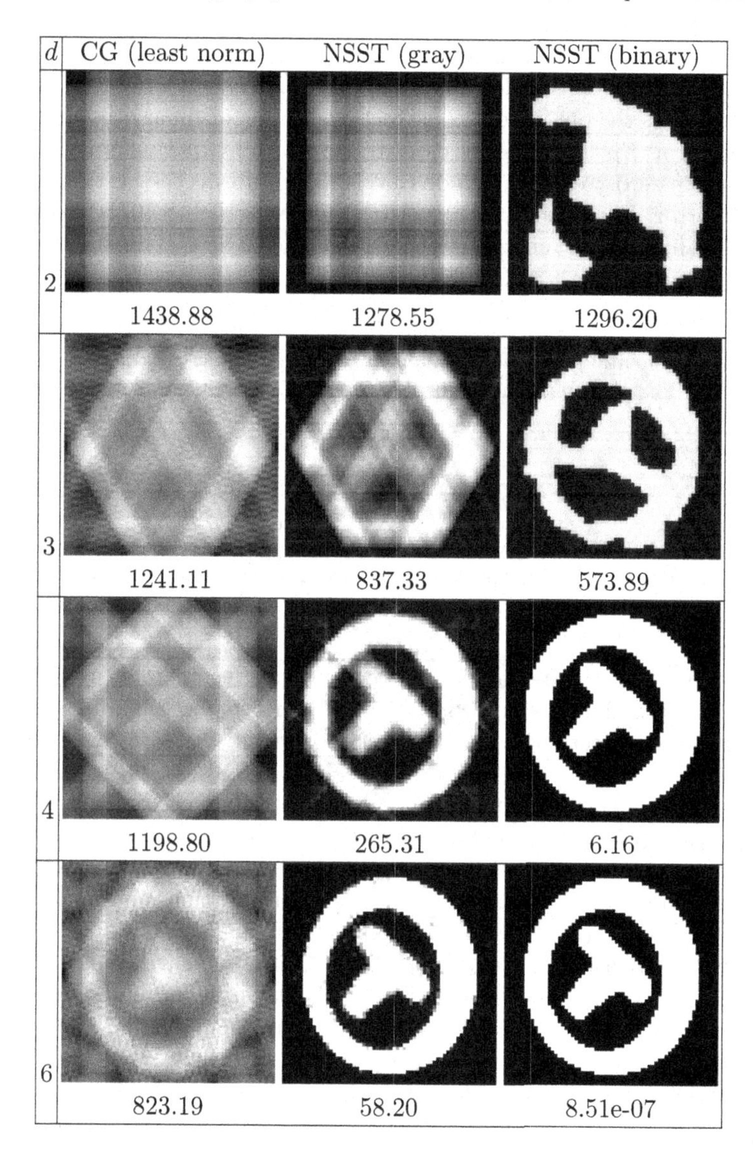

Fig. 5. Reconstructions of the phantom image IM2. The presented values below the reconstructions shows the corresponding reconstruction errors (E_R).

number of misclassified pixels, while rE_R express this number but relative to the size of the object, i.e., number of white pixels.

We note that the running time is not the advantage of the NSST method, mostly due to increased memory consumption and computational costs: the numerous $(n - m)$ basis vectors $\mathbf{b}_i \in \mathbb{R}^n$ of the null space of the projection matrix, see Sect. 3, must be memorized and their linear combinations must be manipulated during the whole reconstruction process.

Table 1 shows the obtained reconstruction results for binary test images. The projection error E_P for NSST is close to zero in all experiments. If we round these values to zero, we can conclude that NSST performs the best regarding the projection error. In terms of the reconstruction error E_R, SPG is the winner in four cases, while in the remaining other four cases NSST gives the best results or shares first place with SPG.

Table 2. Experimental results for IM3 and IM4 images, using two different reconstruction methods. The abbreviation d indicates the number of taken projection directions.

d		IM3				IM4			
		10	15	20	25	10	15	20	25
NSST	E_P	8.71e−08	9.50e−08	9.78e−08	6.90e−08	9.42e−08	9.184e−08	8.85e−08	8.63e−08
	E_R	114.87	77.49	65.36	52.26	191.60	152.31	130.18	100.56
SIRT	E_P	0.29	0.26	0.30	0.31	0.07	0.14	0.25	0.27
	E_R	253.27	197.87	170.24	135.58	212.35	176.25	152.70	125.87

Table 2 summarises the obtained reconstructions results for gray test images. The NSST method has significantly better performance in all experiments, regarding both projection and reconstruction errors, than the "control" SIRT algorithm.

Figure 5 shows three important phases of the proposed NSST reconstruction process for four different projection direction settings (d). First column shows least norm reconstructions, obtained by the CG algorithm. In the next column we can see results of the second phase, where reconstructions are provided by the minimization model (3). The third column shows final results obtained by the regularized binarization model (9). We note, that the binarization process may be not "completed", which means that pixel intensities are not always purely binary, but just close to binary. The effect of this issue you can follow in cases of low amount of projection data, when d is 2 and 3. One of the possible reason for that can by the "highly" non convexity of the used potential function, see Fig. 3.

Figure 6 shows reconstructions of gray images. NSST provides visually most appealing results in all presented cases.

Summarizing all obtained experimental results, we can conclude that the proposed NSST method shows best performance regarding the projection error minimization, and also NSST shows good competence regarding the quality of reconstructions.

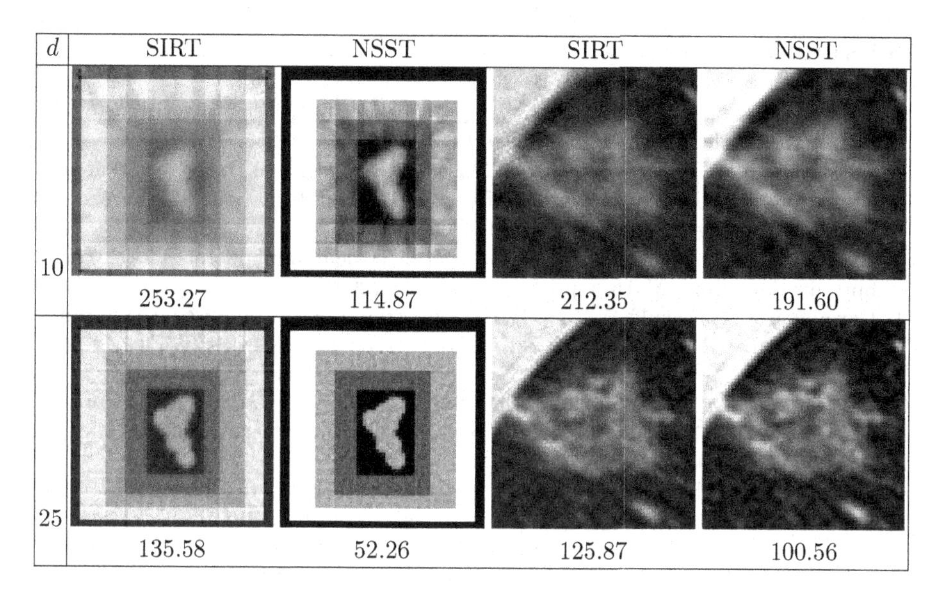

Fig. 6. Reconstructions of test images IM3 and IM4. The presented values below of reconstructions show the corresponding reconstruction errors (E_R).

5 Conclusions

This paper introduces a new deterministic tomography reconstruction approach, called NSST. The proposed method is based on searching through linear combinations of the basis vectors of the null space of the projection matrix. One of the important advantages of the new method is that the projection error of the reconstruction guaranteed to remain at a minimum possible level, practically equal to zero. This is the case even when different regularization terms is involved into the reconstruction process. The minimization problem of the proposed reconstruction model is solved by a gradient based iterative algorithm. The obtained experimental results show good performance competence of the new method in comparison with three well-known reconstruction methods.

Acknowledgement. Authors acknowledge the financial support of Department of Fundamental Sciences, Faculty of Technical Sciences, University of Novi Sad, in the frame of the Project "Primena opštih disciplina u tehničkim i informatičkim naukama". T. Lukić also acknowledges support received from the Hungarian Academy of Sciences through the DOMUS project.

References

1. Batenburg, K.J., Sijbers, J.: DART: a practical reconstruction algorithm for discrete tomography. IEEE Trans. Image Process. **20**, 2542–2553 (2011)
2. Batenburg, K.J., Sijbers, J.: DART: a fast heuristic algebraic reconstruction algorithm for discrete tomography. In: Proceedings of International Conference on Image Processing (ICIP), pp. 133–136 (2007)
3. Birgin, E., Martínez, J.: Spectral conjugate gradient method for unconstrained optimization. Appl. Math. Optim. **43**, 117–128 (2001)
4. Carmignato, S., Dewulf, W., Leach, R.: Industrial X-Ray Computed Tomography. Springer, Heidelberg (2018). https://doi.org/10.1007/978-3-319-59573-3
5. Gilbert, P.: Iterative methods for the three-dimensional reconstruction of an object from projections. J. Theor. Biol. **36**(1), 105–117 (1972). https://doi.org/10.1016/0022-5193(72)90180-4, https://www.sciencedirect.com/science/article/pii/0022519372901804
6. Gordon, R., Bender, R., Herman, G.T.: Algebraic reconstruction techniques (ART) for three-dimensional electron microscopy and x-ray photography. J. Theor. Biol. **29**(3), 471–481 (1970). https://doi.org/10.1016/0022-5193(70)90109-8, https://www.sciencedirect.com/science/article/pii/0022519370901098
7. Herman, G.T.: Image Reconstruction from Projections. Springer, Heidelberg (1980)
8. Herman, G.T., Kuba, A.: Discrete Tomography: Foundations, Algorithms and Applications. Birkhäuser (1999)
9. Herman, G.T., Kuba, A.: Advances in Discrete Tomography and Its Applications. Birkhäuser (2007)
10. Herman, G.T., Kuba, A.: Discrete tomography: Foundations, Algorithms, and Applications. Springer, Heidelberg (2012)
11. Kisner, S.J.: image reconstruction for X-ray computed tomography in security screening applications. Ph.D. thesis, USA (2013)
12. Lukić, T.: Discrete tomography reconstruction based on the multi-well potential. In: Aggarwal, J.K., Barneva, R.P., Brimkov, V.E., Koroutchev, K.N., Korutcheva, E.R. (eds.) IWCIA 2011. LNCS, vol. 6636, pp. 335–345. Springer, Heidelberg (2011). https://doi.org/10.1007/978-3-642-21073-0_30
13. Lukić, T., Balázs, P.: Binary tomography reconstruction based on shape orientation. Pattern Recogn. Lett. **79**, 18–24 (2016)
14. Lukić, T., Balázs, P.: Limited-view binary tomography reconstruction assisted by shape centroid. Vis. Comput. (Springer) **38**, 695–705 (2022)
15. Lukić, T., Lukity, A.: A spectral projected gradient optimization for binary tomography. In: Rudas, I.J., Fodor, J., Kacprzyk, J. (eds.) Computational Intelligence in Engineering. SCI, vol. 313, pp. 263–272. Springer, Heidelberg (2010). https://doi.org/10.1007/978-3-642-15220-7_21
16. Lukić, T., Nagy, B.: Deterministic discrete tomography reconstruction method for images on triangular grid. Pattern Recogn. Lett. **49**, 11–16 (2014)
17. Lukić, T., Nagy, B.: Regularized binary tomography on the hexagonal grid. Phys. Scripta **94**, 025201(9pp) (2019)
18. Lukić, T., Balázs, P.: Shape circularity assisted tomography reconstruction. Phys. Scripta **95**(10), 105211 (2020). https://doi.org/10.1088/1402-4896/abb633
19. Nocedal, J., Wright, S.J.: Numerical Optimization, 2e edn. Springer, New York (2006). https://doi.org/10.1007/978-0-387-40065-5

20. Palenstijn, W.J., Bédorf, J., Sijbers, J., Batenburg, K.J.: A distributed ASTRA toolbox. Adv. Struct. Chem. Imaging **2**(1), 1–13 (2016). https://doi.org/10.1186/s40679-016-0032-z
21. Schüle, T., Schnörr, C., Weber, S., Hornegger, J.: Discrete tomography by convex-concave regularization and D.C. programming. Discrete Appl. Math. **151**, 229–243 (2005)
22. Weber, S., Nagy, A., Schüle, T., Schnörr, C., Kuba, A.: A benchmark evaluation of large-scale optimization approaches to binary tomography. In: Kuba, A., Nyúl, L.G., Palágyi, K. (eds.) DGCI 2006. LNCS, vol. 4245, pp. 146–156. Springer, Heidelberg (2006). https://doi.org/10.1007/11907350_13

Instance Segmentation with BoundaryNet

Teodor Boyadzhiev[1,2](✉) and Krassimira Ivanova[1]

[1] Institute of Mathematics and Informatics at the Bulgarian Academy of Sciences, Sofia, Bulgaria
{t.boadzhiev,kivanova}@math.bas.bg

[2] University of Library Studies and Information Technologies, Sofia, Bulgaria

Abstract. Instance segmentation is one of the key technology in many domains, such as medical image analysis, traffic and critical infrastructures monitoring, understanding of natural scenes. Recent methods for instance segmentation rely on bounding box regression, however the bounding boxes are not a natural representation for many domains.

We address the limitations of the bounding boxes with a new approach called BoundaryNet, in which we train a fully convolutional neural network to draw the boundaries around each object of each class. The boundaries allow for an easy bounding box and mask inference while still providing detailed information about the shape of the object. BoundaryNet avoids the restrictions of YOLO such as the number of bounding boxes, while it is more computationally efficient than the R-CNN methods.

The conducted experiments with the proposed neural network architecture BoundaryNet on the Common Object in Context (COCO) dataset show promising results in improving the instance segmentation process.

Keywords: Instance segmentation · Deep learning · BoundaryNet

1 Introduction

The Instance Segmentation is widely used in various fields such as medical image analysis, traffic monitoring, and remote sensing. The field of medicine has always been a primary source of image analysis tasks. Instance segmentation is extremely useful in histopathology for the detection of nuclei that can be used to diagnose dangerous diseases [9,18] or segmentation of organs or tumors in the organs from CT scans and MRI [1]. The combination between Semantic Segmentation and Instance Segmentation is often used in the recognition of complex street scenarios by self-driving cars [13] or by traffic management systems [24], as well as in the monitoring of critical infrastructures such as stations and airports [20]. The challenging tasks in the sphere of satellite and aerial imagery have also benefited the instance segmentation field. Such tasks include automated artificial object detection and building extraction from satellite images [21], evaluating building damage after a large-scale natural disaster from post-event aerial images [23], extracting geographical features (such as water bodies) from satellite maps using bounding boxes [3]. Of course, these areas do not

R. P. Barneva et al. (Eds.): IWCIA 2022, LNCS 13348, pp. 260–269, 2023.
https://doi.org/10.1007/978-3-031-23612-9_16

exhaust the field of application of instance segmentation - recently its use in more diverse tasks is increasing. A brief reference in Scopus on the keyword "instance segmentation" shows an exponential increase in the number of articles from 2016 so far.

In 2014 Girshick et al. proposed R-CNN for instance segmentation [5]. This approach uses a class agnostic region proposal method, based on a generic objectness score, to propose around 2000 regions for each image. Then from each region, a 4096-dimensional feature vector is extracted by a convolutional neural network, which was trained on the Image-Net challenge. Finally, each of these feature vectors is classified by a SVM.

This approach is later improved by Fast R-CNN for speed and accuracy by sharing the computations for feature extraction between the proposed regions. In the improved solution, feature maps are extracted by a convolutional neural network, then for each region proposal a feature vector is extracted, through a custom max-pool layer. For each of these feature vectors, a class and bounding box are predicted, using fully connected layers [4]. Further improvements in the same direction are made by Faster R-CNN by using a region proposal network and sharing the computations between this network and the feature extraction [17].

Based on these methods is proposed Mask R-CNN [8] which adds a third path to the Fast R-CNN to predict also the semantic mask for each bounding box. Gkioxari et al. [6] replace the mask branch in Mask R-CNN with a branch that predicts a 3D triangular mesh.

A different approach is used by Redmon et al. [14], which is called You Only Look Once (YOLO). YOLO splits the image in $S \times S$ grid and for each cell from the grid it predicts B bounding boxes and C class probabilities. For each bounding box is predicted also a confidence. The input image is processed by 24 convolutional layers, followed by 2 fully connected layers. The shape of the output tensor is $S \times S \times (B * 5 + C)$. This method predicts one category and its bounding box for each cell. YOLO is simpler and works much faster than the R-CNN pipeline, however, the performance is lower at 57.9% mAP. An improved version of YOLO is YOLO9000 which utilizes batch normalization, finer grid, relative to the cell centers bounding box regression, and is capable of detecting over 9000 object categories [15]. Further improvements were made in the third version [16].

Frequently bounding boxes are not a good approximation for the object boundaries in many domains. For instance, to overcome this problem Schmidt et al. [19] use star-convex polygons for the detection of cells in microscope images, while Loncomilla et al. [12] propose replacing bounding boxes with ellipses for detecting rocks. Other methods such as Mask R-CNN could overcome this problem by also predicting masks, however, this might become a problem for overlapping objects of the same category, due to crowding.

Here we propose a different approach to instance segmentation, called BoundaryNet, in which we train a fully convolutional neural network to draw the boundaries around each object of each class. This method is inspired by Yu et al. [22] who use a neural network to draw boundaries around each category in semantic segmentation, to improve the performance of their model. Instead of drawing a boundary around each class, we draw boundaries around each instance of a class.

Drawing the boundaries does not impose hard restrictions on the number of bounding boxes, such as YOLO, does not have a complex pipeline such as the R-CNN architectures, and has great flexibility with respect to the object shape. The boundaries allow for easy bounding box and mask inference while still providing detailed information about the shape of the object.

This paper is organized in 5 sections. Section 2 describes the problem representation, the network architecture, and the error function. Section 3 provides details about dataset size, image resolution, data augmentation, training algorithm parameters, network size, etc. Section 4 shows the results from the experiments and Sect. 5 contains discussion, conclusion, and directions for further research.

2 BoundaryNet

The problem of instance segmentation is addressed by BoundaryNet by predicting the semantic masks for each class as well as the boundaries of each object. The outputs of the network are two tensors, one for the semantic segmentation and one for the boundaries. The semantic tensor has the shape $H \times W \times (C+1)$, where H and W are the height and the width of the image, and C is the number of categories. One more channel is used for the background category, which is considered everything else. The boundary tensor predicts whether each pixel is a part of a boundary or not. It has the shape $H \times W \times 2$. In general, it could be replaced with $H \times W$ and *sigmoid* activation, since it is a binary classification.

2.1 Labelling

For the semantic segmentation the class of each pixel is determined by the category of the object it belongs to. This is a multi-category classification at a pixel level. Therefore, the semantic label is a matrix, $L_s \in \mathbb{L}^{h \times w}$, where $\mathbb{L}$ is the set of the category labels, w is the width and h is the height of the input image. During training each such matrix is converted into one-hot notation, making it a 3D tensor. If there are less than 256 categories, the semantic label can be represented as a gray-scale image.

For the boundary output, the class of each pixel is "background", unless it is on the inside of an edge of an object, in which case it is assigned the label "boundary". The boundary label a matrix, $L_b \in \{0, 1\}^{h \times w}$, where w is the width and h is the height of the input image. The label can be represented as an image containing the edges between the instances of interest and the background, Fig. 1.

Fig. 1. The boundary label is a matrix, where each cell is either 0 or 1 depending on whether the pixel is a part of a boundary of an object of interest.

2.2 Segments Extraction

Once the boundaries and the semantic information is extracted from the network, each object of each class needs to be determined. The method consists of several steps:

1. The boundaries are used to extract several segments of connected background. Each segment is numbered with a different integer, creating a segment mask s.
2. For each category the semantic mask is extracted, c_k, by setting the pixels classified as this category to 1 and the rest to 0.
3. Each semantic mask is multiplied by the segments mask, element-wise

$$s_k = s \odot c_k \tag{1}$$

   where s_k is the segments, belonging to category k.
4. Finally for each category, k, the segments have to be grouped into objects. *This step is outside the scope of this paper and remains for further development.*

2.3 Network Architecture

The architecture of BoundaryNet is based on the architecture of UNet [18], (Fig. 2). It has one encoder and two independent decoders. Each skip connection from the encoders is connected to the corresponding level of both decoders. At each level the network has two convolutions with 3×3 kernel. Each convolution uses batch normalization and has *ReLU* activation. At each level of the encoder a 2×2 MaxPool operation is used. In the decoders UpSize operation with linear resampling is used. After the last convolution of each decoder, a 1×1 convolution is used with *softmax* activation as a classifier for each pixel.

The error for the network is a weighted sum of cross-entropy error for the semantic decoder and focal loss [10] for the boundary decoder

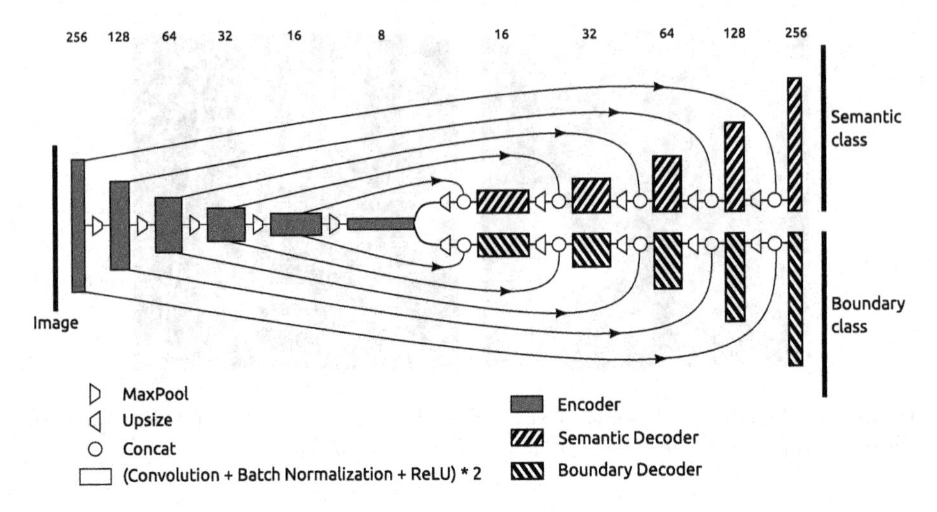

Fig. 2. Framework of BoundaryNet.

$$\mathcal{E} = \alpha\text{Focal} + (1-\alpha)\text{CE} \quad \alpha \in [0,1] \tag{2}$$

where

$$\text{Focal} = \sum_{k=1}^{2}(1-pb_k)^{\gamma} b_k \log(pb_k) \tag{3}$$

and b_k is the one-hot notation for the boundary labeling, pb_k is the probability of category k in the boundary decoder, and γ is a parameter. Since for the boundary a binary classification is used, $k \in 1,2$. This part can be substituted by a *sigmoid* and binary focal loss.

The filters in each level, $f(l)$, of the network are determined by

$$f(l) = \lfloor f2^{\frac{l}{d}} \rfloor \tag{4}$$

where f is the number of frames in level 0, which is before the first MaxPool layer, l is the level number, and d is a divider. For example, if the network has 5 MaxPool layers, it will have 6 levels, numbered from 0 to 5.

3 Methods

The network was trained on the COCO dataset for people only [11]. The images were scaled to a resolution of 256×256. Objects which are composed of less than 256 pixels in total were removed, using image in-painting. In total 38027 images were extracted. The boundary labels were created by using an edge detection algorithm for each of the segments of objects of interest separately and then interpolated on top of each other. Then a dilation of 1 pixel in each direction was used to make them thicker and avoid small discontinuities.

The network was initialized using the Xavier method [7], where the weights were drawn from normal distribution and it was trained with the ADAM algorithm with learning rate 10^{-3}, $\beta_1 = 0.9$, $\beta_2 = 0.999$, and $\epsilon = 10^{-5}$ for 100 iterations. The training algorithm used batch size of 64 images. During the training no L-regularization or dropout was used.

The training and testing set was split randomly, using 80% of the data for training and the rest for testing. The training set was augmented using random zoom, horizontal flip, vertical flip, and rotation. The parameters for these operations are summarized in Table 1.

Table 1. Probability of augmentation and the parameters.

Operation	Probability	Parameters
Zoom	0.2	$\sim U(1, 1.5)$
Rotation	0.2	$\sim U\left(-\frac{\pi}{2}, \frac{\pi}{2}\right)$
Horizontal flip	0.2	
Vertical flip	0.2	

The parameter of the weighted sum of the error function is $\alpha = 0.1$. The parameter for the focal loss is $\gamma = 2$. The filters in the topmost layer are $f = 24$ and the divider is $d = 1.25$. The network has 5 MaxPool layers, meaning that in the lowest level it uses a feature map of 8×8 with 384 channels.

For the initialization and training algorithm were used the implementations provided by Wolfram Mathematica 13.0.

4 Results

Figure 3 shows the training and testing intersection over union (IOU) for the network trained for 100 iterations. The result shows that there is no over-fitting for the first 100 iterations. The semantic IOU reached 67.7% on the testing dataset and 71.6% on the training dataset. The boundary IOU reached 46.5% on the testing dataset and 46.4% on the training dataset. The network could reach better results if it is trained longer time since it has not fully converged.

Using IOU to measure the quality of the boundaries is not very appropriate, since it is too sensitive. For example, if the boundary generated by the network is half the thickness, the IOU would be 0.5, however, if it is unbroken and in the correct place, it can still be used to identify the object correctly.

Figure 4a shows examples from the validation set, where the network managed to find semantic mask and object boundaries with quality sufficient to distinguish between objects. Figure 4b shows examples from the validation set, where the network has made mistakes with the boundaries and the semantic segmentation.

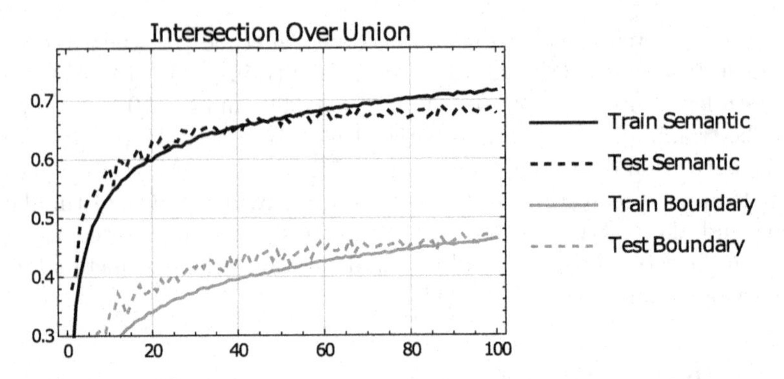

Fig. 3. Training and testing intersection over union (IOU) for the boundaries and the semantic segmentation.

Most of the mistakes for the boundaries are false negative, which will cause incorrect segment merging. For instance, in the first example of the mistakes (the left hand side of Fig. 4b), in the top right corner the contour of the palm of the person is broken, which can cause his hand to be identified as part of the person behind him.

In the second image (the center of Fig 4b), the boundary line of the child's elbow is broken. This will cause the child and the body of the man behind to be identified as the same instance, while the head of the man will become a separate instance.

The haze in the third image (the right hand side of Fig. 4b), causes the boundary between the woman and the child to be broken and they will be merged into the same instance later. In this case the algorithm will identify two people instead of three.

5 Conclusion

In this paper we proposed a new approach for instance segmentation, based on drawing boundaries around each object of interest. We used a deep fully convolutional neural network, named BoundaryNet. The network, which is based on the UNet architecture, has one encoder and two parallel decoders. One of the decoders produces semantic masks and the other produces boundaries around each object of interest.

We demonstrated successful boundary identification for people, however, it is possible to have improvements in the quality of the boundaries. Further developments can include exploring different network architectures, such as channel attention blocks [22] or atrous pooling [2]. Other work in the same area is handling cases when the same object is covered by another object and therefore splitting it into two segments.

The results which we demonstrated here are inferior to other approaches such as YOLO and R-CNN, however, using transfer learning and exploring

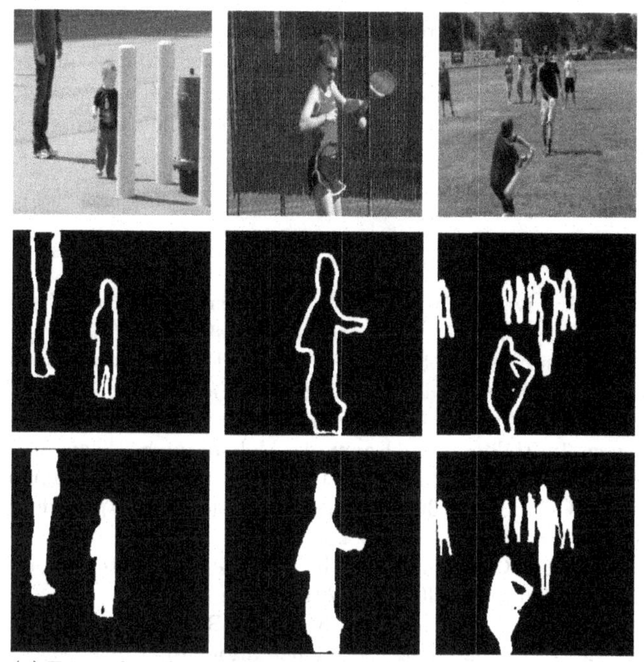

(a) Examples of successful segmentation and finding of object boundaries.

(b) Examples of issues with finding the boundaries between objects.

Fig. 4. Examples for drawing boundaries.

other architectures could improve the quality of the results. Boundaries with improved quality can provide an alternative approach to instance segmentation with greater flexibility concerning the object shape. Such approach can be useful to domains, dealing with objects which are hard to be described by bounding boxes.

References

1. Altini, N., et al.: Liver, kidney and spleen segmentation from CT scans and MRI with deep learning: a survey. Neurocomputing **490**, 30–53 (2022)
2. Chen, L.C., Papandreou, G., Schroff, F., Adam, H.: Rethinking atrous convolution for semantic image segmentation. arXiv preprint arXiv:1706.05587 (2017)
3. Dhyakesh, S., et al.: Mask R-CNN for instance segmentation of water bodies from satellite image. In: Haldorai, A., Ramu, A., Mohanram, S., Chen, M.-Y. (eds.) 2nd EAI International Conference on Big Data Innovation for Sustainable Cognitive Computing. EICC, pp. 301–307. Springer, Cham (2021). https://doi.org/10.1007/978-3-030-47560-4_24
4. Girshick, R.: Fast R-CNN. In: Proceedings of the IEEE International Conference on Computer Vision, pp. 1440–1448 (2015)
5. Girshick, R., Donahue, J., Darrell, T., Malik, J.: Rich feature hierarchies for accurate object detection and semantic segmentation. In: Proceedings of the IEEE Conference on Computer Vision and Pattern Recognition, pp. 580–587 (2014)
6. Gkioxari, G., Malik, J., Johnson, J.: Mesh R-CNN. In: Proceedings of the IEEE/CVF International Conference on Computer Vision, pp. 9785–9795 (2019)
7. Glorot, X., Bengio, Y.: Understanding the difficulty of training deep feedforward neural networks. In: Proceedings of the Thirteenth International Conference on Artificial Intelligence and Statistics, pp. 249–256. JMLR Workshop and Conference Proceedings (2010)
8. He, K., Gkioxari, G., Dollár, P., Girshick, R.: Mask R-CNN. In: Proceedings of the IEEE International Conference on Computer Vision, pp. 2961–2969 (2017)
9. Hou, L., et al.: Sparse autoencoder for unsupervised nucleus detection and representation in histopathology images. Pattern Recogn. **86**, 188–200 (2019)
10. Lin, T.Y., Goyal, P., Girshick, R., He, K., Dollár, P.: Focal loss for dense object detection. In: Proceedings of the IEEE International Conference on Computer Vision, pp. 2980–2988 (2017)
11. Lin, T.-Y., et al.: Microsoft COCO: common objects in context. In: Fleet, D., Pajdla, T., Schiele, B., Tuytelaars, T. (eds.) ECCV 2014. LNCS, vol. 8693, pp. 740–755. Springer, Cham (2014). https://doi.org/10.1007/978-3-319-10602-1_48
12. Loncomilla, P., Samtani, P., Ruiz-del Solar, J.: Detecting rocks in challenging mining environments using convolutional neural networks and ellipses as an alternative to bounding boxes. Expert Syst. Appl. **194**, 116537 (2022)
13. Ojha, A., Sahu, S.P., Dewangan, D.K.: Vehicle detection through instance segmentation using mask R-CNN for intelligent vehicle system. In: 2021 5th International Conference on Intelligent Computing and Control Systems (ICICCS), pp. 954–959. IEEE (2021)
14. Redmon, J., Divvala, S., Girshick, R., Farhadi, A.: You only look once: unified, real-time object detection. In: Proceedings of the IEEE Conference on Computer Vision and Pattern Recognition, pp. 779–788 (2016)

15. Redmon, J., Farhadi, A.: YOLO9000: better, faster, stronger. In: Proceedings of the IEEE Conference on Computer Vision and Pattern Recognition, pp. 7263–7271 (2017)
16. Redmon, J., Farhadi, A.: YOLOv3: an incremental improvement. arXiv preprint arXiv:1804.02767 (2018)
17. Ren, S., He, K., Girshick, R., Sun, J.: Faster R-CNN: towards real-time object detection with region proposal networks. Adv. Neural Inf. Process. Syst. **28** (2015)
18. Ronneberger, O., Fischer, P., Brox, T.: U-net: convolutional networks for biomedical image segmentation. In: Navab, N., Hornegger, J., Wells, W.M., Frangi, A.F. (eds.) MICCAI 2015. LNCS, vol. 9351, pp. 234–241. Springer, Cham (2015). https://doi.org/10.1007/978-3-319-24574-4_28
19. Schmidt, U., Weigert, M., Broaddus, C., Myers, G.: Cell detection with star-convex polygons. In: Frangi, A.F., Schnabel, J.A., Davatzikos, C., Alberola-López, C., Fichtinger, G. (eds.) MICCAI 2018. LNCS, vol. 11071, pp. 265–273. Springer, Cham (2018). https://doi.org/10.1007/978-3-030-00934-2_30
20. Tseng, C.H., Hsieh, C.C., Jwo, D.J., Wu, J.H., Sheu, R.K., Chen, L.C.: Person retrieval in video surveillance using deep learning-based instance segmentation. J. Sens. **2021**, 12, 9566628 (2021). https://doi.org/10.1155/2021/9566628
21. Vakalopoulou, M., Karantzalos, K., Komodakis, N., Paragios, N.: Building detection in very high resolution multispectral data with deep learning features. In: 2015 IEEE International Geoscience and Remote Sensing Symposium (IGARSS), pp. 1873–1876. IEEE (2015)
22. Yu, C., Wang, J., Peng, C., Gao, C., Yu, G., Sang, N.: Learning a discriminative feature network for semantic segmentation. In: Proceedings of the IEEE Conference on Computer Vision and Pattern Recognition, pp. 1857–1866 (2018)
23. Zhan, Y., Liu, W., Maruyama, Y.: Damaged building extraction using modified mask R-CNN model using post-event aerial images of the 2016 kumamoto earthquake. Remote Sens. **14**(4), 1002 (2022)
24. Zhang, X.: A method to estimate position relationship between pedestrian and crosswalk based on YOLCAT++. In: 2021 2nd International Seminar on Artificial Intelligence, Networking and Information Technology (AINIT), pp. 38–42. IEEE (2021)

Curvature-Based Denoising of Vector-Valued Images

Christian Gapp(✉) and Martin Welk

UMIT TIROL – Private University for Health Sciences,
Medical Informatics and Technology, Eduard-Wallnöfer-Zentrum 1,
6060 Hall in Tirol, Austria
{christian.gapp,martin.welk}@umit-tirol.at

Abstract. Salient visual information in images is often concentrated on contours or on regions where edges or curves change their direction abruptly. It is therefore of utmost importance in the processing of images to preserve this kind of information. Recently, a curvature-based denoising method has been proposed which first transforms an image into a level-line tree, then smoothes the level lines, and finally reassembles the image from those. Curvature information generated in this approach has also potential for further applications in image analysis.

Focusing on denoising, we transfer curvature-based smoothing to vector-valued images. We replace level lines by pseudo-level lines (integral curves of the vector field of directions of least vectorial contrast) and design a robust algorithm for their extraction from a vector-valued image. In this context we also propose a modification of the level line extraction from grey-scale images for better rotational invariance. Since intensities along pseudo-level lines are not constant, our method stores this information along the pseudo-level lines, and performs an appropriate smoothing on intensities. Finally we adapt the reconstruction process.

We present experiments on grey-scale and colour images to validate our proposed modification of the original grey-scale method as well as our new vector-valued curvature-based denoising method.

Keywords: Denoising · Curvature · Affine morphological scale space · Pseudo-level lines

1 Introduction

Due to the ubiquity of noise of various sources across image formation processes, denoising continues to be a fundamental task of image processing. The purpose of denoising is to remove noise while at the same time preserving as much as possible the image features needed for further processing of images by humans or computers. Telling apart noise from the relevant features is challenging, such that denoising methods inevitably interfere with image features along with removing noise. Together with the great variability in both noise sources and features that need to be preserved depending on application context, this constitutes a major

R. P. Barneva et al. (Eds.): IWCIA 2022, LNCS 13348, pp. 270–287, 2023.
https://doi.org/10.1007/978-3-031-23612-9_17

Fig. 1. Morphological Cat, constructed by taking 38 points of maximum curvature from the level lines, that represent the contours, and then connecting them with straight lines. From [4].

reason why even after decades of research there is no universal denoising method that suits all kinds of applications. For example, methods that directly smooth the source image by minimising an energy functional struggle with preserving the contrast and sharpness of contours.

A long-standing observation [4] is that salient information in images, especially for human observers, is concentrated along contours as well as feature points like angles or curvature extrema, compare the example in Fig. 1. This has inspired researchers to design denoising methods that specifically focus on this kind of features.

In [5] it is suggested to denoise images by extracting first the curvature image, similar to Fig. 1, then denoising the curvature image, and obtaining the final denoised image by reconstruction from the modified curvature image. It turns out that curvature images are less affected by additive noise n, leading to a better separation of salient image information and noise in the process.

Despite promising results, the method from [5] has so far only been studied for grey-scale images. Our aim in this paper is to extend the approach to vector-valued, such as (RGB) colour images. To this end, several obstacles need to be overcome.

First, the concept of level sets as such is suitable for grey-scale images only, and needs to be replaced by a suitable generalisation in the case of vector-valued images. To this end, we resort to the concept from [6] in which lines of minimal colour/vector contrast are proposed as level lines; for clarity, we will denote these as *pseudo-level lines*. Adapting the level-line extraction procedure from [19] to pseudo-level lines is the first component of our proposed method.

Second, as image intensities along pseudo-level lines are not constant as they are along level lines in grey-scale images, richer intensity information must accompany an extracted pseudo-level line. In the smoothing step, it is therefore necessary to not just smooth the pseudo-level line curves (which can be done essentially by the same affine morphological scale space as for grey-scale images) but also to take care of the intensity information. This is the second component of our proposed method.

Third, we need to provide a way to (approximately) reconstruct the original image from pseudo-level line information, which again is more complex than in the case of grey-scale images.

Our Contributions. Our main contribution is the extension of the curvature-based smoothing algorithm from grey-scale to vector-valued images, which relies on the concept of pseudo-level lines as a replacement for level lines. We spell out the necessary adaptations of, and additions to the algorithm step by step. Moreover, we introduce a modified choice for pixel neighbourhoods for the sake of reducing directional bias, which can also be used beneficially in the base algorithm for grey-scale images.

Structure of the Paper. In Sect. 2 we recall the curvature-based denoising method for grey-scale images from literature, and introduce our modification of pixel neighbourhoods. Our extension of the method to vector-valued images is developed in Sect. 3. Section 4 is devoted to the experimental demonstration of the techniques. A summary and outlook in Sect. 5 conclude the paper.

2 Curvature-Based Denoising of Scalar-Valued Images

Let us recall first the curvature-based denoising method for grey-scale images. We largely follow [7] but introduce a small modification of the pixel neighbourhoods that helps to avoid directional bias.

2.1 Level Line Tree

Level lines in a (space-continuous) grey-scale image are lines of constant intensity. They are closed curves (where curves ending in the image boundary can be closed by suitable boundary segments), and different level lines cannot cross each other; of two level lines, either one encircles the other, or both lie apart. Any finite set of level lines is therefore naturally organised in a Level Line Tree (LLTree), with the image boundary as its root. This intuition also carries over to discrete images, with the only caveat that segments of different discrete level lines can coincide, but still a strict tree-order is established by inclusion and exclusion.

A level line is a closed list of edge elements (edgels). An edge element (edgel) is given by a pair of neighbouring pixels, with the understanding that the space-continuous curve represented by the level line passes between these pixels. One of the pixels making up an edgel, the immediate interior pixel (IIP), is inside the private region (pregion) of the level line. The other one represents the immediate exterior pixel (IEP) outside this region. The sequences of immediate interior pixels (IIPs) and immediate exterior pixels (IEPs) in the list of edgels progress from pixels to neighbouring pixels; repetitions (i.e. subsequent edgels sharing their IIPs or IEPs) are allowed. Closedness of the level line means that the first and last entry of the list of edgels are identical. The pregion is represented by a list of pixels. As for grey-scale images a private value (pvalue) can easily be defined as either being the maximum or minimum intensity of all IIPs along the

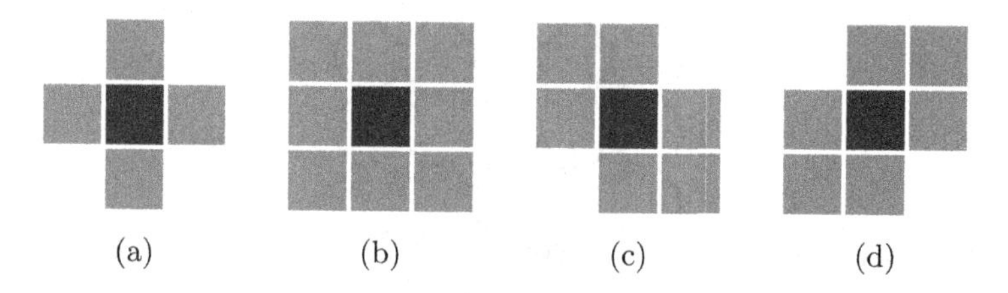

Fig. 2. Neighbourhood types as proposed in [19]; dark centre pixels are shown surrounded by neighbours in grey. (a) 4-neighbourhood, (b) 8-neighbourhood, (c) 6-neighbourhood of type 1, (d) 6-neighbourhood of type 2.

level line, the pregion contains all pixels inside the region with intensity equal to the pvalue. Considering the connectedness graph of an image, i.e. the graph whose vertices are the pixels, and edges connect exactly those pixels which are neighbours (thus, edges represent edgels), one sees that a level line represents a cut of this graph. The pregion of the level line corresponds to one of the connected components into which the connectedness graph is split by the cut.

Defining a curve as a list of pixels, a level line is associated with two curves: the curve-immediate-interior-pixel (curveIIP) that contains all IIPs, and the curve-immediate-exterior-pixel (curveIEP) of all IEPs.

To establish the LLTree, each line is stored in a node together with its pvalue and pregion, and an unordered list of references to their children (child nodes). The LLTree contains all these nodes from the uppermost parent node down to the bottom child node. Here the first node, the uppermost parent, always has a level line that expresses the border of an image. A child of the parent describes a region inside the parent's region. Two or more children are called siblings. They are in a common list of references and on the same level in the LLTree.

Note that a parent and a child node can have common IIPs and IEPs. Two siblings can only have common IEPs.

Neighbourhoods. The set of edgels available in the extraction process depends on a choice of neighbourhoods, see Fig. 2. Admissible edgels are always the pairs (p, q_i) of a centre pixel p as IIP and one of its n neighbours q_i ($i = 0, \dots, n-1$ where n is 4, 8 or 6) as IEP. Each of these choices, however, comes with a downside. Using 8-neighbourhoods, Fig. 2b, on the pixel set of the given image implies a non-planar connectedness graph due to the intersection of diagonal edgels. However, representing level lines, thus cuts of the connectedness graph, by *sequences* of edgels, actually relies on the assumption that the connectedness graph is planar.

For the further discussion we remark that the level line extraction process will be designed in a way that it proceeds from edgel to edgel via the meshes of the connectedness graph. Whenever a level line enters one mesh of the connectedness graph via an edgel, one has to determine by which edgel it leaves the mesh. This will be particularly easy if the meshes of the connectedness graph are triangles: In this case, the exit edge is always adjacent to the entrance edge of each mesh.

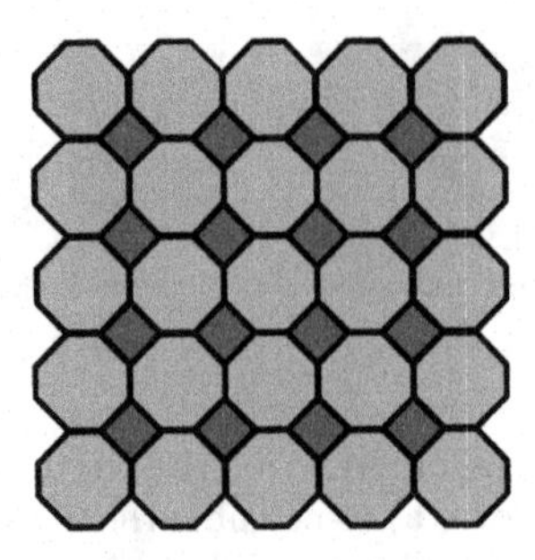

Fig. 3. Image region Ω containing 5×5 original pixels (grey), extended by dummy pixels (orange). Note that original pixels have 8-connectedness, whereas dummy pixels have 4-connectedness. (Color figure online)

It only takes to choose between these to adjacent edges by keeping fixed either the IIP or the IEP. Meshes circumscribed by more than three edges need more complicated case distinctions.

Returning to neighbourhood choices, the 4-neighbourhood, Fig. 2a, leads to a connectedness graph with quadrilateral meshes, which is therefore unfavourable for the level line extraction process. For this reason, the 6-neighbourhoods from Fig. 2c and Fig. 2d have been proposed; they yield planar graphs with triangular cells that are a perfect fit for the extraction algorithm. Unfortunately, this comes at the cost of sacrificing symmetry by preferring one diagonal direction over the other. Thereby they introduce a directional bias which is generally unfavourable in image processing; indeed, it leads to visible artifacts in the smoothed images, cf. Fig. 12 in Sect. 4 where diagonal streaks in the direction of the preferred diagonals are clearly visible.

We therefore favour an alternative approach. We insert dummy pixels located at the common corners of four adjacent pixels of the original image grid, see Fig. 3. By bilinear interpolation, each dummy pixel is assigned the average of the intensities of the four surrounding original pixels as its intensity value. The neighbourhood relation within this extended set of pixels, and thereby the connectedness graph, is defined as follows. Each original pixel has eight neighbours: the four original pixels which are located next to it in vertical and horizontal direction, and the four dummy pixels next to it. In contrast, each dummy pixel has only the four original pixels next to it as neighbours. In Fig. 3 this is visualised by showing original pixels as octagons but dummy pixels as squares. Pixels are considered neighbours if and only if they have a common border in this representation. With this definition, the connectedness graph is planar and consists entirely of triangular meshes. Each mesh is made up by two original pixels and one dummy pixel. Thus, the graph meets the needs of the extraction algorithm, while retaining all symmetries of the regular pixel grid.

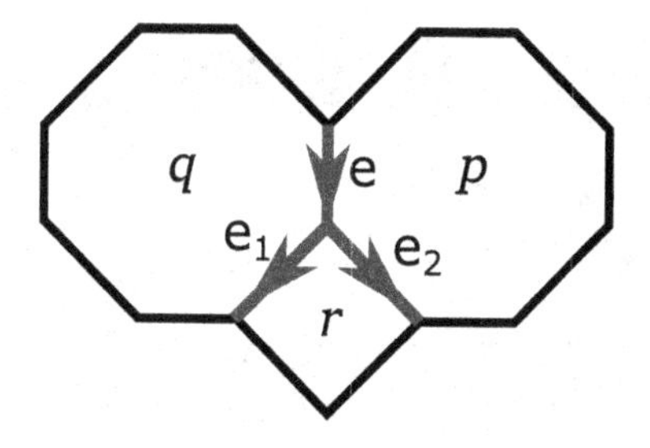

Fig. 4. Two choices for the next edgel: e_1 or e_2. Edgel $e = (p, q)$ is already in the boundary. In case r is inside the region, the next edgel is $e_1 = (r, q)$. Otherwise r becomes IEP and thus the next edgel is $e_2 = (p, r)$.

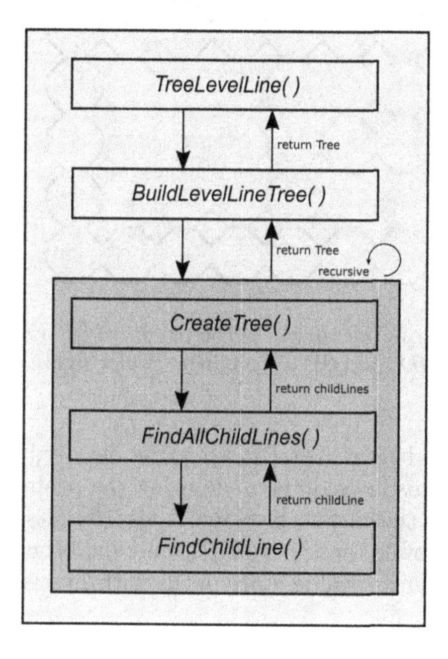

Fig. 5. Routine *TreeLevelLine* for scalar valued images.

Level Line Extraction. A new level line always begins with a start edgel α. Let the actual edgel be $e = (p, q)$. Then the *NextPixel*(p, q) operation returns the pixel r that either is the next IIP or IEP, see Fig. 4.

As soon as the start edgel α is reached again, the level line is closed, and further pixels are investigated. These are either pushed to the pregion or represent IIPs of a child's start edgel, meaning a child line is passing through.

The overall algorithm to build the LLTree is visualised in Fig. 5. In the recursive part (*CreateTree*), all children are successively added to the parent currently processed within the subroutine *FindAllChildLines*.

Within *FindChildLine*, edgels are added successively until the start edgel is reached again. The edgel (p, q) is either followed by (r, q) or (p, r), depending

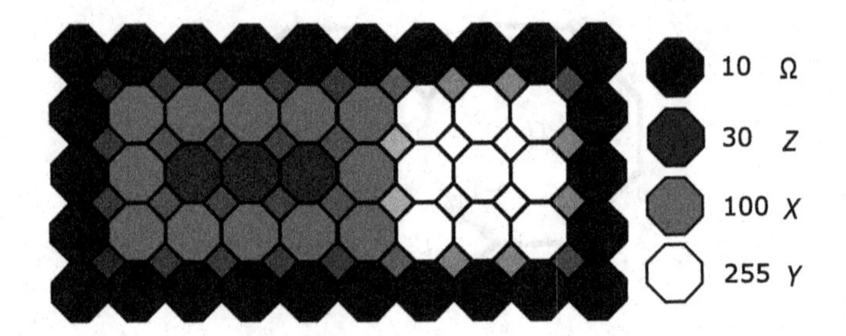

Fig. 6. Example grey-scale image (10 × 5) with dummy pixels inserted. Ω, X, Y, Z are the four regions with identical intensities each.

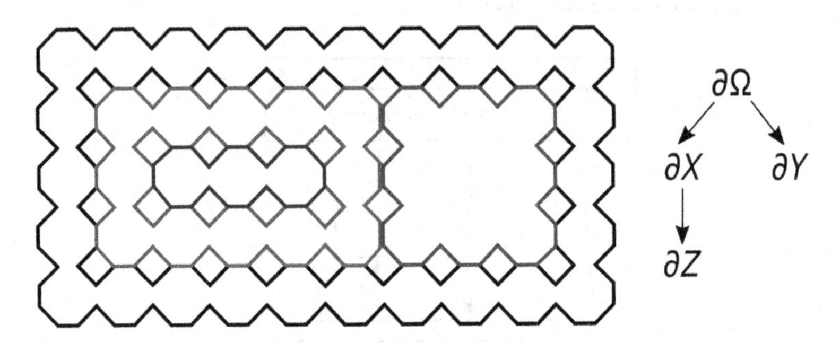

Fig. 7. Level lines in Fig. 6. $\partial\Omega$ (black) is the parent of ∂X (green) and ∂Y (red). ∂Z (blue) is a child of ∂X. ∂X and ∂Y are siblings. (Color figure online)

on whether r (returned from *NextPixel*(p, q)) is declared to be an IIP or IEP. In grey-scale images this decision is made using the pvalue as threshold. Let us assume the pvalue v of the current level line is less than its parent's pvalue. Then pixels with intensity lower than v are IIPs, and the others IEPs. In the case v is higher than the parent's pvalue, only pixels with intensity greater than v are IIPs (Fig. 6 and 7).

2.2 Level Line Shortening

Discrete Curvature of Level Lines. Let $\boldsymbol{\Gamma} = \{\boldsymbol{x}(s) : s \in [0, L], \|\boldsymbol{x}'(s)\| = 1\}$ be a sufficiently smooth (C^2) curve in arc-length parametrisation. The second derivative $\boldsymbol{x}''(s)$ then always points in normal direction, i.e. [12]

$$\boldsymbol{x}''(s) = \kappa(s)\,\boldsymbol{n}(s) \tag{1}$$

with some function $\kappa(s)$ which is called *curvature* of $\boldsymbol{\Gamma}$, and $\boldsymbol{n}(s)$ denoting the unit normal vector $\boldsymbol{n}(s) \perp \boldsymbol{x}'(s)$. Assuming that the moving frame $(\boldsymbol{x}'(s), \boldsymbol{n}(s))$ is positively oriented, $\kappa(s) > 0$ indicates that the curve is locally bent in mathematically positive sense whereas $\kappa(s) < 0$ indicates it turns in the mathematically negative sense. The definition of κ via arc-length parametrisation is transferred to curves in arbitrary parametrisation by reparametrisation, making $\kappa(s) \equiv \kappa(\boldsymbol{x})$

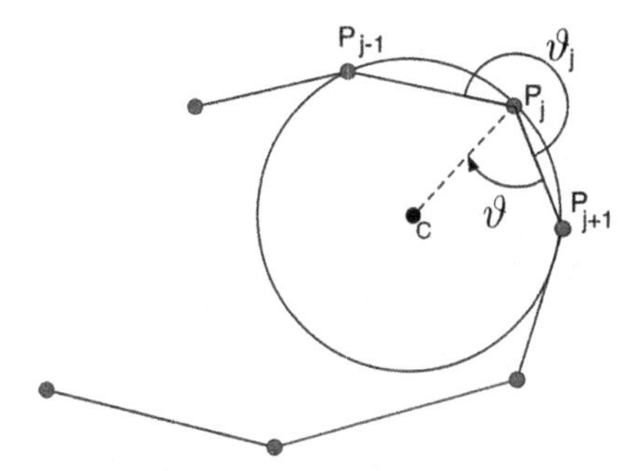

Fig. 8. Definitions and discretisation of the level line to display the computation of the discrete curvature at the vertex $\overline{P_{j-1}P_jP_{j+1}}$, from [7]. The curvature is $\kappa(P_j) = 1/r_j$, with $r_j = \|P_jC\|$.

in fact dependent only on the shape but not the parametrisation of the curve. In fact, for $\kappa(\boldsymbol{x}) \neq 0$, $1/|\kappa(\boldsymbol{x})|$ is the radius of a best-fit (osculating) circle to $\boldsymbol{\Gamma}$ at $\boldsymbol{x}$.

For any non-singular point $\boldsymbol{x}$ of a sufficiently smooth (C^2) image $u : \mathbb{R}^2 \to \mathbb{R}$, the curvature $\kappa(\boldsymbol{x})$ of the level line of u passing through $\boldsymbol{x}$ can be computed as [7,17]

$$\kappa(\boldsymbol{x}) = \frac{u_{xx}u_y^2 - 2u_{xy}u_xu_y + u_{yy}u_x^2}{(u_x^2 + u_y^2)^{3/2}}(\boldsymbol{x}). \tag{2}$$

Regarding the sign of κ, level lines are understood here to be oriented such that the normalised local image gradient vector $\nabla u/\|\nabla u\|$ points to the right of the level line, i.e. $\nabla u/\|\nabla u\| = -\boldsymbol{n}$ locally.

As pointed out by Mondelli and Ciomaga [16], direct implementation of (2) by finite difference scheme (FDS) models as done in [2] and [9] suffers from numerous artifacts. Therefore it is preferable to compute instead the curvature directly on the level lines.

To this end, let $\boldsymbol{\Gamma}$ be a closed discrete curve denoted as $\boldsymbol{\Gamma} = \{P_j(x_j, y_j)\}$, with $j \in \{0, \ldots, N\}$ and $P_0 = P_N$. We assume that $\boldsymbol{\Gamma}$ approximates a (space-continuous) level line; in fact, we will use for $\boldsymbol{\Gamma}$ the curveIIP of a discrete level line. Building on the relation between curvature and osculating circles, the curvature $\kappa(P_j)$ of the discrete curve at P_j can be defined as

$$\kappa(P_j) := \pm 1/r_j, \tag{3}$$

with the radius $r_j = \|P_jC\|$ computed with the three points P_{j-1}, P_j, P_{j+1} (see also Fig. 8) and the sign consistent with (1), compare Fig. 8.

As noted in [7], the discrete curvature at point P_j is

$$\kappa(P_j) = \frac{-2\,\sin(\vartheta_j)}{\|P_{j-1}P_{j+1}\|} = \frac{-2\,\det\left(P_jP_{j-1} \quad P_jP_{j+1}\right)}{\|P_{j-1}P_j\|\,\|P_jP_{j+1}\|\,\|P_{j-1}P_{j+1}\|}, \tag{4}$$

with

$$\det\left(P_jP_{j-1} \quad P_jP_{j+1}\right) := \det\begin{pmatrix} x_{j-1} - x_j & x_{j+1} - x_j \\ y_{j-1} - y_j & y_{j+1} - y_j \end{pmatrix}. \tag{5}$$

AMSS – Affine Morphological Scale Space. Affine morphological scale space is a curvature-driven process that preserves invariance properties such as monotonicity, morphology and affine invariance [1,15].

The affine scale space can be interpreted as an intrinsic heat equation [13]. Let $\sigma \mapsto \boldsymbol{x}(t,\sigma)$ be a Jordan arc (or curve) for each scale t. Then, in any neighbourhood without an inflection point, the affine scale space

$$\frac{\partial \boldsymbol{x}}{\partial t} = \kappa(\boldsymbol{x})^{1/3}\,\boldsymbol{n}(\boldsymbol{x}) \tag{6}$$

is equivalent to the intrinsic heat equation $\partial \boldsymbol{x}/\partial t = \partial^2 \boldsymbol{x}/\partial \sigma^2$ with parametrisation σ (affine length) [13,18].

To implement (6) the geometric scheme proposed by Moisan [15] is used. The latter equation can be interpreted as an alternating filter, switching between affine erosion and dilation in dependence of the scale space parameter σ. This can be realised working with affine erosion on the individual convex and concave parts of the discrete curve $\boldsymbol{\Gamma}$. Therefore first the inflection points $E_i = P_{j(i)}$ must be detected. After the resampling process, where points are added/removed to get good smoothing results, each convex component

$$C_i = (E_i = P_{j(i)}, P_{j(i)+1}, P_{j(i)+2}, \ldots, P_{j(i+1)} = E_{i+1}) \tag{7}$$

is processed by affine erosion resulting in an envelope of σ-chords

$$C_i^\sigma = (E_i, P^\sigma_{j(i)}, P^\sigma_{j(i)+1}, P^\sigma_{j(i)+2}, \ldots, P^\sigma_{j(i+1)}, E_{i+1}), \tag{8}$$

a set of middle points of σ-chords with unchanged inflection points [7]. This is important here because the convex and concave parts are glued together after each iteration. A σ-chord C_i^σ is defined as a segment connecting two points of the (discrete) curve that cuts off a (polygonal) area of size σ between itself and the curve (see Fig. 9).

The larger σ, the less the accuracy of the geometric scheme. Hence the affine shortening process is iterated with a small σ as often as needed to achieve the desired smoothness. The smoothing process can be described with Algorithm 1.

Algorithm 1. *Smooth Curves*

```
1: for all curves c ∈ LLTree do
2:     while (desired scale t not reached) do
3:         split c into convex and concave parts
4:         resample c
5:         affineErosion(c)
6:     resample c
```

For all curves ∈ LLTree the process of splitting, resampling and affine erosion is iterated until the desired scale t is reached (lines 2–5). As a last step the new curve (σ-chord) is resampled again.

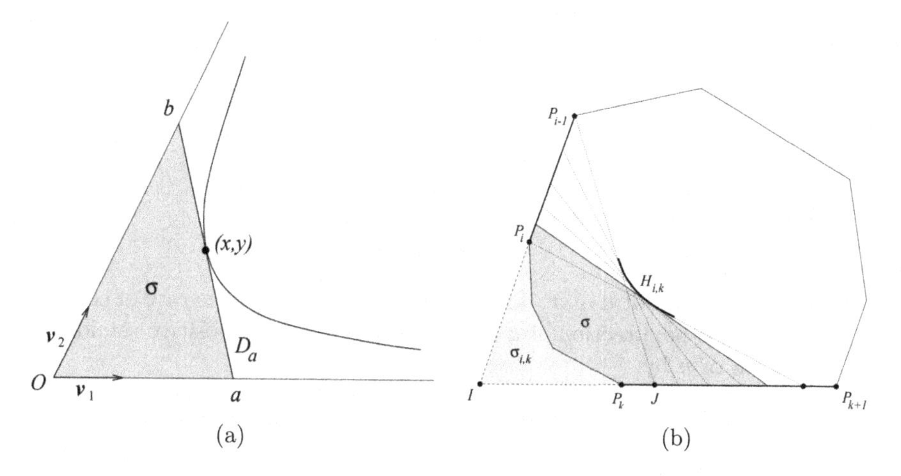

Fig. 9. Affine erosion of a corner, from [15]. (a) The affine erosion of a corner results in a hyperbola. σ displays the area cut after several iterations. (b) Evolution of a hyperbola ($H_{i,k}$) resulting from two edges. $\sigma_{i,k}$ is the area cut after a couple of iterations, whereas σ (note that σ includes $\sigma_{i,k}$) analogously to (a) displays the area trimmed after many iterations.

2.3 Reconstruction

For scalar-valued images the regions can be filled straightforward by iterating the LLTree from top to bottom and printing the inside of the regions with the level lines' pvalue. Herein the children's pregion overprint their parents'. As siblings do not affect each other, it is irrelevant which region is filled first on the same level of the LLTree.

3 Curvature-Based Denoising of Vector-Valued Images

This section is devoted to adapting the denoising method recalled in the previous section to vector-valued images.

3.1 Pseudo-level Lines

Unlike grey-scale images, vector-valued images are not filled by curves of constant intensity, i.e. level lines. As a surrogate for these, it is proposed in [6] to consider the integral curves of the directions of minimal vectorial change, also denoted as level lines there. For a clear distinction, we will use the term *pseudo-level lines* in the following. The first step towards computing pseudo-level lines is the computation of gradients of vector-valued images; following [6] this can be done using standard tools from Riemannian geometry [14].

For a (space-continuous) vector-valued image $\boldsymbol{u}(\boldsymbol{x}) : \mathbb{R}^2 \to \mathbb{R}^n$, at each location $\boldsymbol{x} = (x, y)^\mathsf{T}$ the directional derivative of the vector-valued image $\boldsymbol{u}$ at $\boldsymbol{x}$

in the direction of $\boldsymbol{v}$ then is the vector

$$\partial_{\boldsymbol{v}} \boldsymbol{u}(\boldsymbol{x}) = D\boldsymbol{u}(\boldsymbol{x})\boldsymbol{v} \tag{9}$$

where

$$D\boldsymbol{u}(\boldsymbol{x}) = \begin{pmatrix} \partial_x u_1(\boldsymbol{x}) & \partial_y u_1(\boldsymbol{x}) \\ \vdots & \vdots \\ \partial_x u_n(\boldsymbol{x}) & \partial_y u_n(\boldsymbol{x}) \end{pmatrix} \tag{10}$$

is the Jacobian matrix of $\boldsymbol{u}$ at $\boldsymbol{x}$. The norm $\|\partial_{\boldsymbol{v}} \boldsymbol{u}(\boldsymbol{x})\|$ yields the rate of change of the values of $\boldsymbol{u}$ in the direction of $\boldsymbol{v}$ and can be written as a positive semidefinite quadratic form of $\boldsymbol{v}$ by

$$\|\partial_{\boldsymbol{v}} \boldsymbol{u}(\boldsymbol{x})\|^2 = (D\boldsymbol{u}(\boldsymbol{x})\boldsymbol{v})^\mathsf{T} D\boldsymbol{u}(\boldsymbol{x})\boldsymbol{v} = \boldsymbol{v}^\mathsf{T} \big(D\boldsymbol{u}(\boldsymbol{x})^\mathsf{T} D\boldsymbol{u}(\boldsymbol{x})\big)\boldsymbol{v}. \tag{11}$$

As a result, in each non-singular point $\boldsymbol{x}$ of the image domain there are two mutually orthogonal directions $\boldsymbol{v}_{1,2}(\boldsymbol{x})$ in which the greatest and least rates of change, respectively, are found; $\boldsymbol{v}_{1,2}$ are the eigenvectors of $\boldsymbol{J}(\boldsymbol{x}) := D\boldsymbol{u}(\boldsymbol{x})^\mathsf{T} D\boldsymbol{u}(\boldsymbol{x})$. (The matrix $\boldsymbol{J}(\boldsymbol{x})$ is also known as *structure tensor*.) The vector field $\boldsymbol{v}_1(\boldsymbol{x})$ denoting the directions of fastest change can be understood as surrogate of a gradient vector field of $\boldsymbol{u}$, and the vector field $\boldsymbol{v}_2(\boldsymbol{x})$ of directions of slowest change as surrogate of a level-line direction field; thus the pseudo-level lines are the integral curves of the vector field $\boldsymbol{v}_2(\boldsymbol{x})$. Note that in the continuous case, under regularity conditions, pseudo-level lines are closed curves, see [3,10,11].

The following Subsects. 3.2–3.4 correspond to the three steps of our overall algorithm for curvature-based smoothing of vector-valued images.

3.2 First Step: Construction of the Pseudo-level Line Tree

From here on we assume that the vector-valued image is an RGB colour image with the colour channels R, G, B. Unlike for grey-scale images, a unique pvalue can not be defined for vector-valued images. Nevertheless, we define the mean value of all IIPs as pvalue in order to push a pixel $\hat{p}$, that fulfills the criteria

$$\left(R_{\hat{p}} - R_{\text{nIIP}}\right)^2 + \left(G_{\hat{p}} - G_{\text{nIIP}}\right)^2 + \left(B_{\hat{p}} - B_{\text{nIIP}}\right)^2 < s^2, \tag{12}$$

$$\left(R_{\hat{p}} - R_{\text{nIEP}}\right)^2 + \left(G_{\hat{p}} - G_{\text{nIEP}}\right)^2 + \left(B_{\hat{p}} - B_{\text{nIEP}}\right)^2 < s^2, \tag{13}$$

where nIIP, nIEP represent the nearest IIP, IEP of the actual level line, and $s \in \mathbb{R}$ is a tolerance limit, to a pseudo-level line's pregion. Additionally, $\hat{p}$ must have a pixel $\hat{q_n}$ in its immediate neighbourhood with radius $r = 5$, that is already part of a pseudo-level line.

Pseudo-level Line Extraction. In order to build a discrete pseudo-level line for vector-valued images, we need strict criteria for a pixel r to either be the next IIP or IEP. To this end, we start from the 2×2 structure tensor

$$\boldsymbol{J} = \nabla R \, \nabla R^\mathsf{T} + \nabla G \, \nabla G^\mathsf{T} + \nabla B \, \nabla B^\mathsf{T} \tag{14}$$

where $\nabla c = (\partial_x c, \partial_y c)^\mathsf{T}$ for $c = R, G, B$ can be computed using central differences involving the four immediate neighbours of the dummy pixel. (In each step, only one of the three pixels p, q, r represents a dummy pixel, the others are integer.)

The spectral decomposition of the structure tensor $\boldsymbol{J} = \lambda_1 \boldsymbol{v}_1 \boldsymbol{v}_1^\mathsf{T} + \lambda_2 \boldsymbol{v}_2 \boldsymbol{v}_2^\mathsf{T}$ with the eigenvalues $\lambda_1 \geq \lambda_2 \geq 0$ and eigenvectors $\boldsymbol{v}_1 \perp \boldsymbol{v}_2$ yields the (pseudo-) gradient direction $\boldsymbol{v}_1$ and pseudo-level line direction $\boldsymbol{v}_2$. The projection matrix

$$\boldsymbol{Z} = \begin{pmatrix} \langle \nabla R, \boldsymbol{v}_1 \rangle \\ \langle \nabla G, \boldsymbol{v}_1 \rangle \\ \langle \nabla B, \boldsymbol{v}_1 \rangle \end{pmatrix}, \tag{15}$$

is a 3×1 matrix that projects a 3×1 Red-Green-Blue (RGB)-vector (the intensity of a pixel) onto the gradient in the colour space. Thus,

$$p(p) = p_p = \left\langle \boldsymbol{Z}, \left(R_p \;\; G_p \;\; B_p\right)^\mathsf{T} \right\rangle \tag{16}$$

is the projection of the pixel p onto this gradient; p_q and p_r are computed analogously. To give a certain criterion for r to be the next IIP or IEP, let $\hat{\alpha} \in [0, 1]$ be the division ratio that splits p_p and p_q into two parts, then

$$p_{\hat{\alpha}} = (1 - \hat{\alpha}) \cdot p_p + \hat{\alpha} \cdot p_q. \tag{17}$$

If $p_p > p_q$, the eigenvector $\boldsymbol{v}_1$ is replaced with $-\boldsymbol{v}_1$ such that $p_p \leq p_{\hat{\alpha}} \leq p_q$ is applicable. The pixel r can now certainly be chosen as next IIP if

$$p_r \leq p_{\hat{\alpha}}, \tag{18}$$

and as IEP otherwise.

For the first edgel after the start edgel, $\hat{\alpha}$ is initialised with 0.5. Every time a new edgel is added to the level line, $\hat{\alpha}$ is updated for the next step via

$$\hat{\alpha} = \frac{p_{\hat{\alpha}} - p_r}{p_q - p_r} \quad \text{if} \;\; r = \text{IIP}, \qquad \text{or} \qquad \hat{\alpha} = \frac{p_{\hat{\alpha}} - p_p}{p_r - p_p} \quad \text{if} \;\; r = \text{IEP}. \tag{19}$$

Crash Handling. Although in the continuous domain pseudo-level lines are closed curves, the discrete algorithm described so far can fail to yield a closed level line. This is essentially due to accumulated errors in the estimation of colour gradient projection matrices $\boldsymbol{Z}$ in the course of the computation. This means that the sequence of pseudo-level line edgels might not return to the start edgel exactly. If this is the case, the pseudo-level line LL crsahed into itself and must be modified. Crashes can both happen from the inside and the outside. An inside crash occurs if some IIP $\in LL$ is picked as a new IEP. Crashes from the outside are detected if some IEP $\in LL$ is selected as new IIP. Another possibility is, that a whole edgel $\hat{\beta}$ already $\in LL$ is found again. If the IIP($\hat{\beta}$) is found first, LL crashed into itself from the inside. If IEP($\hat{\beta}$) is reached first, the crash occurred from the outside.

In both cases, one edgel $\in LL$ needs to be changed: $(p, r) \leftrightarrow (r, q)$. Starting from this edgel the pseudo-level line is recomputed. In the case of an inside crash, candidate edgels $i \in \{1, \ldots, l-1\}$ – with $l = \text{length}(LL)$ – for a change are of the type

$$(p_i, r_i) \rightarrow (r_i, q_{i-1}), \tag{20}$$

forcing LL to make a turn to the outside. In the other case, LL is forced to make a turn to the inside, meaning admissible edgels are of the type

$$(r_i, q_i) \rightarrow (p_{i-1}, r_i). \tag{21}$$

In most cases, more than one edgel lends itself as a candidate. To implement a reliable rule to determine the best candidate, we assign alternative edgels as described in (20) and (21) with costs ψ that measure how expensive it is for LL to take the alternative direction. The more clearly the decision is for r to be IEP or IIP (18), the higher the costs for the edgel to take r *wrongly* as IIP (20) or IEP (21), respectively. Depending on the decision for r, we compute

$$\psi = \frac{p_{\hat{\alpha}} - p_r}{p_q - p_p} \quad \text{if} \;\; r = \text{IIP}, \qquad \text{or} \qquad \psi = \frac{p_r - p_{\hat{\alpha}}}{p_q - p_p} \quad \text{if} \;\; r = \text{IEP}. \tag{22}$$

When LL is approximated, from all potential candidates the one with the lowest costs is modified. Each level line modified has a handicap Ψ, initialised with the costs ψ_i of the first changed edgel e_i. Let the second crash occur from the same side with e_j with $i \neq j$, then Ψ is either set to ψ_j, if $j < i$, or ψ_j is added to Ψ ($\Psi = \psi_i + \psi_j$), if $j > i$. Note that Ψ is added to all costs $\psi_{i+1}, \ldots, \psi_{l-1}$ of the edgels $\in$ alternative path $(\mathrm{e}_{i+1}, \ldots, \mathrm{e}_{l-1})$ within each crash.

Intensity Handling. For vector-valued images the intensities must be carried along the pseudo-level lines for each node in order to reconstruct a clean image after having applied affine morphological scale space (AMSS) smoothing. Special attention is paid to IIPs shared by parents and children: In such a case $\text{RGB}(\text{IIP}_{\text{parent}})$ is replaced with $\text{RGB}(\text{IEP}_{\text{child}})$ in order to have a consistent treatment of which RGB values are associated to the inner and outer sides of both pseudo-level lines, respectively.

Similar modifications regarding coincidences between IIP/IEP pixels of siblings are under investigation but currently not part of our implementation.

3.3 Second Step: Smoothing

Smoothing the Pseudo-level Lines: AMSS. Affine morphological scale space works equally to scalar-valued images with respect to process of curve evolution itself. Additionally, for vector-valued images the curveIEP is evolved too.

Furthermore, the intensities of each subpixel carried along the pseudo-level line must be stored and, if necessary, modified correctly. This is only possible with huge expense, because the number of subpixels $\in$ c changes within the smoothing process. New points inserted to the curve get the arithmetic mean value, computed with the intensities of the immediate former and immediate next subpixel, associated.

Smoothing the Intensities Along Curves. The second step of denoising is to smooth the vectorial intensities along the curves. As these intensities affect the quality of the denoised image, a smooth colour gradient is desired. Therefore linear explicit diffusion is applied to all curves.

Denoting by v_i the RGB value of the pixel $p_i \in$ curve c with $i \in \{0, \dots, l\}$, $l = \text{length}(c)$, we smooth the discrete 1-D signal $(v_0, \dots, v_l)$ by linear diffusion [21, Chap. 1]. In doing so, we approximate the diffusion PDE $v_t = v_{xx}$ by the standard explicit finite-difference scheme

$$v_i^{k+1} = v_i^k + \tau(v_{i+1}^k - 2v_i^k + v_{i-1}^k). \tag{23}$$

for iteration numbers $k \geq 0$, starting with the given signal in step $k = 0$ and assuming a spatial step size of 1. This explicit scheme is stable for time step sizes $\tau \leq 1/2$. In the present paper, we run three iterations, amounting to a diffusion time $t = 1.5$.

3.4 Third Step: Reconstruction

Given the curvature image – shortened level lines printed equipped correct intensities – it is necessary to reconstruct, finally, a clean denoised image.

To this end, the intensities (RGB) of IIPs and IEPs from the curvature image are fixed. Using these as Dirichlet boundary conditions, intensities on the remaining pixels can be inpainted by linear diffusion [21, Chap. 1] which should in principle be computed until numerical convergence to a steady state. Using a standard explicit finite-difference scheme for 2D diffusion, one computes

$$u_{i,j}^k = \left(1 - 4\frac{\tau}{h^2}\right) u_{i,j}^{k-1} + \frac{\tau}{h^2} \left(u_{i-1,j}^{k-1} + u_{i,j-1}^{k-1} + u_{i,j+1}^{k-1} + u_{i+1,j}^{k-1}\right) \tag{24}$$

for all non-IIP/IEP pixels (i, j) and iterations $k = 1, 2, \dots$ until

$$|u_{i,j}^k - u_{i,j}^{k-1}| < \varepsilon \qquad \text{for all } i, j. \tag{25}$$

In (24), τ denotes the time step size and h the spatial grid step size of the image; assuming $h = 1$ the scheme is stable for $\tau \leq 1/4$.

The number of iterations until the stopping condition (25) is met can be reduced by a suitable initialisation; to this end, the pregion of each node nd_l with $l = 0, \dots, \text{number(nodes)} - 1$ can be prefilled line by line by colouring the pixels $u_{i,j}^0 \in \text{pregion}(nd_l)$ with the intensity of the last met $\text{IIP}(nd_l)$ in this line.

4 Experimental Demonstration

In this section we will illustrate curvature-based denoising of grey-scale and colour images with some example images. All methods were implemented entirely in C++ on the basis of the standard library, some components being adapted from the published implementation of [7].

Before we turn to show actual image smoothing examples, we discuss the visualisation of curvature maps.

Fig. 10. Viridis colour bar. (Color figure online)

Fig. 11. Rescaled Viridis colour bar. (Color figure online)

(a) Noisy source (b) SC2, E_6^1 (c) SC2, E_6^2 (d) SC2, $E_4 + E_8$

Fig. 12. The level lines in the source image *camera40.pgm* [256 × 256] (a) noisy with Gaussian noise ($\sigma = 40$) are extracted using different edgel types (cf. Fig. 2 in Sect. 2.1). (b)–(d) Denoised images after having applied AMSS smoothing with SC2. (b) 6-connectedness of type 1 (E_6^1), (c) 6-connectedness of type 2 (E_6^2) used. (b) shows stripes 45° in lower right, (c) 45° in upper right direction. The 4- (for dummy pixels) and 8-connectedness edgels (for integer pixels) – $E_4 + E_8$ – used in (d) remove the artifacts.

4.1 Curvature Maps and Visualisation

Curvature maps give a coloured information about the curvatures present in the denoised image. As not all pixels are part of a level line, we compute first by (4) the curvatures in those pixels that are IIPs of the shortened level lines. Fixing these as Dirichlet boundary conditions, we inpaint the curvature map to the remaining pixels by running linear diffusion until numerical convergence to a steady state is reached, analogous to Sect. 3.4.

For visualisation, the so obtained dense curvature field $(c_{i,j})$ with values in $[-1, 1]$ is coloured on a modification of the Viridis colour scale [20] (see also [8]) reaching from dark blue/purple, $(R, G, B) = (68.0, 1.0, 84.0)$, for $c_{i,j} = -1$, via blue/green, $(R, G, B) = (32.0, 146.0, 140.0)$, for $c_{i,j} = 0$, to yellow/orange, $(R, G, B) = (253.0, 231.0, 37.0)$, for $c_{i,j} = 1$.

Whereas the original Viridis colour scale, Fig. 10, ensures linear contrast between different values visualised, we use a modification in which instead of $c_{i,j}$ itself the quantity $\tanh(10\, c_{i,j})$ is coloured by the original Viridis scale, resulting in enhanced colour contrast for curvatures around zero as shown in Fig. 11.

(a) Source 402 × 302 (b) CurvImg. SC5 (c) Smoothed SC5 (d) CurvMap(b)

Fig. 13. Smoothing a colour image. (a) Image *flowers*. (b)–(d): Result of AMSS smoothing with SC5 applied to the 96 744 level lines. (Image source: https://cs.colby.edu/courses/S19/cs151-labs/labs/lab04/Flowers.png, accessed 2022-02-02. Author: Colby)

(a) Noisy source 502 × 207 (b) Smoothed SC2

Fig. 14. Denoising of a colour image with Gaussian noise. (a) Image *HoheMunde40* degraded by Gaussian noise ($\sigma = 40$). – (b) Denoised by applying AMSS smoothing with SC2 to the 206 842 level lines. (Image source: https://www.telfs.at/files/user_upload/915x375/wohnen-leben-hohe-munde-hausberg-telfs-02.jpg, accessed: 2022-02-02.)

4.2 Image Smoothing Experiments

First, we show an experiment on a grey-scale image, Fig. 12, to demonstrate the effect of the modified neighbourhood setting with dummy pixels as introduced in Sect. 2.1. Note that the denoising results in Fig. 12b and Fig. 12c are visibly biased to the respective diagonal directions of the chosen 6-neighbourhoods.

For vector-valued images, first the effect of AMSS is highlighted with Fig. 13. Further, in Fig. 14 we show a denoising result for a colour image with Gaussian noise. In Fig. 15 we demonstrate the effect for impulse noise.

With our non-optimised implementation, run times ranged from about 2 min (grey-scale experiment, Fig. 12) to about 10 min (noise-free colour experiment, Fig. 13); surprisingly, the noisy colour images in Fig. 14 and Fig. 15 were processed much faster than Fig. 13, probably due to the dominance of much shorter pseudo-level lines. At any rate, we expect that run times can be significantly reduced by future algorithmic optimisations.

5 Summary and Outlook

In this work, we have extended the curvature-based denoising algorithm for grey-scale images from [7] to vector-valued, such as RGB colour, images. In

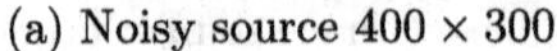

(a) Noisy source 400 × 300 (b) Smoothed SC1

Fig. 15. Denoising of a colour image with impulse noise. (a) Image *MTBIN20_cbc* degraded by synthetic impulse noise where 20% of all pixels are replaced with a random RGB value. – (b) Denoised by applying AMSS smoothing with SC1 to the 206 726 level lines.

the course of this extension, we have designed a robust extraction algorithm for pseudo-level lines. Due to the absence of a usable pvalue in vector-valued images, the intensities must be carried along each line in a properly manner. Special handling is required in certain configurations where nested discrete pseudo-level lines touch each other.

The smoothing process for level lines using AMSS could be transferred verbatim to pseudo-level lines. However, along with smoothing the pseudo-level lines their attached intensity information needs to be smoothed as well.

Finally, the reconstruction step required again an adaptation because for vector-valued images it is no longer sufficient to fill private regions with constant values. Diffusion inpainting was used to overcome this difficulty.

By experiments the viability of the approach was demonstrated. We presented processed colour images as well as exemplary curvature maps. Ongoing work is, on one hand, directed at algorithmic optimisations.

On the other hand, as already pointed out in [7], denoising is just one application of the curvature-based image processing paradigm underlying this work. The sub-pixel localised curvature information extracted in the course of the method bears potential for a range of further applications like image registration, segmentation, sharpening, or feature extraction for computer vision applications; note that points with extremal curvature such as corners are, in different formulations, long-established features in computer vision. With our curvature-based analysis method for vector-valued images the new paradigm can also be made available for colour images.

References

1. Alvarez, L., Morales, F.: Affine morphological multiscale analysis of corners and multiple junctions. Int. J. Comput. Vis. **25**(2) (1997)
2. Alvarez, L., Morel, J.M.: Formalization and computational aspects of image analysis. Acta Numer. **3**, 1–59 (1994). https://doi.org/10.1017/S0962492900002415

3. Ambrosio, L., Caselles, V., Masnou, S., Morel, J.M.: Connected components of sets of finite perimeter and applications to image processing. J. Eur. Math. Soc. **3**, 39–92 (2001). https://doi.org/10.1007/PL00011302
4. Attneave, F.: Some informational aspects of visual perception. Psychol. Rev. **61**(3), 183–193 (1954)
5. Bertalmío, M., Levine, S.: Denoising an image by denoising its curvature image. SIAM J. Imaging Sci. **7**(1), 187–211 (2014). https://doi.org/10.1137/120901246
6. Chung, D.H., Sapiro, G.: On the level lines and geometry of vector-valued images. IEEE Signal Process. Lett. **7**(9), 241–243 (2000). https://doi.org/10.1109/97.863143
7. Ciomaga, A., Monasse, P., Morel, J.M.: The image curvature microscope: accurate curvature computation at subpixel resolution. Image Process. On Line **7**, 197–217 (2017). https://doi.org/10.5201/ipol.2017.212
8. Crameri, F., Shephard, G.E., Heron, P.J.: The misuse of colour in science communication. Nat. Commun. **11** (2020). https://doi.org/10.1038/s41467-020-19160-7
9. Crandall, M.G., Lions, P.L.: Convergent difference schemes for nonlinear parabolic equations and mean curvature motion. Numer. Math. **75**(1), 17–41 (1996)
10. Cumani, A.: Edge detection in multispectral images. CVGIP: Graph. Models Image Process. **53**(1), 40–51 (1991). https://doi.org/10.1016/1049-9652(91)90018-F
11. Di Zenzo, S.: A note on the gradient of a multi-image. Comput. Vis. Graph. Image Process. **33**(1), 116–125 (1986). https://doi.org/10.1016/0734-189X(86)90223-9
12. Guggenheimer, H.W.: Differential Geometry. McGrawHill, New York (1963)
13. Guichard, F., Morel, J.M., Ryan, R.: Contrast invariant image analysis and PDE's. Technical report, Image Processing On Line (2004). http://dev.ipol.im/~morel/LivreGMR/MMBookOct04.ps
14. Kreyszig, E.: Differential Geometry. University of Toronto Press, Toronto (2019). https://doi.org/10.3138/9781487589455
15. Moisan, L.: Affine plane curve evolution: a fully consistent scheme. IEEE Trans. Image Process. **7**(3), 411–420 (1998). https://doi.org/10.1109/83.661191
16. Mondelli, M., Ciomaga, A.: Finite difference schemes for MCM and AMSS. Image Process. On Line **1**, 127–177 (2011). https://doi.org/10.5201/ipol.2011.cm_fds
17. Sapiro, G.: Geometric Partial Differential Equations and Image Analysis. Cambridge University Press, Cambridge (2001)
18. Sapiro, G., Tannenbaum, A.: On affine plane curve evolution. J. Funct. Anal. **119**, 79–120 (1994)
19. Song, Y.: A topdown algorithm for computation of level line trees. IEEE Trans. Image Process. **16**(8), 2107–2116 (2007). https://doi.org/10.1109/TIP.2007.899616
20. van der Walt, S., Smith, N.: MPL colour maps. Online Resource (2020). https://bids.github.io/colormap. Accessed 12 Mar 2022
21. Weickert, J.: Anisotropic Diffusion in Image Processing. Teubner, Stuttgart (1998)

Face Characterization Using Convex Surface Decomposition

Somrita Saha(✉) and Arindam Biswas

Department of Information Technolgy, Indian Institute of Engineering Science and Technology, Howrah, Shibpur, WB, India
somrita.besu@gmail.com, barindam@gmail.com

Abstract. In 2-dimensions, the analysis of face images has been done in a detailed way. The characterization of faces from 3D inputs may be interesting and discerning with respect to the uniqueness of a face. In this work, we have attempted to capture the surface curvature of a 3D face and thereof derive its characteristic features. The approach is based on the convex surface decomposition of the face models. Experimental results are encouraging and amenable to further treatise towards better realization of the characterization of a face.

Keywords: Face characterization · Convex decomposition · Surface decomposition · Face recognition

1 Introduction

The early research works in the area of human face characterization can be found almost half a century ago. Since then it has remained a very active field of research and there are a lot of innovative works going on even now. This field has an increasing demand in today's digital world where human face recognition is a very important aspect for information security, biometrics, smart cards, access control, law enforcement, and surveillance system. Human faces are unique, hence, most significant with respect to biometric traits.

Jafri et al. presented a survey on the face recognition techniques in 2009 in [4]. A 2D Gabor filter based face recognition system has been presented by Barbu in [1]. Kar et al. proposed an automatic attendance system based on face recognition [5]. Parmar et al. presented a detailed discussion on the existing works on face recognition in [6]. Another detailed discussion on the evolution of 2D face recognition techniques and their comparative study can be found in [2]. In 2018, Yang et al. came up with an emotion recognition model based on facial recognition in virtual learning environment [10]. Interest point based face recognition system using adaptive neuro fuzzy interface system was proposed by Rejeesh et al. [8]. A novel technique for spontaneous facial micro-expression recognition was proposed by Reddy et al. using 3D spatiotemporal convolutional neural networks in 2019 [7]. Watson et al. presented another method of data-driven face characterization of natural facial expressions when giving good and

R. P. Barneva et al. (Eds.): IWCIA 2022, LNCS 13348, pp. 288–300, 2023.
https://doi.org/10.1007/978-3-031-23612-9_18

bad news [9]. A 2D human face recognition technique was proposed by Gupta et al. in 2021 using SIFT and SURF descriptors of the feature regions of a face [3].

The methodology presented in this work is based on the decomposition of the human face model into a set of convex surface regions. The various parts of human face are oriented with different convexity and can be identified separately through convex surface decomposition. The convexity checking criterion is based on scalar triple product of vectors. The outputs produced by the algorithm reveal characteristic features of different faces. A comparative study of some related works are given in Table 1.

The rest of the paper is organized in the following manner. Section 2 contains the basic definitions and mathematical background of the presented work. In Sect. 3, the proposed method is discussed. Section 4 contains the algorithm and explanations of the steps of the algorithm. Section 5 contains the complexity analysis. In Sect. 6, the results are given. Finally, concluding remarks are given in Sect. 7.

The main contribution of this work is a simple method which exploits the mutual orientation of adjacent faces to derive convex regions present in a 3D face with which the face can be characterized. The runtime complexity of the algorithm corresponding to this method is $\mathcal{O}(n)$ where n is the total number of face triangles in the input face object.

2 Definitions and Preliminaries

Definition 1. *Convex Decomposition: It is the process by which a large and complicated digital object is decomposed into a number of convex subsets, union of which results in the original object.*

Exact Convex Decomposition (ECD) of a 3D input dataset generates a huge number of convex subsets which may not be useful for the further analysis of the object. Instead of ECD, some approximation can be adapted to produce lesser number of convex regions which are comparable with the visual impression of the original input object.

Figure 1 illustrates an example where the proposed convex surface decomposition algorithm has been applied to an input face model, given in Fig. 1(a), and the output produced is given in Fig. 1(b).

Definition 2. *Surface Decomposition: It is the process by which the surface of a 3D digital object is decomposed into a number of convex or concave subsets, union of which results in the surface of the original object.*

In this work, we have presented a method of face characterization using convex surface decomposition where the convexity checking criterion is governed by the volume of the tetrahedron formed by two edge-adjacent face triangles of the 3D face input dataset. This volume can be calculated using the scalar triple product or the mixed product of the vectors corresponding to the tetrahedron. The volume of the tetrahedron, described by the two edge-adjacent face triangles

Table 1. Related works and their features

Algorithm	Remarks
Gupta et al., 2021 [3] Approach: Speeded up robust features (SURF) and scale-invariant feature transform (SIFT)	An integration of feature extraction using SURF and SIFT, the algorithm has a high rate of recognition accuracy
Reddy et al., 2019 [7] Approach: Based on Convolutional Neural Network	Proposes two 3D-CNN methods: MicroExpSTCNN and MicroExpFuseNet, for spontaneous facial micro-expression recognition utilizing the spatiotemporal information
M. Rejeesh, 2019 [8] Approach: An Adaptive Genetic Algorithm and ANFIS-ABC based algorithm	Interest points in the face objects are determined using AGA and then classified using ANFIS
Yang et al., 2018 [10] Approach: Haar Cascades method to detect input image and Neural Network classifier training to detect different emotions	Method facilitates emotion recognition during distance education. Helps teacher to change teaching strategies as required
Kar et al., 2012 [5] Approach: Personal Component Analysis (PCA) algorithm	Implements an automatic attendance system which can be deployed in a classroom environment
Proposed method Approach: BFS Complexity: $\mathcal{O}(n)$	Convexity determined by the volume of the tetrahedron generated by each two edge-adjacent face triangles of the input face object with both time and space complexity of $\mathcal{O}(n)$, n being the number of faces in input. Decomposed face helps in detection of distinct features of the face

is negative when the faces are oriented in a convex manner w.r.t. each other. Conversely, if this volume is positive, then the corresponding face triangles are concavely oriented. A zero volume indicates that the face triangles are coplanar. Figure 2 demonstrates the cases of convex and concave surfaces. The volume, v, of the tetrahedron is the scalar triple products of the vectors $\boldsymbol{p}, \boldsymbol{q}$, and $\boldsymbol{r}$. This product is mathematically denoted by

$$v = [\boldsymbol{p}, \boldsymbol{q}, \boldsymbol{r}] = \boldsymbol{p} \cdot (\boldsymbol{q} \times \boldsymbol{r})$$

Specific Volume: Specific volume, here, volume per area, v_s, is denoted by the ratio of the volume, v, of the tetrahedron generated by two edge-adjacent face triangles and the sum of the areas, a, of these edge-adjacent face triangles. Here, in this work, we will consider the volume, v, as the volume of the tetrahedron

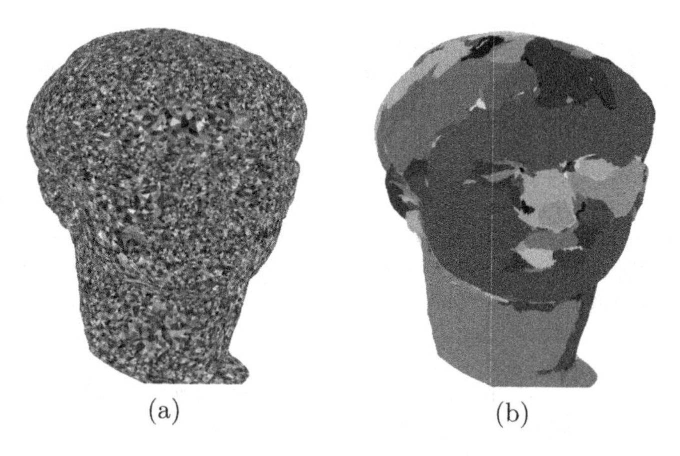

Fig. 1. (a) Original input object and (b) After the decomposition. (Color figure online)

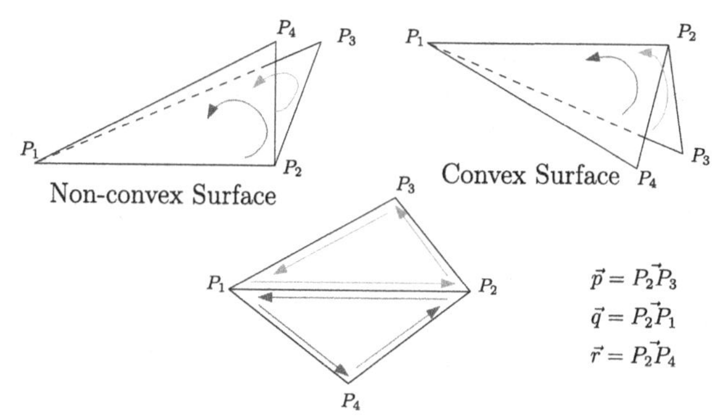

Fig. 2. Scalar triple product. (Color figure online)

formed by the two edge-adjacent triangles $\triangle P_1P_2P_3$ and $\triangle P_1P_2P_4$, and the area, a, as the sum of their areas a_1 and a_2, respectively.

Definition 3. *ρ-convexity: Two edge-adjacent face triangles are said to be ρ-convex with respect to each other, if the specific volume, v_s, of the tetrahedron formed by them is less than a predefined threshold, ρ.*

Let the area of the edge-adjacent triangles ($\triangle P_1P_2P_3$ and $\triangle P_1P_2P_4$, as in Fig. 2) are a_1 and a_2, respectively, then the condition for ρ-convexity is

$$\frac{v}{(a_1 + a_2)} \leqslant \rho \tag{1}$$

where ρ is a threshold on the specific volume of the tetrahedron formed by the corresponding triangles.

The convex surface decomposition method for face objects is implemented using Doubly Connected Edge List (DCEL) data structure. Apart from the original structure of DCEL, some additional information are maintained in the DCEL for this work. For the vertex object, additional information on whether it is a part of the boundary of any convex region, number of region boundaries it is part of, are stored. For the edge object, additional information on destination vertex, twin half edge, new previous half edge, new next half edge, whether the edge is dropped, and the convex region it is a part of, are maintained. For the face object, additional information normal, color code of the convex region it belongs to, area, and the base face corresponding to the convex region it is part of, are stored. Finally, a new object structure for the convex regions is added with some attributes, such as region *id*, base face, area, total number of faces under this region, color, outer region *id*, first edge on boundary, etc.

3 Face Characterization Method

Two face triangles are said to be convexly oriented if the volume of the tetrahedron generated by them is negative, otherwise oriented in a concave way. The proposed method can process watertight as well as not-watertight 3D face objects. By the term 'watertight', it is meant that the 3D input object does not have holes, cracks, or missing faces. A face has multiple surface regions with varying curvature and convexity. Our objective is to decompose a human face into some visually meaningful convex subsets based on the convexity checking criterion given in Eq. 1.

The face list of the 3D face object is sorted based on descending order of area. This sorting is performed as a preprocessing work before starting the actual method of decomposition. The largest area triangle is selected as the base face for the first convex surface region to be grown. The three edge-adjacent faces of the base face are taken into consideration, one by one, for checking the convexity. Each edge-adjacent face forms a tetrahedron with the base face. The volume, v, of this tetrahedron can be computed by the scalar triple product of the vectors defining the tetrahedron. Now, in our approach, the convexity criterion is defined such that if the specific volume, v_s, defined by the ratio of the volume, v, of the tetrahedron and the sum of the areas of the concerning faces, is less than a predefined threshold, ρ, then the tetrahedron is ρ-convex and the faces describing the tetrahedron are convexly oriented. If the edge-adjacent face satisfies the convexity criterion, then it is enqueued for further investigation of its edge-adjacent faces. Once, all the three edge-adjacent faces are checked, a face is dequeued from the queue and each of its edge-adjacent faces are checked for convexity in turn. This process continues till the queue is exhausted. Once the queue is exhausted, the first convex surface region of the face object is produced. For the next region to be grown, the next uncolored face is chosen as the base face from the sorted face list.

The above mentioned process continues till the last face in the sorted face list. The algorithm generates a set of convex surface regions, once all the faces of the sorted face list are checked for convexity. The convex regions generated by the algorithm vary widely with respect to their area. The regions with a very small area are merged with one of the adjacent convex regions. These small regions are insignificant with respect to the visual impression of the face, therefore, may not be considered as distinct convex surface regions.

4 Algorithm

The steps of the proposed algorithm are given in Algorithm 1. It takes a triangulated face object in the form of DCEL(D) and the threshold, ρ, for specific volume as input and produces modified DCEL(D) for the decomposed convex surface of the face object. In Steps (1–4), initialization tasks are performed. F is initialized to the already sorted face list in descending order of area (Step 1). In Step 2, base face, b, is initialized to the first face from F. Then, in Step 3, b is enqueued to a queue, Q and its color is initialized to *Current Color*, a color which will be uniquely assigned to all the faces under the current region to be grown. The base face, b, is added to the current ρ-convex set, S, of face triangles.

Then, in Steps (5–17), each face from the face list is checked for ρ-convexity until F is empty. In Steps (6–11), each edge-adjacent face, t, of b is investigated. If t is not colored and also ρ-convex with respect to b (Step 7), then it is enqueued to Q and also added to the set S (Step 8). Also, t is assigned the *Current Color* (Step 9). Else, in Step 10, if t already has the *Current Color*, then the shared half-edges of b and t are marked as dropped (Step 11). In Steps (12–16), if the Q is empty, it indicates that no more triangular faces can be added to this convex region and the algorithm produces the convex region for the current iteration. The ith convex surface region is formed with the set, S, of ρ-convex faces (Step 13). The *Current Color* is reset in the next step, Step 14. Then some reset operations are performed for the next region to be grown. S is set to NULL and i is incremented in Step 15. In Step 16, the next uncoloured face from F is identified and enqueued to Q and added to S. In Step 17, Q is dequeued to get the new face, b, to be checked.

When Steps (5–17) are executed till the last face of F, each of the significantly small convex surface regions is merged with one of its adjacent convex surface region which is bigger in size. Finally, the modified DCEL, (D), is returned as the output of the algorithm.

Algorithm 1. FACE-DECOMPOSITION

Require: Triangulated face object in the form of DCEL(D), ρ
Ensure: Modified DCEL (D) for the decomposed convex surface of the face

```
1: F ← Sorted face triangle list in descending order of area.
2: Base face, b ← first face from F, i ← 1
3: Enqueue b to a queue, Q. Set the color of b as the Current Color.
4: Add b to the current ρ-convex set, S, of face triangles.
5: while (F ≠ ∅) do
6:     for each edge-adjacent face t of b do
7:         if t is Not Colored and ρ-convex w.r.t. b then
8:             ENQUEUE(Q, t) and add t to S
9:             Assign t the Current Color.
10:        else if t has Current Color then
11:            Drop the shared half-edges of t and b.
12:    if Q = ∅ then
13:        CR_i (i^th convex region) constructed with the faces in S.
14:        Reset Current Color.
15:        S ← ∅, i ← i + 1
16:        Next uncolored face from F is identified, enqueued to Q, and added to S.
17:    b ← DEQUEUE(Q)
18: for each CR do
19:    if CR.area < ε then
20:        Merge CR with adjacent bigger region.
21: return D
```

5 Complexity

The Algorithm 1 investigates the convexity of all the n face triangles in the face list, F. The face list of the input dataset is sorted in descending order of area, as a part of preprocessing, which takes $\mathcal{O}(n \log n)$ time. Though, the time and space complexity of the main algorithm are bounded by $\mathcal{O}(n)$.

5.1 Time Complexity

Each face triangle of the original input object is investigated a maximum of three times as the triangle is edge-adjacent to only three other triangles. If the convexity criterion satisfies for a face triangle, it is enqueued and dequeued only once and is never investigated thereafter. In the loop Steps (5–17), the operations involved are mainly enqueue, dequeue, convexity checking, and some assignment operations which take constant amount of time. The merging operation in Step 20 is performed for a very small number of regions. The region boundary is traversed to identify an adjacent bigger region. Once identified, the smaller region is merged with the bigger one. As the region is small, its boundary is small as well. Moreover, boundary traversal stops the moment a bigger region is identified. The number of regions with smaller area, n_1, is much less compared to n, i.e., $n_1 \ll n$. Each edge of such a small region is considered only once. Hence, the complexity of merging smaller regions is well within $\mathcal{O}(n)$. Therefore, the overall time complexity of the algorithm is bounded by $\mathcal{O}(n)$, n being the number of face triangles in the original input object.

5.2 Space Complexity

The algorithm is based on DCEL data structure which has a space complexity of $\mathcal{O}(n)$. The original DCEL data structure is augmented by adding some more attributes for vertex, edge, and face objects. A new object structure for the convex regions is also added. All these augmentations are done keeping the overall space complexity same as the original data structure. Hence, the space complexity remains $\mathcal{O}(n)$.

6 Results

The proposed algorithm has been applied on some standard human face models represented by 3D digital object datasets. These inputs are *.obj* files which are processed by the programmatic implementation of the Algorithm 1 in C. The output files are plotted in Python to produce the final images. Figure 3, Fig. 4, Fig. 5, and Fig. 6 contain the outputs of four face objects for different convexity thresholds (ρ) for specific volume of tetrahedrons. Each figure represents four different levels of decomposition. The ρ values are chosen based on the overall convexity of the input object. The number of convex regions, $\#CR$, depends on the value of ρ.

The bottom left images in all the figures (namely, Fig. 3, 4, 5, and 6) depict the output for a ρ value which best justifies the original input data of the face model. Looking at the bottom left image of the aforementioned figures, it can be clearly seen that the different convex regions of a human face are being captured distinctly in the results provided. For example, from the output image of Augustus, in Fig. 5c, the forehead, cheeks, both the lips, and the nose can be discerned separately. Also, it reveals the curly texture of his hair. For the male and female heads, given in Fig. 3 and Fig. 4 respectively, the different parts of the faces, such as forehead, eyes, nose, lips, ears, cheeks, and chin, can be identified. The input datasets does not contain data on the hair texture. Therefore, the hair

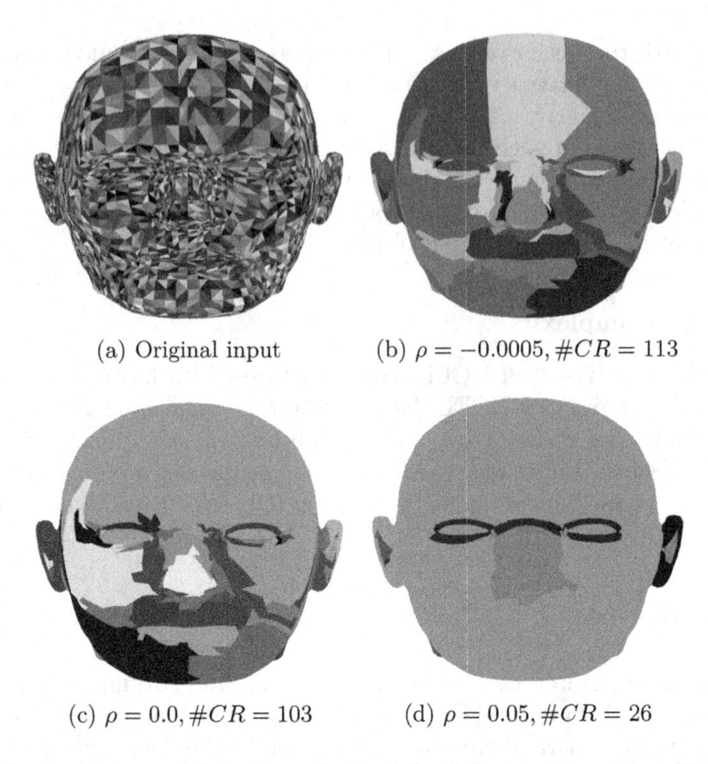

(a) Original input (b) $\rho = -0.0005, \#CR = 113$

(c) $\rho = 0.0, \#CR = 103$ (d) $\rho = 0.05, \#CR = 26$

Fig. 3. Convex decomposition of a male head for different values of ρ. (Color figure online)

texture is not discernible from the output images as well. Finally, the output image given in Fig. 6c reveals the hair texture and elevated distinct portions just above the two eyes along with other facial parts.

The algorithm is capable of identifying different parts of a human face model. Therefore, it can be extended for the reading of facial expression and comparative study of similar faces. Also, the boundary lines between two adjacent convex regions can be smoothened for a better visual quality of the output.

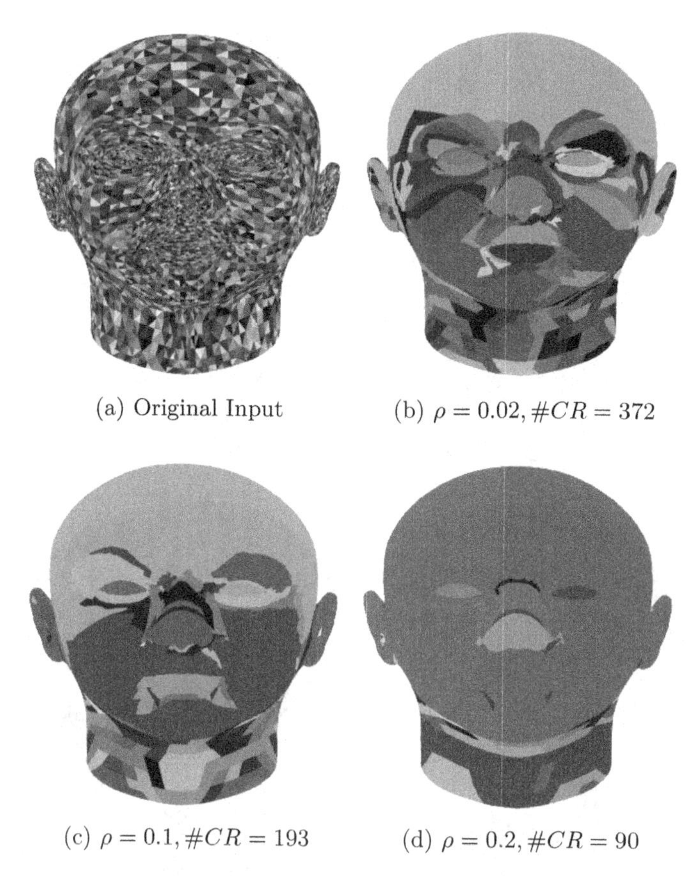

Fig. 4. Convex decomposition of a female head for different values of ρ. (Color figure online)

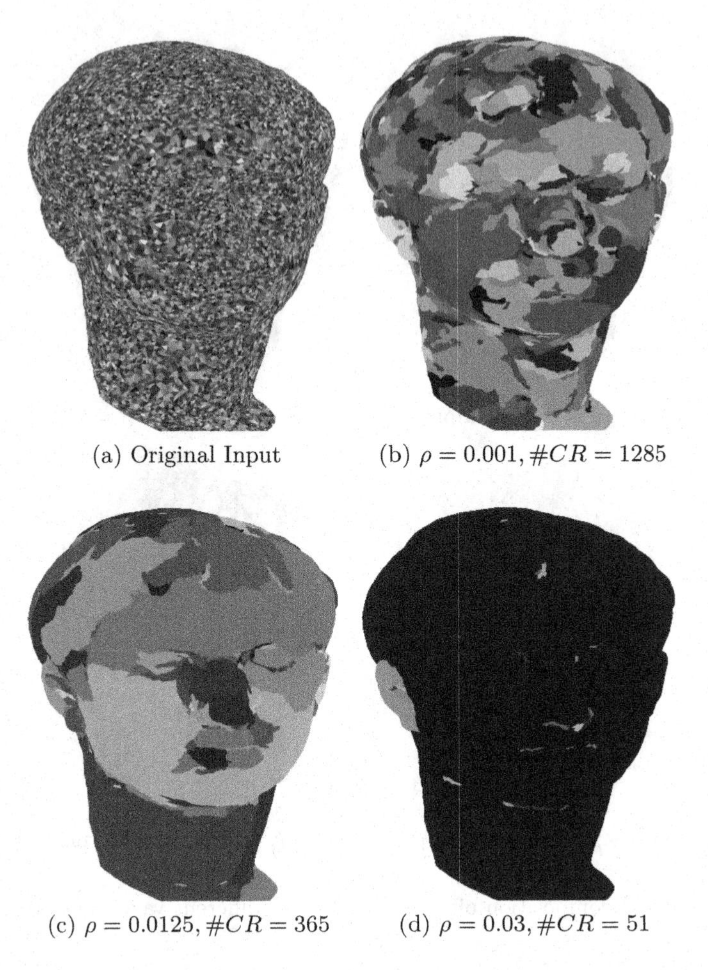

Fig. 5. Convex decomposition of augustus for different values of ρ. (Color figure online)

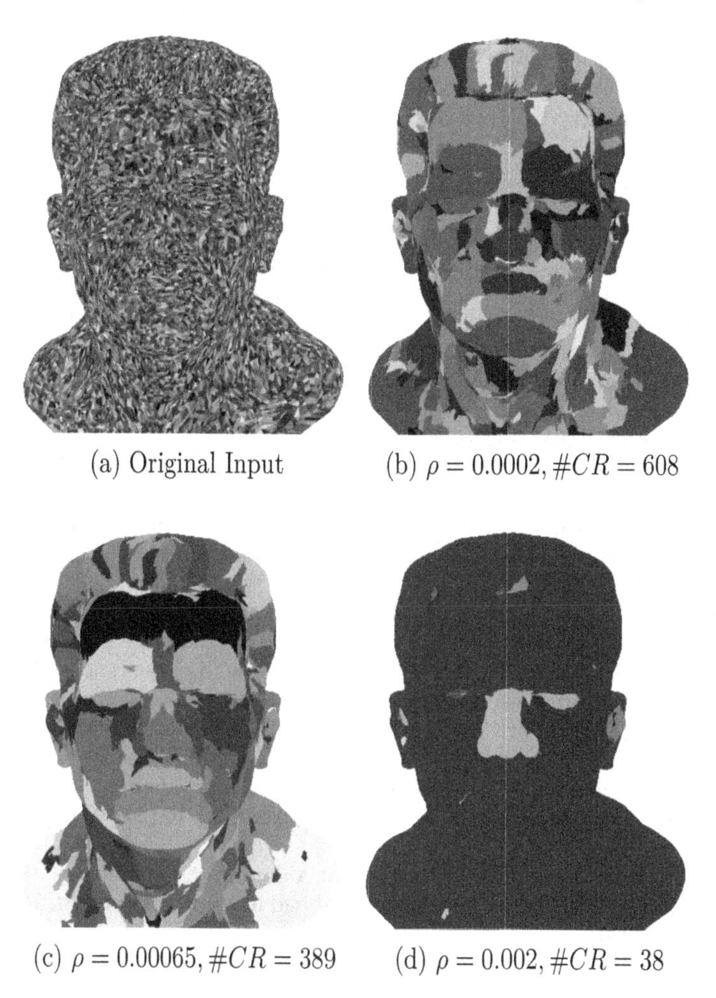

(a) Original Input (b) $\rho = 0.0002, \#CR = 608$

(c) $\rho = 0.00065, \#CR = 389$ (d) $\rho = 0.002, \#CR = 38$

Fig. 6. Convex decomposition of a male bust sculpture for different values of ρ. (Color figure online)

7 Conclusion

The method proposed in this work is suitable for the easy characterization of different parts of a human face in $\mathcal{O}(n)$ time. The algorithm has scope of practical utility considering the application of human face recognition in various fields such as biometrics, access control, smart cards, law enforcement, information security, and surveillance system. The method can be used in extracting the important convex regions which can be used as characterization of a face. Such characterization may be useful in recognition of human faces in a real time end-to-end application. Our approach is simple as it is based on an elementary vector mathematics, called scalar triple product of vectors. As of now, various parts of human face and some distinct features are identifiable through our method. We intend to improvise the work as to identify various facial expressions with clear and smooth boundary between adjacent regions.

References

1. Barbu, T.: Gabor filter-based face recognition technique. Proc. Rom. Acad. **11**(3), 277–283 (2010)
2. Chihaoui, M., Elkefi, A., Bellil, W., Ben Amar, C.: A survey of 2D face recognition techniques. Computers **5**(4), 21 (2016)
3. Gupta, S., Thakur, K., Kumar, M.: 2D-human face recognition using sift and surf descriptors of face's feature regions. Vis. Comput. **37**(3), 447–456 (2021). https://doi.org/10.1007/s00371-020-01814-8
4. Jafri, R., Arabnia, H.R.: A survey of face recognition techniques. J. Inf. Process. Syst. **5**(2), 41–68 (2009)
5. Kar, N., Debbarma, M.K., Saha, A., Pal, D.R.: Study of implementing automated attendance system using face recognition technique. Int. J. Comput. Commun. Eng. **1**(2), 100 (2012)
6. Parmar, D.N., Mehta, B.B.: Face recognition methods & applications. arXiv preprint arXiv:1403.0485 (2014)
7. Reddy, S.P.T., Karri, S.T., Dubey, S.R., Mukherjee, S.: Spontaneous facial micro-expression recognition using 3D spatiotemporal convolutional neural networks. In: 2019 International Joint Conference on Neural Networks (IJCNN), pp. 1–8. IEEE (2019)
8. Rejeesh, M.: Interest point based face recognition using adaptive neuro fuzzy inference system. Multimed. Tools Appl. **78**(16), 22691–22710 (2019). https://doi.org/10.1007/s11042-019-7577-5
9. Watson, D.M., Brown, B.B., Johnston, A.: A data-driven characterisation of natural facial expressions when giving good and bad news. PLoS Comput. Biol. **16**(10), e1008335 (2020)
10. Yang, D., Alsadoon, A., Prasad, P.C., Singh, A.K., Elchouemi, A.: An emotion recognition model based on facial recognition in virtual learning environment. Procedia Comput. Sci. **125**, 2–10 (2018)

Characterization and Reconstruction of Hypergraphic Pattern Sequences

Michela Ascolese(✉) and Andrea Frosini

University of Florence, Florence, Italy
{michela.ascolese,andrea.frosini}@unifi.it

Abstract. The notion of hypergraph has been introduced as a generalization of graphs so that each hyperedge is a subset of the set of vertices, without constraints on its cardinality. Our study focuses on 3-uniform hypergraphs, i.e., those hypergraphs whose (hyper)edges have three as common cardinality. A widely investigated problem related both to graphs and to hypergraphs concerns their characterization and reconstruction from their degree sequences. Concerning graphs, this problem has been efficiently solved in 1960 by Erdös and Gallai, while no efficient solutions are possible in the case of hypergraphs, even in the simple case of 3-uniform hypergraphs, as shown in 2018 by Deza et al. [4]. These problems are among the most studied in the field of Discrete Tomography (see [11,12] for a complete survey) and, in a more general fashion, of Image Analysis. So, to reduce the NP-hard core of the hypergraph reconstruction problem, we consider a class of degree sequences defined in [4] that show interesting properties. Here, in particular, we characterize the subclass $\mathcal{P}$ by using the new notion of pattern and pattern sequence. First, we focus on t-pattern sequences, i.e., sequences with constant pattern $t \geq 1$, and we study the remarkable behaviour of their last elements, called *tails*. In particular, for any fixed t, we show that the tails tend to a fixed point when increasing the sequences' lengths. The elements of these fixed points, on varying t, are the same and form the sequence $A002620$ in [13], generalizing the results in [6]. Finally, we provide a fast algorithm to reconstruct the hypergraphs that realize the sequences in $\mathcal{P}$ by iteratively discovering the elements of the characterizing pattern sequence.

Keywords: Discrete tomography · Reconstruction problem · Image analysis · Hypergraph · Degree sequences

AMS classification: 05C60 · 05C65 · 05C85 · 05C99

1 Introduction

Since their introduction, one of the most challenging problems related to (simple) hypergraphs has been their characterization and reconstruction starting from the degree sequences' knowledge. Formally, we define the following problems: Given a non-increasing sequence of positive integers π

R. P. Barneva et al. (Eds.): IWCIA 2022, LNCS 13348, pp. 301–316, 2023.
https://doi.org/10.1007/978-3-031-23612-9_19

Consistency: does there exist a simple hypergraph having π as degree sequence?

Reconstruction: is it possible to reconstruct in polynomial time a hypergraph whose degree sequence is π?

In case of graphs, in 1960, Erdös and Gallai completely characterized the degree sequences related to simple graphs [5], and later many algorithms were defined to provide a polynomial time solution for their reconstruction too (see [9,10]). Moving to hypergraphs, the studies mainly focused on k-*uniform* hypergraphs, that can be seen as a first, simple generalization of the concept of graph. In literature, we can find the characterization [2,3,8] and the efficient reconstruction [7] of their degree sequences in case of classes of specific instances.

Due to the fact that each k-uniform hypergraph can be represented by its incidence matrix, so that each row sums to k, and the i-th column sums to the degree of the i-th vertex, we observe that those problems admit a general formulation in the field of Discrete Tomography. In fact, it is required there to characterize the couples of row and column sums that admit at least one binary matrix realizing them (*Consistency problem* in [11]) and, in case of positive answer, to reconstruct one (*Reconstruction problem* in [11]).

Coming back to hypergraphs, both the problems, in their general form, remained open for many years. In 2018, Deza et al. [4] proved that the reconstruction problem is NP-hard even in the simplest case of 3-uniform hypergraphs, then it becomes relevant to define classes of hypergraphs that admit a polynomial time solution for their reconstruction, in order to limit the NP-complete core of the problem.

The NP-completeness proof in [4] relies on the definition, starting from a generic integer sequence s, of a gadget 3-hypergraph where the instances of the chosen NP-complete problem set down. Relying on this, in [6], the authors defined a class of 3-hypergraphs whose degree sequences show very interesting properties. Our studies here, deepen and generalize these properties to a wider class $\mathcal{P}$ of instances. Then, we define a polynomial time reconstruction strategy that, provided a degree sequence π computed from s, produces the related 3-uniform hypergraph in $\mathcal{P}$ shrinking the NP-hard core of the general problem. In particular, our strategy proceeds in recovering the mutual differences between the positive elements of s, obtaining a new and yet unexplored way of facing the reconstruction of the whole class $\mathcal{D}$ (defined in [1]).

2 Definitions and Previous Results

In this section we recall basic notions concerning hypergraphs and fix the notation we are going to use. Hypergraphs generalize the notion of graphs, as an edge can join any number of vertices. Formally, a hypergraph H is a pair of sets (V, E) where $V = \{v_1, \ldots, v_n\}$ is the set of vertices and $E = \{e_1, \ldots, e_m\} \subset \mathcal{P}(V) \backslash \emptyset$ is the set of hyperedges, briefly edges, with $\mathcal{P}(V)$ the power set of V. A hypergraph is *simple* if it contains neither singleton nor repeated edges, and it is called k-uniform, or simply k-hypergraph, if each edge has the same cardinality k. It is clear that if $k = 2$ we get back the definition of graph. One of the standard representations of a hypergraph is through its *incidence matrix*, that is a $m \times n$ binary matrix in which each column corresponds to a vertex and each row to an edge, such that the element in position (i, j) is set to 1 if and only if the edge e_i contains the vertex v_j. We can observe that if H is simple we do not have repeated rows in its incidence matrix, and that considering the rows' sum we get the edges' cardinalities; trivially, when H is k-uniform the sum of each row is k. In the sequel, by H we will equally refer both to a hypergraph and to its incidence matrix.

The *degree* of a vertex v is the number of edges that contain v, and the *degree sequence* of a hypergraph is the list of the degrees of all its vertices, usually arranged in non-increasing order. If we consider the columns' sum of its incidence matrix, we get the degree sequence of the hypergraph. One of the main problems related to hypergraphs is the so called *reconstruction problem*, that is the reconstruction of a hypergraph, if it exists, having a given degree sequence. In [4] the authors proved the NP-hardness of this problem even in the simple case of 3-uniform hypergraphs. In Sect. 3 and 4, we define a class of 3-uniform degree sequences that show interesting properties, and that reveal to be useful to perform the related reconstruction problem in polynomial time.

More precisely, in their proof, Deza et al. reduced an instance of the NP-complete problem *3-partition* to an instance of the decision problem of the existence of a 3-uniform hypergraph having a given degree sequence. In an intermediate step of the proof, it has been defined a class of 3-uniform hypergraphs that show very strong properties, later generalized to the class $\mathcal{D}$ in [1]. So, let us briefly recall the definition of the class $\mathcal{D}$ that will be the focus of our study.

Starting from a non-increasing integer sequence $s = \big(s(1), \ldots, s(n)\big)$, with $n \geq 3$, we define the 3-uniform hypergraph $H = (V, E)$ as follows: the set of vertices is $V = \{v_1, \ldots, v_n\}$, and the set of edges is composed by the triplets (v_i, v_j, v_k) such that $s(i) + s(j) + s(k) > 0$. This definition allows the incidence matrix of H to be easily computed by (the pseudo-code of) Algorithm 1, *Gen*(s).

Algorithm 1. *Gen(s)*

```
  Input: A weakly-decreasing integer sequence s of length n
1 Initialize an empty matrix H;
2 for each 1 ≤ i < j < k ≤ n do
3 |   if s(i) + s(j) + s(k) > 0 then
4 |   |   append to the matrix H the row corresponding to the edge (v_i, v_j, v_k);
5 |   end
6 end
  Output: H
```

The degree sequences of all the 3-hypergraphs thus obtained, on varying of n, form the class $\mathcal{D}$. One of the main properties of the degree sequences in $\mathcal{D}$ is their *uniqueness*, i.e., if $\pi \in \mathcal{D}$ then there exists one only 3-hypergraph (up to isomorphism) having π as degree sequence (see [1]).

Example 1. *The degree sequence* $\pi = (6, 4, 3, 2, 2, 1) \in \mathcal{D}$ *can be obtained from the integer sequence* $s = (3, 1, 0, -1, -2, -3)$. *Here is shown the incidence matrix of the related 3-uniform hypergraph.*

$$H = \begin{pmatrix} 1\,1\,1\,0\,0\,0 \\ 1\,1\,0\,1\,0\,0 \\ 1\,1\,0\,0\,1\,0 \\ 1\,1\,0\,0\,0\,1 \\ 1\,0\,1\,1\,0\,0 \\ 1\,0\,1\,0\,1\,0 \end{pmatrix}$$

We underline the following remarkable

Property 1. *Given* π *the degree sequence obtained from* $s = (s(1), \dots, s(n))$ *and* H *the related hypergraph, if* (v_i, v_j, v_k) *is an edge of* H, *then* $(v_i, v_j, v_{k'})$ *is an edge of* H *for all* $j + 1 \leq k' \leq k$.

This property is clear from the construction of H, since s is non-increasing by definition.

3 Definition and Properties of the Pattern Sequences

When the integer sequence s has a specific shape, some properties arise in the degree sequence π of the constructed hypergraph. An interesting example is studied in [6], where the authors considered the n-length sequences s such that $s(i+1) - s(i) = 1$ with $i = 1, \dots, n-1$, and they found a link between their number on varying n and some combinatorial structures enumerated by the same sequence (sequence $A002620$ in [13]).

Our aim is to generalize these sequences s by relaxing the differences between two consecutive, non-negative elements to a generic pattern, and keeping the value 1 only if $s(i) < 0$. So, we just define a *pattern* as an array $\mathbf{p} = (t_1, \dots, t_{n-1})$ of non-negative integer values that store the differences between two consecutive elements in the sequence s. In general, $s(i) - s(i+1) = t_i$. Each pattern can be regarded as the concatenation of two arrays, $\mathbf{p} = (\mathbf{p}^+, \mathbf{p}^-)$, where $\mathbf{p}^+$ is the pattern concerning the differences between the non-negative elements in s, while $\mathbf{p}^-$ refers to the negative part of the sequence. We underline that this decomposition is unique for each pattern $\mathbf{p}$ and strictly depends from the sequence s. In this paper, the pattern defined for the negative values of s will be fixed as $\mathbf{p}^- = (1, \dots, 1)$. With abuse of language, from now on, if not differently specified, with the term pattern we will refer only to the array $\mathbf{p}^+$.

Example 2. *The integer sequence* $s = (6, 3, 1, 1, 0, 0, -1, -2, -3, -4, -5, -6, -7, -8)$ *has pattern*

$$\mathbf{p} = (\underbrace{3,2,0,1,0}_{\mathbf{p}^+},\underbrace{1,1,1,1,1,1,1,1}_{\mathbf{p}^-}).$$

Our aim is to study the class of the *pattern degree sequences*, briefly *pattern sequences*, defined as the set of all the 3-degree sequences obtained on varying of the pattern $\mathbf{p}^+$. At first, we will consider a constant pattern with value t. We call the value t as the *step* of the sequence s.

Formally, for a given integer $t \geq 2$ we iteratively define

$$\begin{cases} s^{1,t} = (t, 0, -1, -2, \ldots, -(t-1)), \\ s^{i,t} = (it, (i-1)t, \ldots, 0, -1, -2, \ldots, -[2(i-1)t-1]) \quad \text{for all } i \geq 2 \end{cases} \tag{1}$$

and denote $\pi^{i,t}$ the degree sequence of the 3-hypergraph $H_i = Gen(s^{i,t})$, whose length is $n_i = (2t+1)i - t$ (see Proposition 1). In general, the sequence $s^{i,t}$ is iteratively obtained from $s^{i-1,t}$ by adding a first and $2t$ last elements according to the pattern rules previous described. We observe that for $t = 1$ and $i \geq 2$, we get the sequences defined and studied in [6]. We underline that all the sequences $s^{i,t}$ are such that $s^{i,t}(1) + s^{i,t}(2) + s^{i,t}(n_i) = 1$, with $s^{i,t}(n_i)$ its last element. This choice is such that $s^{i,t}(n_i)$ is the minimum allowed value to avoid singleton vertices, that is $\pi^{i,t}_{n_i} \neq 0$.

Lemma 1. *Given a step $t \geq 2$ and the integer sequence $s^{i,t}$, we have that $s^{i,t}(i+1) = 0$ for all $i \geq 1$. Moreover, it holds that*

$$s^{i,t}(1) + s^{i,t}(k) + s^{i,t}(i + t(2i - k + 1)) = 1$$

for $k = 2, \ldots, i+1$.

Proof. It directly follows from the construction of the sequences $s^{i,t}$. Indeed, we have that $s^{i,t}(1) = it$ and $s^{i,t}(k) = it - (k-1)t$ for all $k = 2, \ldots, i+1$. The value $s^{i,t}(j) = -[(2i-k+1)t - 1]$ is required in the equality, and, by construction, it lies in position $j = i + 1 + (2i - k + 1)t - 1$. □

Corollary 1. *Keeping the notation of Lemma 1, it holds*

$$s^{i,t}(1) + s^{i,t}(k) + s^{i,t}(j) \leq 0 \quad \text{for all } j \geq i + 1 + t(2i - k + 1)$$

for $k = 2, \ldots, i+1$.

It directly follows by the non-increasing of the sequence $s^{i,t}$.

For a fixed value of t, we define $\mathcal{P}_t \subset \mathcal{D}$ the class of the *t-pattern sequences* as the set of all the degree sequences $\pi^{i,t}$, with $i \geq 1$. Since t is fixed, from now on, to simplify the notation, we omit it when its value is clear from the context.

Example 3. *If $t = 2$, the first integer sequences are*

$$\begin{aligned} s^1 &= (2, 0, -1) \\ s^2 &= (4, 2, 0, -1, -2, -3, -4, -5) \\ s^3 &= (6, 4, 2, 0, -1, -2, -3, -4, -5, -6, -7, -8, -9) \\ s^4 &= (8, 6, 4, 2, 0, -1, -2, -3, -4, -5, -6, -7, -8, -9, -10, -11, -12, -13). \end{aligned}$$

The corresponding 2-pattern sequences are

$$\begin{aligned}
\pi^1 &= (1, 1, 1) \\
\pi^2 &= (10, 7, 5, 4, 3, 2, 1, 1) \\
\pi^3 &= (28, 21, 16, 12, 10, 8, 7, 5, 4, 2, 2, 1, 1) \\
\pi^4 &= (55, 44, 35, 28, 22, 19, 16, 14, 12, 10, 7, 6, 4, 4, 2, 2, 1, 1).
\end{aligned}$$

Proposition 1. *Given $t \geq 2$, the length of the integer sequence s^i is $n_i = (2t+1)i - t$ for all $i \geq 1$.*

Proof. By induction on the index i. For $i = 1$, the sequence s^1 has t as first entry and then all the negative integers from 0 to $-(t-1)$, that is $n_1 = t+1$. The generic sequence s^i is obtained by adding $2t+1$ elements to the previous one, s^{i-1}, that is $n_i = n_{i-1} + 2t + 1$. It follows that $n_i = n_1 + (i-1) \cdot (2t+1) = (2t+1)i - t$, and so the thesis. □

If we iteratively generate the t-pattern sequences π^i, we observe an interesting property on the lower values of their degrees: in Example 3, the last entries of the array π^i are preserved both as values and positions, from right to left, in the last elements of π^{i+1}. We are going to prove that they generate a numerical sequence that generalizes the sequence indexed as *A*002620 in [13], defined and studied in [6]. At each iteration, new elements of this numerical sequence are revealed in the last part of the degree sequence. Formally: for a fixed step t and $i \geq 1$, we define the *tail* of π^i as the array composed by its $t(i-1)+1$ last entries, that is

$$T(i,t) = \left(\pi^i(n_i - t(i-1)), \ldots, \pi^i(n_i)\right),$$

being n_i the length of π^i. It turns out that the tails of the 2-pattern sequences in Example 3 are

$$\begin{aligned}
T(1) &= (1) \\
T(2) &= (2, 1, 1) \\
T(3) &= (4, 2, 2, 1, 1) \\
T(4) &= (6, 4, 4, 2, 2, 1, 1).
\end{aligned}$$

In order to define a reconstruction algorithm for degree sequences having fixed step, we investigate the behaviour of their tails on varying i.

Definition 1. *For all $i \geq 1$, we define e_{2i} as the sum of the first* even *numbers from 2 to $2i$ (included) and o_{2i-1} as the sum of the first* odd *numbers from 1 to $2i-1$ (included).*

In the sequel, we use the standard exponential notation a^t to indicate the sequence of t entries equal to a.

Theorem 1. *Let $\mathcal{P}_t$ be the class of the t-pattern sequences π^i iteratively generated for a fixed value $t \geq 2$, and let $T(i)$ be the tail of the i-th sequence. For all $i \geq 1$, we have that*

$$\begin{aligned}
T(i) &= \left(o_i, e_{i-1}^t, o_{i-2}^t, e_{i-3}^t, \ldots, 2^t, 1^t\right) \text{ if } i \text{ is odd,} \\
T(i) &= \left(e_i, o_{i-1}^t, e_{i-2}^t, o_{i-3}^t, \ldots, 2^t, 1^t\right) \text{ otherwise.}
\end{aligned}$$

Proof. By induction on i.

If $i = 1$, $T(1)$ has length 1. By definition $s(1) + s(2) + s(n_1) = 1$ holds, then $T(1) = 1 = o_1$.

If $i = 2$, $T(2)$ has length $t + 1$. By Lemma 1 and Property 1, we have that

$$\begin{aligned} &s^2(1) + s^2(2) + s^2(k) > 0 \text{ for } 3 \le k \le n_2, \\ &s^2(1) + s^2(3) + s^2(k) > 0 \text{ for } 4 \le k \le n_2 - t, \\ &s^2(1) + s^2(4) + s^2(k) \le 0 \text{ for } k \ge 5. \end{aligned}$$

It follows that the last $t + 1$ entries of π^2 are $(2, 1, \ldots, 1)$, i.e. $T(2) = (e_2, o_1^t)$. Analogously we get $T(3) = (4, 2, \ldots, 2, 1, \ldots, 1) = (o_3, e_2^t, o_1^t)$.

Let us now consider the generic iteration from i to $i + 1$. We remind that $T(i)$ and $T(i + 1)$ have lengths $t(i - 1) + 1$ and $ti + 1$ respectively. Two cases arise: if i is even, then considering only the hyperedges added to H_i to obtain H_{i+1}, by Lemma 1 it holds that the degrees of the last $ti + 1$ vertices are

$$D = (i + 1, i^t, (i - 1)^t, \ldots, 3^t, 2^t, 1^t).$$

Moreover, by the iterative construction of the integer sequences s^i and s^{i+1}, we have

$$T(i + 1) = \left(\widetilde{T}(i), 0^{2t}\right) + D,$$

where $\widetilde{T}(i)$ stands for the tail $T(i)$ up to its first t elements. We get rid of the first t elements of $T(i)$ since they overcame the required length of $T(i + 1)$. Furthermore, an easy check reveals that those first elements will be also changed in the next iteration $i + 2$ of π^i.

By induction hypothesis, we get

$$T(i{+}1) = \left(o_{i-1}{+}(i{+}1), e_{i-2}^t{+}(i)^t, o_{i-3}^t{+}(i{-}1)^t, \ldots, e_2^t{+}4^t, o_1^t{+}3^t, 0^t{+}2^t, 0^t{+}1^t\right)$$

that gives the thesis,

$$T(i + 1) = (o_{i+1}, e_i^t, o_{i-1}^t, \ldots, 2^t, 1^t).$$

When i is odd, the proof is similar. □

Corollary 2. *For a fixed step $t \ge 2$ and index $i \ge 1$, the tail related to the i-th t-pattern sequence is*

$$T(i) = (a_i, a_{i-1}^t, \ldots, a_2^t, a_1^t),$$

with $\{a_n\}_{n\ge1}$ the numerical sequence whose generic element is

$$a_n = \left\lceil \frac{n+1}{2} \right\rceil \cdot \left\lfloor \frac{n+1}{2} \right\rfloor \qquad \textit{for } n \ge 1.$$

Proof. If n is even, i.e., $n = 2k$, we get $a_n = k(k+1)$, that is the sum of the first k even numbers, e_{2k}. On the other hand if n is odd, i.e., $n = 2k+1$, we get $a_n = (k+1)^2$, that is the sum of the first $k+1$ odd numbers, o_{2k+1}. □

We again underline that $\{a_n\}_{n\geq 1}$ is the *A*002620 numerical sequence in [13].

Definition 2. *Given a generic* $t \geq 1$*, we define the sequence*

$$\boldsymbol{\pi} = (\ldots\ldots, a_{i+1}^t, a_i^t, \ldots\ldots, a_3^t, a_2^t, a_1^t)$$

as the fixed point *of the class* $\mathcal{P}_t$.

The sequence $\boldsymbol{\pi}$ can be read as the limit of π^i as i grows. We highlight that at each iteration, t new elements of the fixed point appear in the tail $T(i)$ of π^i. This is clear by Theorem 1.

The case $t = 1$

In this section we briefly point out a further property of the tails $T(i)$ of 1-pattern sequences already presented in [6]. Since by Eq. (1), s^1 gives the empty hypergraph, in this case we will consider the sequences s^i starting from $i = 2$.

By definition, the tail $T(i)$ of the i-th degree sequence has length i. Let us now consider the case of i even index, and observe $\pi^i(2i-1)$: it is the first element on the left of the tail of π^i (we remind that π^i has length $n_i = 3i - 1$). An easy check reveals that in case of $t = 1$ and i even, the equality of Lemma 1 also holds for $k = i + 2$, that is

$$s^i(1) + s^i(i+2) + s^i(2i-1) = 1.$$

Following the same argument used in the proof of Theorem 1, we can conclude that $\pi^i(2i-1) = o_{2i+1}$. On the other hand, if i is odd, the same equality of Lemma 1 can not be extended till $k = i + 2$. We finally conclude that

Proposition 2. *Let* $t = 1$ *and* $k \geq 1$*. The following statements hold:*

1. $T(2k+1)$ *reveals* one *new element of the fixed point* $\boldsymbol{\pi}$,
2. $(\pi^{2k}(4k-1), T(2k))$ *reveals* two *new elements of the fixed point* $\boldsymbol{\pi}$.

Proposition 2 is in accordance with the result obtained in [6]. The same result holds if we iteratively generate degree sequences starting from an integer sequence $s = (s(1), \ldots, s(n))$ such that $s(i+1) - s(i) = t$ for all $i = 1, \ldots, n$, with $t \geq 1$. Indeed, if we choose the constant pattern $\mathbf{p} = (t, \ldots, t)$, we get again the class of 1-pattern sequences $\mathcal{P}_1$, since s and $t \cdot s = (ts(1), \ldots, ts(n))$ generate by construction the same degree sequence π for all $t \geq 1$ (see [1]).

Example 4. *Let us consider the integer sequence* $s = (20, 15, 10, 5, 0, -5, -10, -15, -20, -25, -30)$*, whose pattern is* $\mathbf{p} = (5, 5, 5, 5, 5, 5, 5, 5, 5, 5)$*. The degree sequence of the corresponding* 3*-hypergraph is*

$$\pi = (25, 21, 18, 15, 12, 10, \mathbf{9}, \mathbf{6}, \mathbf{4}, \mathbf{2}, \mathbf{1}).$$

We observe that π corresponds to $\pi^{4,1}$, since $s = 5 \cdot (4, 3, 2, 1, 0, -1, -2, -3, -4, -5, -6)$.

Moreover, the array $(\pi(7), T(4))$ in boldface consists of the first five elements of the numerical sequence $\{a_n\}_{n \geq 1}$, according to Proposition 2.

4 Two Reconstruction Algorithms for Pattern Sequences

Due to the particular behaviour of their tails, the reconstruction problem is polynomially solvable for the class $\mathcal{P}_t$. As a matter of fact, if the smallest elements of a given degree sequence $\pi \in \mathcal{P}_t$ constitute the first elements (from right to left) of the fixed point $\boldsymbol{\pi}$, we can immediately compute the values of the step t and the i-th iteration of the integer sequence s that realizes it. Then, it is sufficient to construct $s^{i,t}$ and generate the corresponding hypergraph. The uniqueness property of the degree sequences in the class $\mathcal{P}_t$ assures that the correspondence between the degree sequence of the generated hypergraph and the input one holds if and only if $\pi \in \mathcal{P}_t$. The described procedure is provided in pseudo-code in Algorithm 2, *RecStep*.

Algorithm 2. *RecStep(π)*

Input: A weakly-decreasing integer sequence π of length n
1 Compute t the number of 1 elements in π;
2 set $i = 2$;
3 **while** $\pi(n - (i-1)t) = \cdots = \pi(n - (i-2)t + 1) = a_i$ **do**
4 | $i = i + 1$;
5 **end**
6 **if** $\pi(n - (i-1)t) = a_i$ **then**
7 | compute the integer sequence $s^{i,t}$;
8 | compute $\pi^{i,t}$ the degree sequence generated by $s^{i,t}$;
9 | **if** $\pi^{i,t} = \pi$ **then**
10 | | **return** success;
11 | **else**
12 | | **return** failure;
13 | **end**
14 **else**
15 | **return** failure;
16 **end**
Output: $s^{i,t}$

Example 5. *Let us consider the following degree sequences (for brevity sake, we omit the incidence matrices of the 3-hypergraphs that realize them):*

$\pi_1 =$ *(27,20,14,11,10,8,6,5,4,2,2,1,1): the entries $\pi_1(13), \ldots, \pi_1(9)$ equals $T(3)$, so they allow us to conjecture the inclusion of π_1 in $\mathcal{P}_2$.* RecStep(π_1) *computes $s^{3,2}$ and successively $\pi^{3,2}$. Since $\pi_1 \neq \pi^{3,2}$, then it holds that $\pi_1 \notin \mathcal{P}_2$.*

$\pi_2 = (28,21,16,12,10,8,7,5,4,2,2,1,1)$: *acting as in case of* π_1, *we note that* RecStep(π_2) *ends with success and returns* $s^{3,2}$. *A final run of* $Gen(s^{3,2})$ *generates in polynomial time the hypergraph having* π_2 *as degree sequence.*

In this section, we further generalize the concept of pattern sequences and introduce an algorithm that extends the classes $\mathcal{P}_t$ of instances that are polinomially reconstructable.

We define $\mathcal{P}$ as the class of the *pattern degree sequences*, weakening the hypothesis on the integer sequences s that generate them and allowing any possible pattern $\mathbf{p}^+$. Obviously, this class is such that $\mathcal{P}_t \subset \mathcal{P} \subset \mathcal{D}$, with $t \geq 1$.

Definition 3. *A degree sequence* $\pi \in \mathcal{D}$ *is in* $\mathcal{P}$ *if and only if the integer sequence* s *of length* n *that realizes it satisfies the following conditions:*

1. *there exists at least one index* i *s.t.* $s(i) = 0$, *with* $2 \leq i \leq n-1$;
2. $s(i) - s(i+1) = 1$ *for all* $z \leq i \leq n-1$, *where* $z = \max\{1 < z < n \mid s(z) = 0\}$ *denotes the position of the last entry equal to zero in* s;
3. $s(1) + s(2) + s(n) = 1$.

As one can argue from the previous definition, the integer sequences s preserve the same negative pattern $\mathbf{p}^- = (1, \ldots, 1)$ as the elements in $\mathcal{P}_t$. On the other hand, we do not impose any restriction on its positive part, allowing a generic pattern $\mathbf{p}^+$.

Example 6. *The integer sequence* $s = (9, 3, 1, 0, -1, -2, -3, -4, -5, -6, -7, -8$ $-9, -10, -11)$ *realizes the degree sequence* $\pi \in \mathcal{P}$, *with* $\pi = (43, 19, 15, 13, 11, 10, 8, 6, 6, 5, 4, 3, 2, 1, 1)$. *In this example, the distance between positive elements in* s *is not fixed, and the pattern is* $\mathbf{p}^+ = (6, 2, 1)$.

From Example 6, we can note that the characterization of the tail provided in Theorem 1 concerning the elements of $\mathcal{P}_t$ is lost. In words, this is ascribable to the lack of a constant gap between the non-negative elements of s in the sense that the iterative procedure to generate the tails at step $i+1$ does not allow sequences of the same values to pack together in $\widetilde{T}(i)$ and D.

Despite the fewer information directly provided by the degree sequence, the strong hypothesis on the pattern of the negative elements of s allow again to provide a polynomial time algorithm, say *Rec*, for the reconstruction of the hypergraphs related to pattern sequences. The pseudo-code of *Rec* is described in Algorithm 3, and its behavior sketched below.

Starting from a degree sequence π of length n, *Rec* computes the integer sequence s that generates the relative hypergraph. If $\pi \notin \mathcal{P}$, then the reconstruction fails.

Without loss of generality, we assume to have as prior knowledge both the index z i.e., the position of the last null entry in the array s, and the first value of the pattern, $t_1 = s(1) - s(2)$. These hypothesis are not restrictive, since we can suppose to start n parallel computations, one for each position of z, and then consider all the possible related values of t_1, however keeping the polynomiality of the process. Indeed, by the structure of s, the value t_1 may vary from 0 to $-\left\lfloor \frac{s(n)}{2} \right\rfloor$, with $s(n)$ that can assume the value $-n+2$ at most.

Rec is an iterative algorithm that progressively computes the elements of the sequence s, revealing its pattern from the degree sequence π. The negative entries, $s(z), \ldots, s(n)$, are still known by the knowledge of z and $\mathbf{p}^-$. The algorithm computes the missing values with the following iterative steps:

1. Let q be the number of 1 elements of π. It is easy to check that $q = s(2) - s(3)$, since if a vertex v_k of a hypergraph in $\mathcal{P}$ has degree 1, then the inequality $s(i) + s(j) + s(k) > 0$ is satisfied for one only couple of indices, that is necessarily $i = 1$ and $j = 2$ by the non-increasing assumption on the elements in s;
2. By the prior knowledge of t_1 and $q = t_2$, the algorithm uniquely computes $s(1), s(2)$ and $s(3)$;
3. *Rec* computes the indexes i, j, and k such that $s(i) + s(j) + s(k) > 0$, with the three indexes ranging on all the known values of s. Then, it inserts in the hypergraph the computed edges (v_i, v_j, v_k). The degree sequence π is then updated by subtracting the degree sequence of the added hyperedges;
4. The algorithm computes the number of vertices that reached degree equal to zero after the update of the sequence π. This value provides the difference $t_3 = s(3) - s(4)$ and $s(4)$ is determined;
5. The procedure described in points 3 and 4 is iteratively repeated until all the entries of s are detected.

The proof of Theorem 2 will provide the correctness of the reconstruction procedure.

Algorithm 3. *Rec*(π,z,t_1)

Input: A weakly-decreasing integer sequence π of length n
1 Initialize s a null array of length n;
2 set $[s(z), \ldots, s(n)] = [0, -1, \ldots, -(n-z)]$;
3 compute q the number of elements in π that are equal to 1;
4 set $s(2) = \frac{1-s(n)-t_1}{2}$, $s(1) = 1 - s(n) - s(2)$ and $s(3) = s(2) - q$;
5 **for** $i, j, k \in \{1, 2, 3, z, z+1, \ldots, n\}$ *and* $i < j < k$ **do**
6 | **if** $s(i) + s(j) + s(k) > 0$ **then**
7 | | $\pi(i) = \pi(i) - 1$;
8 | | $\pi(j) = \pi(j) - 1$;
9 | | $\pi(k) = \pi(k) - 1$;
10 | **end**
11 **end**
12 $\pi^4 = \pi$;
13 **for** $l = 4 : z - 1$ **do**
14 | compute w the number of elements in π^l that are equal to 0;
15 | set $t = w - q$; $s(l) = s(l-1) - t$; $q = q + t$; $\pi^{l+1} = \pi^l$;
16 | **for** $i, j, k \in \{1, \ldots, l, z, z+1, \ldots, n\}$ *and* $i < j < k$ **do**
17 | | **if** $s(i) + s(j) + s(k) > 0$ *and was not previously computed* **then**
18 | | | $\pi^{l+1}(i) = \pi^{l+1}(i) - 1$;
19 | | | $\pi^{l+1}(j) = \pi^{l+1}(j) - 1$;
20 | | | $\pi^{l+1}(k) = \pi^{l+1}(k) - 1$;
21 | | **end**
22 | **end**
23 **end**
24 compute π^* the degree sequence generated by s;
25 **if** $\pi^* = \pi$ **then**
26 | **return** success;
27 **else**
28 | **return** failure;
29 **end**
Output: s

Theorem 2. *Given an integer sequence π of length n, if $\pi \in \mathcal{P}$ then the algorithm* Rec(π, z, t_1) *reconstructs in polynomial time the integer sequence s that generates π.*

Proof. We proceed by induction, proving that at each step of the *For* loop on l (Algorithm 3, line 13) the algorithm correctly computes a new element of the sequence s.

First, we observe that the values $s(1)$, $s(2)$ and $s(3)$ are correctly computed. In particular, $s(1)$ and $s(2)$ are uniquely determined by the conditions $s(1) + s(2) + s(n) = 1$ and $s(1) - s(2) = t_1$. Furthermore, the number q of elements equal to 1 in the degree sequence corresponds to the difference between $s(2)$ and $s(3)$. Indeed, v_k has degree 1 if and only if the edge (v_1, v_2, v_k) is part of the hypergraph, and reaches degree at least 2 if and only if the edge (v_1, v_3, v_k) is in the hypergraph too. This holds since $s(1)+s(2)$ and $s(1)+s(3)$ are the greatest sums we can get with couples of elements in s. We can conclude that q is the number of vertices v_k such that $s(1)+s(2)+s(k)>0$ and $s(1)+s(3)+s(k)\leq 0$ hold at the same time, i.e. $q=t_2=s(2)-s(3)$.

As a matter of fact, the values $s(i)$ for $z\leq i\leq n$ are correctly individuated by the index z and the fixed pattern $\mathbf{p}^-$.

Basis, $l=4$.

After the insertion of all the possible edges (v_i, v_j, v_k), with i, j, and k different indexes in $\{1,2,3,z,\ldots n\}$, we update π to $\pi^4=(*,\ldots,*,0^{u_4},0^q)$, with $\pi^4(n-q-u_4)\neq 0$. Stars stand for non-null values. The following statements hold by construction:

i) The edge (v_1, v_4, v_{n-q-u_4}) is part of the hypergraph, since, considering the previous insertions we made, $s(1)+s(4)$ is now the greatest sum we can reach summing two elements of the sequence s;
ii) $\pi(n-q+1),\ldots,\pi(n)$ became null due to the insertion of edges of type (v_1, v_2, v_x);
iii) $\pi(n-q-u_4+1),\ldots,\pi(n-q)$ became null also due to the insertion of edges of type (v_1, v_3, v_x), since they had degree ≥ 2.

It follows that the value u_4 is exactly the number of vertices v_x such that (v_1, v_3, v_x) is an edge of the hypergraph but (v_1, v_4, v_x) is not, i.e. the difference between $s(1)+s(3)$ and $s(1)+s(4)$. It directly follows that $u_4=t_3=s(3)-s(4)$, and the new value $s(4)$ of the sequence is correctly computed by the algorithm.

Induction step, from l *to* $l+1$.

The values of the integer sequence s up to $s(l)$ are known.

After the insertion of the edges (v_i, v_j, v_k), with $i<j<k\in\{1,2,\ldots,l,z,\ldots,n\}$, we get the updated degree sequence

$$\pi^{l+1}=(*,\ldots,*,0^{u_{l+1}},0^{q_{l+1}}).$$

By induction hypothesis, $q_{l+1}=\sum_{i=2}^{l-1}s(i)-s(i+1)$, and the elements $\pi^{l+1}(n-u_{l+1}-q_{l+1}+1),\ldots,\pi^{l+1}(n-q_{l+1})$ became null due to the insertion of edges (v_1, v_l, v_k). On the other hand, $\pi(n-u_{l+1}-q_{l+1})\neq 0$ implies that $(v_1, v_{l+1}, v_{n-u_{l+1}-q_{l+1}})$ is an edge of the hypergraph, since $s(1)+s(l+1)$ is now the greatest sum we can obtain in s. It follows that u_{l+1} is the number of vertices v_k such that (v_1, v_l, v_k) is an edge of the hypergraph while (v_1, v_{l+1}, v_k) is not, i.e. $u_{l+1}=t_l=s(l)-s(l+1)$. Then the value $s(l+1)$ can be correctly computed.

At the end of the *For* loop in Algorithm 3, line 13, it is revealed the whole pattern $\mathbf{p}^+=(t_1,\ldots,t_{z-1})$ of the sequence, and all the elements of s are correctly detected. □

Corollary 3. *If* $\pi\in\mathcal{P}$, *then the (unique) 3-hypergraph having* π *as degree sequence can be reconstructed in polynomial time.*

Proof. The algorithm *Rec* allows to compute in polynomial time the integer sequence s that generates the hypergraph H related to π. Starting from s, the incidence matrix of H can be computed by *Gen* in polynomial time. □

To clarify the action of *Rec*, we provide the following detailed example

Example 7. *Let us consider* $\text{Rec}(\pi_{input}, z, t_1)$*, with* $z = 7$*,* $t_1 = 4$ *and*

$$\pi_{input} = (110, 76, 56, 50, 40, 36, 36, 31, 29, 24, 20, 17, 16, 12, 10, 9, 7, 4, 3, 3, 2, 1, 1, 1).$$

First, Rec *creates a* 24 *length sequence* s *and initializes its last* 18 *elements to* $[s(7), \ldots, s(24)] = [0, -1, \ldots, -17]$*. Then the value* $q = 3$ *is computed and* $s(1) = 11$*,* $s(2) = 7$*,* $s(3) = 4$ *are obtained from*

$$\begin{cases} s(1) + s(2) - 17 = 1 \\ s(1) - s(2) = 4 \\ s(2) - s(3) = 3. \end{cases}$$

The algorithm proceeds performing the insertion of the following edges:

(v_1, v_2, v_i) *for* $i = 3, \ldots, 24$,
(v_1, v_3, v_i) *for* $i = 4, \ldots, 21$*, since* $s(1) + s(3) + s(22) = 0$,
(v_2, v_3, v_i) *for* $i = 4, \ldots, 17$*, since* $s(2) + s(3) + s(18) = 0$,
(v_i, v_j, v_k) *such that* $i = 1, 2, 3$*,* $j = 7, \ldots, 23$*,* $k = j + 1, \ldots, 24$ *and* $s(i) + s(j) + s(k) > 0$,

getting the updated degree sequence

$$\pi^4 = (46, 34, 25, 50, 40, 36, 14, 12, 12, 10, 8, 7, 7, 5, 4, 4, 3, 2, 1, 1, \mathbf{0}, 0, 0, 0).$$

The value $t_3 = 1$ *is computed by subtracting* q *to the number of* 0 *elements in* π^4 *(in boldface), getting* $s(4) = 3$.

The insertions of the edges involving v_4 *and the vertices* v_i*, with* $i \in \{1, 2, 3, 7, 8, \ldots, 24\}$*, are performed, updating* π^4 *to the following*

$$\pi^5 = (30, 22, 16, 14, 40, 36, 9, 8, 8, 7, 5, 4, 4, 3, 2, 2, 2, 1, \mathbf{0}, \mathbf{0}, 0, 0, 0, 0).$$

The difference between $s(4)$ *and* $s(5)$ *is obtained by the new* 0 *elements (in boldface) that appear updating* π^4 *to* π^5*, i.e.,* $t_4 = 6 - 4$*. The entry* $s(5) = 1$ *is also updated.*

Acting similarly, the algorithm computes

$$\pi^6 = (15, 11, 8, 7, 5, 36, 5, 4, 4, 3, 2, 2, 2, 1, 1, 1, \mathbf{0}, 0, 0, 0, 0, 0, 0, 0)$$

and $t_5 = 1$*, that allows to get* $s(6) = 0$*. The procedure stops since all the unknown values of* s *have been detected. We can check that* Rec *terminates with success and that the output sequence* $s = (11, 7, 4, 3, 1, 0, 0, -1, -2, \ldots, -16, -17)$ *corresponds to the integer sequence that generates* π_{input}.

Finally, Gen(s) *constructs the incidence matrix of the* 3*-uniform hypergraph that realizes* π_{input}.

5 Conclusion and Open Problems

The present study focuses on the characterization and reconstruction of an interesting class of hypergraphic degree sequences, say $\mathcal{D}$, that held appeal after its introduction by Deza et al. in [4]. In particular, those sequences appear as a gadget in the NP-completeness proof of the characterization of 3-uniform hypergraphs by degree sequences, and immediately show interesting properties. To study how far we can push in their characterization, we investigate a subclass of $\mathcal{D}$ having a prior knowledge about part of their entries.

We rely on a previous study in [6] about Saind degree sequences and we extend those results to a wider class $\mathcal{P}$ by introducing the notion of pattern and pattern sequence. We show that the last entries of a t-pattern sequence, with t a constant integer, provide a fixed point on growing in length. Moreover, the sequence of the elements of those fixed points, when varying t, are similar and equal the sequence found in [6], where $t = 1$.

Furthermore, we provide an algorithm that performs in polynomial time the reconstruction of the elements of $\mathcal{P}$ (with generic pattern) by discovering the different elements of the pattern and, consequently, the hyperedges of the related 3-uniform hypergraph.

By using the notion of pattern, we provide a different perspective for the reconstruction of subclasses of uniform hypergraphs, one for all the class $\mathcal{D}$, with the aim of shrinking the hard-to-compute core of their reconstruction problem.

References

1. Ascolese, M., Frosini, A., Kocay, W.L., Tarsissi, L.: Properties of unique degree sequences of 3-uniform hypergraphs. In: Lindblad, J., Malmberg, F., Sladoje, N. (eds.) DGMM 2021. LNCS, vol. 12708, pp. 312–324. Springer, Cham (2021). https://doi.org/10.1007/978-3-030-76657-3_22
2. Behrens, S., et al.: New results on degree sequences of uniform hypergraphs. Electron. J. Comb. **20**(4), P14 (2013)
3. Brlek, S., Frosini, A.: A tomographical interpretation of a sufficient condition on h-graphical sequences. In: Normand, N., Guédon, J., Autrusseau, F. (eds.) DGCI 2016. LNCS, vol. 9647, pp. 95–104. Springer, Cham (2016). https://doi.org/10.1007/978-3-319-32360-2_7
4. Deza, A., Levin, A., Meesum, S.M., Onn, S.: Optimization over degree sequences. SIAM J. Discret. Math. **32**(3), 2067–2079 (2018)
5. Erdös, P., Gallai, T.: Graphs with prescribed degrees of vertices (in Hungarian). Mat. Lapok (N.S.) **11**, 264–274 (1960)
6. Frosini, A., Palma, G., Rinaldi, S.: Combinatorial properties of degree sequences of 3-uniform hypergraphs arising from Saind arrays. In: Anselmo, M., Della Vedova, G., Manea, F., Pauly, A. (eds.) CiE 2020. LNCS, vol. 12098, pp. 228–238. Springer, Cham (2020). https://doi.org/10.1007/978-3-030-51466-2_20
7. Frosini, A., Picouleau, C., Rinaldi, S.: On the degree sequences of uniform hypergraphs. In: Gonzalez-Diaz, R., Jimenez, M.-J., Medrano, B. (eds.) DGCI 2013. LNCS, vol. 7749, pp. 300–310. Springer, Heidelberg (2013). https://doi.org/10.1007/978-3-642-37067-0_26

8. Frosini, A., Picouleau, C., Rinaldi, S.: New sufficient conditions on the degree sequences of uniform hypergraphs. Theoret. Comput. Sci. **868**, 97–111 (2021)
9. Hakimi, S.L.: On realizability of a set of integers as degrees of the vertices of a linear graph. J. Soc. Ind. Appl. Math. **10**, 496–506 (1962)
10. Havel, V.: A remark on the existence of finite graphs (in Czech). Časopis pro pěstování matematiky **80**, 477–480 (1955)
11. Herman, G.T., Kuba, A.: Discrete Tomography: Foundations, Algorithms, and Applications. Birkhauser, Boston (1999)
12. Herman, G.T., Kuba, A.: Advances in Discrete Tomography and Its Applications. Birkhauser, Boston (2007)
13. The On-Line Encyclopedia of Integer Sequences. http://oeis.org

The Generalized Microscopic Image Reconstruction Problem for Hypergraphs

Niccolò Di Marco(✉) and Andrea Frosini

University of Florence, Florence, Italy
{niccolo.dimarco,andrea.frosini}@unifi.it

Abstract. In this paper we study a particular case of the *microscopic image reconstruction* problem, first introduced in [6,10] and then extended to undirected unweighted graphs in [2]. We consider a general hypergraph $H = (V, E)$ such that each node v has assigned a physical value l_v that we would determine. Since in many applications it may be difficult or almost impossible to directly extract these values, we study how to retrieve them starting from the set of probes $P_v = \sum_{w \in N(w)} l_w$, i.e. the sum of labels of v's neighbors. In particular, we prove that the values l_v can be found in polynomial time using linear algebra tools and that the problem can be shifted to undirected weighted graphs trough the concept of 2-section of a hypergraph. Finally, we provide some classes of hypergraphs whose 2-intersection graphs have a specific form (a line or a s-tree) and whose related reconstruction problem from the probes can be performed with the minimum number of zero or one surgical probe.

Keywords: Image analysis · Discrete tomography · Reconstruction problem · Hypergraph · Graph spectra

AMS Classification: 52C99 · 05C65 · 05C85 · 05C99

1 Introduction

In this paper we consider the problem of retrieving information and, at its best, reconstructing, a physical discrete object from quantitative measurements of the neighbours of each point. To each point x of the unknown object is assumed to be assigned a value l_x and our goal is to determine these values. As a matter of fact, in many applications it is not easy or even not possible, to obtain them trough a precise inspection (called *surgical probe*), since it may damage the structure or may alter these values. A common alternative proposes the use of aggregate measuring techniques, whereby measurements are taken over a larger area and the values at each point are subsequently extracted by computational methods.

This general problem is often referred to as the *Discrete Tomography Reconstruction* problem (DTR) (for a survey on the topic and the related problems see [7,8]). The *Microscopic Image Reconstruction* problem (MIR) has been introduced in [6,10] as a natural extension of DTR. In both problems the object is

R. P. Barneva et al. (Eds.): IWCIA 2022, LNCS 13348, pp. 317–331, 2023.
https://doi.org/10.1007/978-3-031-23612-9_20

represented by a subset $U \subseteq \mathbb{Z} \times \mathbb{Z}$ whose points have assigned non-negative integer values. In the DTR problem, the window of a probe is typically an entire row or column, and they are called probes. In contrast, in the MIR problem it is assumed that the microscope's scanning window is a subset of the plane (see [3,6] for some examples). In [2] the authors extend this framework considering a generalized setting where the inspected object is represented by an undirected unweighted connected graph $G = (V, E)$. In this case, the vector $\mathbf{l} \in \mathbb{R}^n$ is an assignment of values l_v to each node $v \in V$. Moreover, in this context, a probe centered in v captures the sum of its neighbours' labels. Since in general this problem may have multiple solutions, the authors of [2] studied the so-called *Minimum Surgical Probing* problem (MSP) in which we ask to find the vector $\mathbf{l}$ with as few as possible surgical probes, where a surgical probe is intended as the exact knowledge of a single label. In the same paper the authors show that the problem can be solved in polynomial time using linear algebra tools.

Here, we extend the MSP to hypergraphs, and we prove that it can be related to the same problem on weighted graphs. Moreover, the problem can be solved in polynomial time using tools similar to those used in [2]. Then, we consider some classes of hypergraphs whose 2-intersection graphs (as defined in [5]) have a specific form. In particular, we focus on line and s-trees 2-intersection graphs, and we investigate the MSP problem of the related hypergraphs. We prove that the values assigned to the hypergraphs' nodes can be retrieved (in polynomial time) by using zero or one surgical probes, and we show how to detect them.

2 Preliminaries and Previous Results

Recall that a *graph* G is composed by two set (V, E), where $V = \{v_1, \dots, v_n\}$ is called *set of nodes* and $E \subseteq V \times V$ is called *set of edges*. The definition of *hypergraph* is a generalization of that of graph obtained by relaxing the condition on the edges to $E \subseteq \mathcal{P}(V)$, where $\mathcal{P}(V)$ is the power set of V. From now on, we will indifferently indicate each node v_i with its integer index i. If it holds that every edge in E has the same cardinality k, we say that H is a k-uniform hypergraph, simply k-hypergraph.

Given a graph (or hypergraph) G, the vector $\mathbf{l} \in \mathbb{R}^n$ is an assignment of values l_v to each node v and we call it *label* vector of the nodes.

For a node v we define its probes as

$$P_v = \sum_{w \in V} F(v, w) l_w \tag{1}$$

where $F(v, w)$ is the *number of hyperedges that contain both v and w* (we suppose that $F(v, v) = 0$ for every node v). We stress that the probes of a node v considers each neighbour's label (neighbours indicates a node that share at least one hyperedge with v) with a coefficient counting its occurrences in different hyperedges (provided by $F(\cdot, \cdot)$). On the other hand, the choice of computing the probes of v without such coefficients allows the MSP problem to go back to the (underlying hypergraph neighbour's) graph case. We define the *neighborhood matrix* F of a hypergraph as the matrix whose generic entry $f_{i,j}$ equals $F(i, j)$.

Note that in [1,2] the authors consider v itself among its neighbours. However, in [1] they inspect the links between the two cases. These few definitions are enough to state two main questions:

Question 1. *Is it possible to compute vector* $\boldsymbol{l}$ *from the knowledge of* $\mathbf{P}$*, the vector of probes* P_v*, for each* $v \in V$*?*

So, from the knowledge of the whole set of probes of a given weighted (hyper)graph, we are asking if it is possible to find the exact label assigned to each node. Since, in general, it may exist several vectors $\mathbf{l}$ that would yield to the same probes $\mathbf{P}$, we study a generalization of this problem described in [2]. Let us define *surgical probe* at node v to be the knowledge of its label l_v:

Question 2. *What is the minimal number of surgical probes needed for a unique reconstruction of the label vector* $\boldsymbol{l}$ *from the knowledge of* $\mathbf{P}$*?*

We address this problem as the *Minimum Surgical Probing* problem (MSP). Given a hypergraph G and a vector $\mathbf{P}$, we aim at finding the vector $\mathbf{l}$ that generates $\mathbf{P}$, using as few surgical probes as possible.

In [2], the authors prove that MSP problem on graphs can be solved in polynomial time and present an efficient algorithm to perform the task. In this paper, we extend their study to hypergraphs.

So, our interest in Question 2 concerns the study of the solutions of the linear system:

$$F\mathbf{l} = \mathbf{P}. \tag{2}$$

Some linear algebra notions need to be recalled: given a matrix A, denote with $rank(A)$ its rank and with $\phi_\lambda(A)$ the geometric multiplicity of each $\lambda \in \Lambda(A)$, where Λ is the set of the eigenvalues of A.

We note that F is a non-negative $n \times n$ symmetric matrix, and therefore $\phi_0(F) = n - rank(F)$ is the kernel dimension of F.

Theorem 1. *Let us consider a hypergraph* H *and let* $\boldsymbol{P}$ *be its probes vector and* F *be its neighborhood matrix. It holds that*

1. *if* F *has full rank, then* $\boldsymbol{l}$ *can be found in polynomial time without surgical probes;*
2. *otherwise the minimum number of surgical probes needed to compute* $\boldsymbol{l}$ *is* $s = n - rank(F) = \phi_0(F)$.

The proof of Theorem 1 can be easily obtained from Theorem 2.2 in [2]. We stress that its proof only relies on the the symmetry of the matrix, without imposing any constraint on its coefficients.

Observation 1. *The minimum number of probes pointed out in Theorem 1 holds for a generic vector of labels* $\boldsymbol{l}$*, while it can be considered as a lower bound in case we require* $\boldsymbol{l}$ *to be an integer vector.*

Observation 2. *In the framework we are setting up, we implicitly assume that the probes vector $\mathbf{P}$ is obtained by the machinery scanning of a real object. This implies that system (2) has always at least one solution. However, from a mathematical point of view, it may happen that $rank(F) < |V|$, preventing the problem from having a solution, in general.*

3 The *MSP* Problem on Classes of Hypergraphs

Let $H = (V, E)$ be a hypergraph, we define the graph $G_H = (V_H, E_H)$ such that $V_H = V$ and $E_H = \{\{v, w\} :$ there exists $e \in E$ such that $\{v, w\} \subseteq e\}$. Furthermore, we define a *weight function* W_H on E_H such that $W_H(v, w) = F(v, w)$, being F the neighborhood matrix of H.

In words, starting from H, we compute the graph G_H by replacing each hyperedge $e \in E$ by a complete graph on the same set of nodes of e. The weight function $W(\cdot, \cdot)$ indicates, for each edge (v, w) of G_H, the number of hyperedges of H including the two nodes v and w (see Fig. 1).

We indicate G_H as the *weighted* 2*-section* of H. This notion has been introduced in [4] and later studied in [9], where the authors consider the reconstruction problem of a hypergraph starting from its weighted 2-section. It is worthwhile that the MSP problem on hypergraphs H can be equivalently shifted to the same problem on the weighted graph G_H, where the probes are computed using the edges' weights, so that F turns out to be the weighted adjacency matrix of G_H.

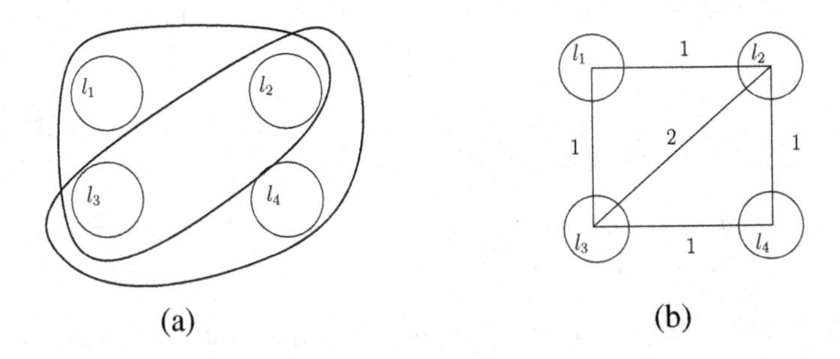

Fig. 1. The hypergraph $H = \{\{1, 2, 3\}\{2, 3, 4\}\}$ is shown in (a). (b) represents the related G_H weighted graph.

Example 1. *Consider the hypergraph H in Fig. 1 (a), whose hyperedges are $\{\{1, 2, 3\}\{2, 3, 4\}\}$. Its probes are $P_1 = l_2+l_3$, $P_2 = l_1+2l_3+l_4$, $P_3 = l_1+2l_2+l_4$, and $P_4 = l_2 + l_3$.*

Consider the 2-section graph G_H depicted in Fig. 1, (b). An easy check reveals that G_H satisfies the same probes $P_1 \ldots P_4$ as H.

Let us recall the definition of t-intersection graph.

Definition 1. *Let $H = (V, E)$ be a hypergraph. The t-intersection graph $I_t(H) = (V', E')$ of H is a graph such that $V' = E$ and $e' = \{e_1, e_2\} \in E'$ if and only if in H it happens that $|e_1 \cap e_2| \geq t$.*

So far, the intersection graph has been a valuable tool to inspect structural properties of uniform hypergraphs, such as the existence of a null labelling [5]. Similarly, in what follows, we use the intersection graph to solve the MSP problem for some relevant classes of hypergraphs.

Observation 3. *Let H be a hypergraph. If its 2-intersection graph $I_2(H)$ has no edges, then no two hyperedges of H share two nodes. So, it holds that the edges' weights of G_H have the same value 1, and the solution of the related MSP problem (on unweighted graph) has already been studied in [2].*

So, we focus on three classes of 3-hypergraphs whose 2-intersection graphs have specific, non trivial, properties of regularity.

3.1 3-Hypergraphs Whose 2-Intersection Graph is a Line

Let us consider a 3-hypergraph H whose $I_2(H)$ intersection graph is a line. Among them, we distinguish two types of hypergraphs: the *cluster hypergraphs* and the *path hypergraphs.*

We begin our study from the former.

The MSP problem on cluster hypergraphs

Definition 2. *A k-cluster hypergraph is a 3-hypergraph such that $|V| = k + 2, |E| = k$ and its hyperedges are defined, up to isomorphism, as*

$$\begin{cases} e_1 = \{1, 2, 3\} \\ e_i = \{1, i+1, i+2\} \quad i = 2 \ldots k. \end{cases} \tag{3}$$

It is worthwhile that the 2-intersection graph of a k-cluster is a line of length k. Figure 2 (a) depicts a 4-cluster hypergraph, and its four-length line 2-intersection graph is shown in (b).

Denote by F_k the neighborhood matrix of a k-cluster. Note that if $|E| = k$, then $|V| = k + 2$.

In Table 1, we explicitly compute the neighborhood matrices of the 2-cluster and 3-cluster hypergraphs $H_2 = \{\{1, 2, 3\}, \{1, 3, 4\}\}$ and $H_3 = \{\{1, 2, 3\}, \{1, 3, 4\} \{1, 4, 5\}\}$.

The elements that are in common are in boldface, according to the successive Lemma 4 that highlights the relation between the generic matrices F_{n-1} and F_n.

Lemma 1. *Let F_{n-1} and F_n be two neighborhood matrices of the $(n-1)$-cluster and n-cluster, respectively. The following recursive equations define F_{n+1}*

$$F_{n+1}(i, j) = F_n(i, j) \textit{ if } 1 \leq i \leq n+1 \textit{ and } 1 \leq j \leq n+1 \tag{4}$$

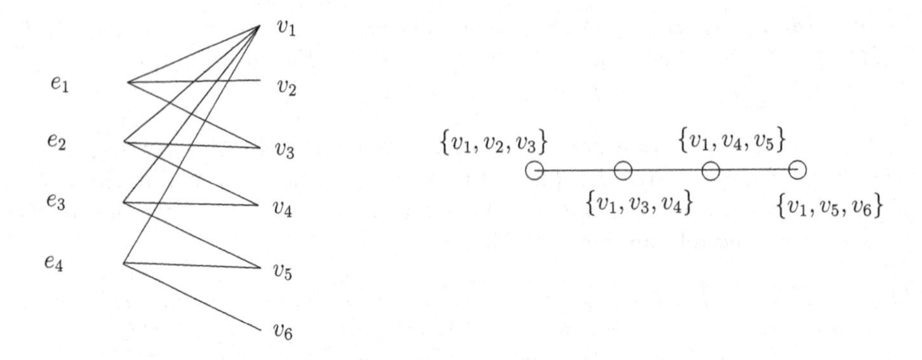

Fig. 2. A 4-cluster hypergraph (left) and its 2-intersection (right)

Table 1. The neighborhood matrices associated to the clusters of dimensions 2 and 3.

$$F_2 = \begin{pmatrix} \mathbf{0} & \mathbf{1} & \mathbf{2} & 1 \\ \mathbf{1} & \mathbf{0} & \mathbf{1} & 0 \\ \mathbf{2} & \mathbf{1} & \mathbf{0} & 1 \\ 1 & 0 & 1 & 0 \end{pmatrix} \qquad F_3 = \begin{pmatrix} \mathbf{0} & \mathbf{1} & \mathbf{2} & 2 & 1 \\ \mathbf{1} & \mathbf{0} & \mathbf{1} & 0 & 0 \\ \mathbf{2} & \mathbf{1} & \mathbf{0} & 1 & 0 \\ 2 & 0 & 1 & 0 & 1 \\ 1 & 0 & 0 & 1 & 0 \end{pmatrix}$$

$$F_{n+1}(i, n+2) = F_{n+1}(n+2, i) = \begin{cases} 2 \text{ if } i = 1 \\ 1 \text{ if } i = n+1 \\ 0 \text{ otherwise} \end{cases} \tag{5}$$

$$F_{n+1}(i, n+3) = F_{n+1}(n+3, i) = \begin{cases} 1 \text{ if } i = 1 \ \vee \ i = n+2 \\ 0 \text{ otherwise.} \end{cases} \tag{6}$$

Proof. Equation (4) states that the multiplicity of the edges $\{i, j\}$ of F_n is not affected by the addition of the new node.

Equation (5) states that adding node $n+3$ implies, by definition, the inclusion of the hyperedge $\{1, n+2, n+3\}$. Therefore, $\{1, n+2\}$ shares two hyperedges and $\{n+2, n+1\}$ shares one hyperedge.

Equation (6) set the values of the row and the column of the new node $n+3$. Since the new hyperedge is $\{1, n+2, n+3\}$ the equation holds. □

From Theorem 1, it follows that the MSP problem on cluster hypergraphs can be solved by studying the rank of the associated F_n matrices.

Theorem 2. *Let $n \geq 4$ and consider n-cluster hypergraph. If n is even, then the MSP problem can be solved with one surgical probes while, if n is odd, then no surgical probes are needed.*

Proof. Let us study $rank(F_{n-2})$ by inspecting its associated homogeneous system:

$$\begin{cases} x_2 + 2x_3 + \ldots + 2x_{n-1} + x_n = 0 \\ x_1 + x_3 = 0 \\ 2x_1 + x_{i-1} + x_{i+1} = 0 \text{ if } i = 3 \ldots n-1 \\ x_1 + x_{n-1} = 0. \end{cases}$$

From the second equation we get $x_3 = -x_1$, and from the last one, $x_{n-1} = -x_1$. By substituting the computed variables in the third equations, with i ranging from 3 to $n-1$, we obtain, for $i = 4$, the equation $x_5 = -x_1$ and, in general, the equation $x_i = -x_1$, with i odd. On the other hand, $x_{n-1} = -x_1$ implies $x_{n-3} = -x_1$ when $i = n-2$ and, in general $x_{(n-1-2k)} = -x_1$. We have two cases:

1. if n is odd we obtain each $x_i = -x_1$. Since the first equation is a sum of all the variables except x_1, we obtain $x_1 = 0$ and therefore $x_i = 0$. So F_{n-2} has maximum rank;
2. if n is even we obtain $x_i = -x_1$, with i odd, and then we have an additional equation that gives $0 = 0$ (i.e., x_1 is a free variable). Note that other equations involving even index i are of the form

$$x_{i-2} + x_i = -x_1$$

Therefore, assuming $i = 2k$, we obtain x_{2k} in terms of x_1 and x_2, so

$$x_{2k} = -x_1 - x_{2k-2} = \begin{cases} x_2 & \text{if } k \equiv_2 0 \\ -x_1 - x_2 & \text{if } k \equiv_2 1 \end{cases} \tag{7}$$

Since the first equation is only a sum of all the other variables except x_1, $rank(F_{n-2}) = n-1$. Therefore, from Theorem 1, we need one surgical probe to determine l. □

The MSP problem on path hypergraphs

Definition 3. *A n-path hypergraph C_n, with $n \geq 1$, is a 3-hypergraph H in which $|V| = n+2$, $|E| = n$ and whose hyperedges are*

$$\begin{cases} e_1 = \{1,2,3\} \\ e_i = \{i, i+1, i+2\} \quad i = 2, \ldots n \end{cases} \tag{8}$$

Also in this case, w.l.g. we suppose that 1 is the starting node. See Fig. 3 for an example.

Call F_n the neighborhood matrix of a n-path. The following lemma holds.

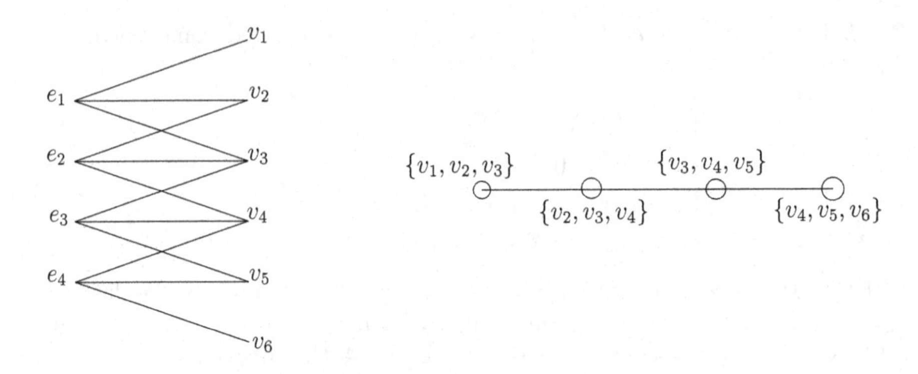

Fig. 3. (a) shows a 4-path hypergraph. (b) shows its 2-intersection graph.

Lemma 2. *Let H be C_n hypergraph. Then F_n is a symmetric pentadiagonal matrix of dimension $(n+2) \times (n+2)$ such that:*

$$\begin{cases} diag(F_n) = (0, \dots, 0) \in \mathbb{R}^{n+2} \\ diag_{\pm 1}(F_n) = (1, 2, 2, \dots, 2, 2, 1) \in \mathbb{R}^{n+1} \\ diag_{\pm 2}(F_n) = (1, \dots, 1) \in \mathbb{R}^{n} \end{cases} \tag{9}$$

where $diag_{\pm k}(F_n)$ indicates the k-th diagonals of F_n above and below the main diagonal.

Proof. The first equation immediately follows by definition of neighborhood matrix. Then, note that each node i different from 1 and $n+2$ is connected twice with $i+1$ and $i-1$. Moreover, nodes 1 and $n+2$ are connected once with node 2 by the hyperedge $\{1, 2, 3\}$, and with $n+2$ by the hyperedge $\{n, n+1, n+2\}$. Therefore the second equation holds. Finally, each node i is connected only once with $i+2$ and $i-2$, obtaining the third equation. $\square$

In Table 2, we explicitly compute the neighborhood matrices F_4 and F_5 of the 4-path and 5-path hypergraphs, respectively. Again the sub-matrices with equal elements are in boldface.

Table 2. The neighborhood matrices associated to the paths of dimensions 4 and 5.

$$F_2 = \begin{pmatrix} \mathbf{0} & \mathbf{1} & \mathbf{1} & 0 \\ \mathbf{1} & \mathbf{0} & \mathbf{2} & 1 \\ \mathbf{1} & \mathbf{2} & \mathbf{0} & 1 \\ 0 & 1 & 1 & 0 \end{pmatrix} \qquad F_3 = \begin{pmatrix} \mathbf{0} & \mathbf{1} & \mathbf{1} & 0 & 0 \\ \mathbf{1} & \mathbf{0} & \mathbf{2} & 1 & 0 \\ \mathbf{1} & \mathbf{2} & \mathbf{0} & 2 & 1 \\ 0 & 1 & 2 & 0 & 1 \\ 0 & 0 & 1 & 1 & 0 \end{pmatrix}$$

The following result holds.

Theorem 3. *Consider a n-path. The MSP problem can be solved without surgical probes if and only if $n \geq 3$. In particular, if $n = 2$, a surgical probe is needed.*

Proof. The case $n = 2$ is obtained after noticing that $rank(F_2) = 3$, so one surgical probe is needed to recover **l**. Consider a generic F_n matrix, with $n \geq 3$. By Theorem 1 the statement is proved if $rank(F_n) = n + 2$. Following the same strategy as in the proof of Theorem 3.1, let us inspect the homogeneous linear system associated with F_{n-2}:

$$\begin{cases} x_2 + x_3 = 0 \\ x_1 + 2x_3 + x_4 = 0 \\ x_{i-2} + 2x_{i-1} + 2x_{i+1} + x_{i+2} = 0 \quad i = 3 \ldots n-2 \\ x_{n-3} + 2x_{n-2} + x_n = 0 \\ x_{n-2} + x_{n-1} = 0 \end{cases} \tag{10}$$

It follows:

$$\begin{cases} x_3 = -x_2 \\ x_4 = -x_1 + 2x_2 \\ x_{i+2} = -x_{i-2} - 2x_{i-1} - 2x_{i+1} \quad i = 3, \ldots n-2 \\ x_n = -x_{n-3} - 2x_{n-2} \\ x_{n-2} = -x_{n-1} \end{cases} \tag{11}$$

The previous equations show that any variable can be expressed by a linear combination of x_1 and x_2. In the equation defining x_i, we indicate α_i the coefficient of x_1 and β_i the coefficient of x_2.

Furthermore, the following initial conditions hold:

$$\begin{cases} \alpha_1 = 1 \\ \alpha_2 = 0 \\ \alpha_3 = 0 \\ \alpha_4 = -1 \end{cases} \quad \begin{cases} \beta_1 = 0 \\ \beta_2 = 1 \\ \beta_3 = -1 \\ \beta_4 = 2 \end{cases} \tag{12}$$

The next equations provide the recursive description of α_i and β_i for a generic index i:

$$\begin{pmatrix} \alpha_i \\ \beta_i \end{pmatrix} = \begin{pmatrix} -\alpha_{i-4} - 2\alpha_{i-3} - 2\alpha_{i-1} \\ -\beta_{i-4} - 2\beta_{i-3} - 2\beta_{i-1} \end{pmatrix}. \tag{13}$$

Note that the two coefficients follow the same equation with different initial conditions. Let us consider α_n (we can act similarly when considering β_n); using basic tools of finite differences equations' theory we obtain that (13) has the following general solution:

$$\alpha_n = \sum_{k=1}^{4} c_k z_k^{n-1}, \; n = 1, 2, \ldots \tag{14}$$

where c_k is a constant determined by the initial conditions and z_k is a root of the characteristic polynomial $z^4+2z^3+2z+1=0$. Moreover, the solutions $\{z_k^n\}_{i=1}^4$ are linearly independent. In particular we have

$$z_1=\frac{1}{2}(-1-\sqrt{2}\sqrt[4]{3}-\sqrt{3}), \qquad z_2=\frac{1}{2}(-1+\sqrt{2}\sqrt[4]{3}-\sqrt{3}),$$

$$z_3=-\frac{1}{2}-\frac{i\sqrt[4]{3}}{\sqrt{2}}+\frac{\sqrt{3}}{2}, \qquad z_4=-\frac{1}{2}+\frac{i\sqrt[4]{3}}{\sqrt{2}}+\frac{\sqrt{3}}{2}.$$

Note that $|z_3|=|z_4|=1$, $|z_2|>1$ and $|z_1|<1$. So, (14) is the general expression of the α_i succession values and we use it in (11)). In particular from its last two equations we obtain:

$$\begin{cases} 0=x_{n-2}+x_{n-1}=x_1(\alpha_{n-2}+\alpha_{n-1})+x_2(\beta_{n-2}+\beta_{n-1}) \\ 0=x_{n-3}+2x_{n-2}+x_n=x_1(\alpha_{n-3}+4\alpha_{n-1}+\alpha_{n-4})+x_2(\beta_{n-3}+4\beta_{n-1}+\beta_{n-4}) \end{cases} \tag{15}$$

These equations have a unique solution $x_1=x_2=0$ since $\alpha_{n-2}+\alpha_{n-1}=\sum_{i=1}^4 c_i(1+z_i)z_i^{n-3}$, i.e. it is a linear combination of the solutions $\{z_i\}_{i=1\dots4}$. Since they are linearly independent, then it holds $\alpha_{n-2}+\alpha_{n-1}\neq 0$, with $n\geq 5$. The same holds for $\beta_{n-1}+\beta_{n-2}$.

Therefore, from the first equation of (15) we obtain $x_2=x_1\frac{(\alpha_{n-2}+\alpha_{n-1})}{(\beta_{n-2}+\beta_{n-1})}$. Substituting the value of x_2 in the second equation of (15), we finally obtain

$$x_1\underbrace{\left(\alpha_n+2\alpha_{n-2}+\alpha_{n-3}+\frac{\alpha_{n-1}+\alpha_{n-2}}{\beta_{n-1}+\beta_{n-2}}(\beta_n+2\beta_{n-2}+\beta_{n-3})\right)}_{=c_n}=0 \tag{16}$$

The proof ends if $c_n\neq 0$. Since $|z_2|$ is the only root greater than 1, then its exponential behavior overwhelms the other terms. In particular, the computation of the values of c_n that we provide up to $n=100$, shows that the condition is verified. As a consequence, the only solution of (16) is $x=0$ obtaining that $rank(F_n)=n+2$. □

As previous observed, both for cluster and path hypergraphs the 2-intersection graph is a line. Conversely, a hypergraph that has a line 2-intersection graph can contain both clusters and paths at the same time. In such cases, it becomes quite hard to find a general expression of the related neighborhood matrices.

3.2 3-Hypergraphs Whose 2-Intersection Graph is a Tree

Consider a 3-hypergraph T_s whose 2-intersection graph $I_2(T_s)$ is a perfectly height-balanced tree of height s and each node, but for the leaves, has degree

3. We indicate it as *s-tree*. Suppose that the edge e of T_s appears (as node) in $I_2(T_s)$ at level r. Then we say that e has *level* $p(e) = r$; if e is a leaf, then we call it *leaf hyperedge*, and it holds $p(e) = s$. The set of all leaf hyperedges is indicated with L.

Consider an internal node $e = \{x, y, z\}$ in $I_2(T_s)$; its children are $\{x, z, k_1\}, \{y, z, k_2\}$. Up to renaming of the nodes of T_s, we suppose that k_1 and k_2 appear only in the hyperedges that are nodes of the subtree of $I_2(T_s)$ having root e. So, at each level of $I_2(T_s)$, we have a sequence of nodes related to hyperedges of T_s of the form

$$\{x_1, z_1, k_1\}, \{y_1, z_1, k_2\}, \ldots, \{x_n, z_n, k_{2n-1}\}, \{y_n, z_n, k_{2n}\}.$$

An order can be set on them according to the $k_1, \ldots, k_{2n}$ numbering. Furthermore, at level s of the s-tree, the new nodes $k_1, \ldots k_{2n}$ of the leaf hyperedges are indicated as *leaf nodes.*

Observation 4. *Since each internal node* $e = \{x, y, z\}$ *of* $I_2(T_s)$ *has degree 3 and* T_s *is a perfectly height-balanced tree,* e *uses all the three couples* $\{x, y\}, \{y, z\}, \{x, z\}$ *to be connected to its neighbourhoods nodes. On the other hand, if* e *is a leaf node, then it is connected with one only couple to its neighbours.*

We assume w.l.g. that the root of $I_2(T_s)$ is the hyperedge $\{1, 2, 3\}$ of T_s. So, by definition, it follows that there exists a unique s-tree hypergraph for each $s \in \mathbb{N}$ (up to isomorphism). Figure 4 shows the s-tree T_2.

Lemma 3. *Let* $s \geq 2$. *The following statements hold:*

1. *given a s-tree hypergraph* T_s, *the corresponding* $I_2(T_s)$ *has* $m = 3 \cdot 2^{s-2}$ *leaves;*
2. *for each node* k *of* T_s, *if* k *is not a leaf node, then there exist exactly two paths in* $I_2(T_s)$ *that connect the first occurrence of* k *in a node to two corresponding leaves. Furthermore, all the nodes in this two paths contains the node* k*;*
3. *an s-tree hypergraph* T_s *has* $6 \cdot 2^{s-2}$ *nodes.*

Proof. 1. Let $s = 2$ and consider T_2. Without loss of generality we have

$$T_2 = \{\{1, 2, 3\}, \{1, 2, 4\}, \{2, 3, 5\}, \{1, 3, 6\}\}$$

(see Fig. 4). Note that it has $3 = 3 \cdot 2^{2-2}$ leaves. Consider a general T_{s+1} and suppose the thesis is true for T_s. Since every leaf of T_s generates two new leaves, T_{s+1} has $2(3 \cdot 2^{s-2}) = 3 \cdot 2^{s-1}$ leaves.

2. let z be a node of T_s and $e = \{x, y, z\}$ (one of its hyperedges), the first internal node of $I_2(T_s)$ in which z appears. By definition of s-tree its two children are $\{x, z, k_1\}$ and $\{y, z, k_2\}$. Note that only one child of each successive node will contain z in its relative edge until reaching the leaves.
3. We proceed by a simple induction. Basis: the T_2 hypergraph has $6 \cdot 2^{2-2} = 6$ nodes. Suppose that the thesis holds for T_s and let us compute the number of nodes of the successive s-tree T_{s+1}. When we add a new level, each leaf of the s-tree hypergraph generates two children and therefore two new nodes. Therefore it holds that the nodes of T_{s+1} double those of T_s, so $2(6 \cdot 2^{s-2}) = 6 \cdot 2^{s-1}$.

□

Example 2. *The T_2 s-tree of Fig. 4 has* $\{1,2,4\}$, $\{2,3,5\}$, $\{1,3,6\}$ *leaf hyperedges, and* 4, 5, *and* 6 *are the leaf nodes.*

Moreover, the neighborhood matrix of T_2 *is* F_2, *i.e., the following* 6×6 *square matrix*

$$F_2 = \begin{pmatrix} \mathbf{0} & \mathbf{2} & \mathbf{2} & \mathit{1} & \mathit{0} & \mathit{1} \\ \mathbf{2} & \mathbf{0} & \mathbf{2} & \mathit{1} & \mathit{1} & \mathit{0} \\ \mathbf{2} & \mathbf{2} & \mathbf{0} & \mathit{0} & \mathit{1} & \mathit{1} \\ \mathit{1} & \mathit{1} & \mathit{0} & 0 & 0 & 0 \\ \mathit{0} & \mathit{1} & \mathit{1} & 0 & 0 & 0 \\ \mathit{1} & \mathit{0} & \mathit{1} & 0 & 0 & 0 \end{pmatrix} = \begin{pmatrix} \mathbf{A_2} & B^t \\ B & \overline{0} \end{pmatrix} \tag{17}$$

Note that the neighborhood matrix F_2 of Example 3.2 can be decomposed into four matrices of dimension 3×3. An easy check reveals that the boldface entries refers to the matrix A_2 that contains the values of the internal nodes' links of $I_2(T_2)$. The italic entries refers to the matrices B and B^t that contains the connections between the internal nodes and the leaves of T_2. The fourth matrix is the zero matrix $\overline{0}$.

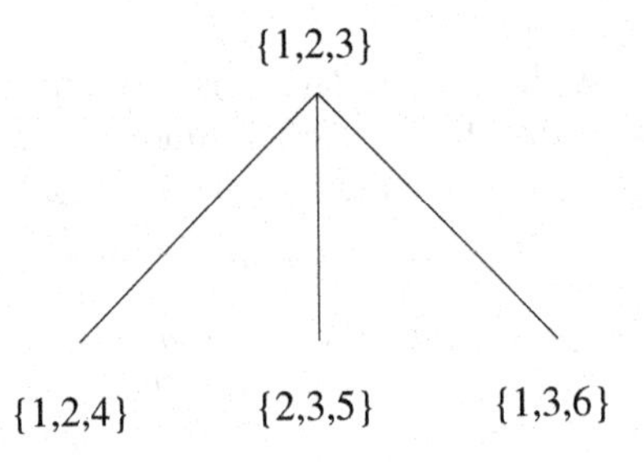

Fig. 4. The figure shows $I_2(T_2)$ hypergraph. We have $L = \{\{1,2,4\},\{2,3,5\},\{1,3,6\}\}$. Moreover nodes 4, 5, 6 are leaf nodes.

Let us consider the linear system associated to F_2. From the last three rows related to the matrix B we get

$$\begin{cases} x_1 = -x_2 \\ x_1 = x_3 \\ 2x_1 = 0. \end{cases} \tag{18}$$

Therefore we obtain $x_1 = x_2 = x_3 = 0$ and $rank(B) = 3$. Since $rank(B^t) = rank(B) = 3$ we obtain also $x_4 = x_5 = x_6 = 0$. Therefore F_2 has maximum rank and the MSP problem on T_2 can be solved without surgical probes.

The previous example shows a general property of the neighbourhood matrix of a T_s hypergraph. We underline that an immediate consequence of Lemma 3 is that the dimensions of F_s, for a generic t-tree hypergraph T_s, is always even.

Lemma 4. *Let T_s be a s-tree hypergraph with $s \geq 3$ and let $n = 6 \cdot 2^{s-2}$. Then F_s can be decomposed into four $\frac{n}{2} \times \frac{n}{2}$ square matrices:*

$$F_s = \begin{pmatrix} A_s & B^t \\ B & \overline{0} \end{pmatrix} \tag{19}$$

Moreover, the entries of A_s can be characterized as follows:

$$A_s(i,j) = \begin{cases} 2 & \text{if } F_{s-1}(i,j) \neq 0 \\ 0 & \text{otherwise.} \end{cases}$$

The matrix B (and its transpose B^t) contains exactly two non-zero elements in each row and in each column. In particular

$$B_{ij} = 1 \text{ if and only if there exists } k > \frac{n}{2} \text{ such that } (i,j,k) \text{ is an edge of } T_s.$$

Finally, $\overline{0}$ is the null matrix.

Proof. Let us focus on the matrix A_s: we note that Lemma 3 assures that A_s has the same dimension of F_{s-1}. By Observation 3.2, and since A_s contains the connections between the first $\frac{n}{2}$ (i.e. non-leaf) nodes, then it has the same non-zero elements of F_{s-1}. Furthermore, since $I_2(T_s)$ has one layer more than $I_2(T_{s-1})$, i.e., the last one, then each 1 entry of F_{s-1} changes in a 2 entry of A_s in the same position.

Concerning matrix B, its characterization follows from the definition of neighborhood matrix. Note that B represents the last $\frac{n}{2}$ (i.e. leaf) nodes of T_s. By Observation 3.2, every row of B has exactly two 1 entries and no two rows can be equal.

Moreover, Lemma 3 states that every non-leaf node appears in exactly two leaf hyperedges, therefore every column contains exactly two 1 entries and again all the columns are different.

Finally, the leaf nodes are not connected so the bottom-right part of matrix F_s is null. □

Lemma 5. *Let us consider the $2 \times \frac{n}{2}$ matrix π_B whose generic elements in positions $(1,i)$ and $(2,i)$, with $1 \leq i \leq \frac{n}{2}$, are the row indexes of the two only first and second non-zero entries in column i of B (as defined in Lemma 4. It holds that matrix π_B is a permutation forming one single cycle of maximal length $\frac{n}{2}$.*

The proof is a direct consequence of Point 2 in Lemma 3 and Lemma 4.

Theorem 4. *Let $s \geq 3$. The MSP problem on T_s can be solved with one surgical probe.*

Proof. Let F_s be the neighborhood matrix associated to T_s and decomposed according to Lemma 4. The linear system associate to the matrix B, considering the property states in Lemma 5, turns out to be, for each $j = 1 \ldots \frac{n}{2}$:

$$\begin{cases} x_j = -x_1 & \text{if } j \text{ is a leaf node of } T_{s-1} \\ x_j = x_1 & \text{otherwise.} \end{cases} \tag{20}$$

By Lemma 3, B has an even number of rows when $s \geq 3$, so the last equation of this system is the identity $0 = 0$, leading to $rank(B) = \frac{n}{2} - 1$. So, each variable $x_2, \ldots, x_j$ is expressed in terms of x_1 and it can be substituted in the variables of A_s obtaining $\mathbf{v}\, x_1$, with $\mathbf{v}$ being the integer column vector of length $\frac{n}{2}$ of the x_1 coefficients in A_s. The variable column $\mathbf{v}\, x_1$ is then concatenated with the part of the linear system related to B^t obtaining:

$$\left(\mathbf{v}\ B^t\right)\mathbf{x} = 0. \tag{21}$$

Note that, since A_s has the same form of F_{s-1}, we have $\mathbf{v} \neq \mathbf{0}$ (the rows of the leaf nodes of T_{s-1} do not sum to 0 since all the associated variables are equal to $-x_1$). With a similar procedure we obtain that $rank(F_s) = n - 1$. So, by Theorem 1 it follows that one only surgical probe is required to solve the MSP related problem. □

4 Conclusion and Future Perspectives

In this paper we study the microscopic image reconstruction problem, a generalization of the standard tomographic reconstruction problem, by extending the notion of projection to a generic-shape probe. In particular, we focus our attention on specific configuration of points that are arranged as hyperedges of a given labelled hypergraph and we consider, as projections of a node, the sums of the labels of its neighbors. These projections are indicated as probes.

We investigate the minimum surgical probing problem (MSP), i.e., the minimum number of labels of the hypergraph required to retrieve all the remaining ones from the probes' knowledge. This problem has been considered in [1,2] with respect to graphs. After generalizing their results to hypergraphs, we provide two classes of hypergraphs whose 2-intersection graph has a specific form (a line and s-tree) and such that they admit zero or one surgical probe to be fully recovered from probes. However, the problem for hypergraphs which admit 2-intersection that is a line still remains open, in its general formulation, since they reveal to be a non yet characterized mix of paths and clusters, as here defined. Therefore, it would be interesting to obtain a general result for the whole class and for super-classes admitting circular inclusions.

On the other hand, since the notion of 2-intersection graph has been useful in the detection of null label, as witnessed by [5], it would be worth studying MSP problem for 3-hypergraphs that admit one. Finally, the result of Theorem 1 obviously holds for k-uniform hypergraph, however it could be interesting to detect differences, if any, on the number of surgical probes needed to recover the uniform hypergraphs' labels according to their uniformity degrees.

References

1. Bar-Noy, A., Böhnlein, T., Lotker, Z., Peleg, D., Rawitz, D.: Weighted microscopic image reconstruction. In: Bureš, T., et al. (eds.) SOFSEM 2021. LNCS, vol. 12607, pp. 373–386. Springer, Cham (2021). https://doi.org/10.1007/978-3-030-67731-2_27
2. Bar-Noy, A., Böhnlein, T., Lotker, Z., Peleg, D., Rawitz, D.: The generalized microscopic image reconstruction problem. Leibniz Int. Proc. Inform. **149**, 42.1–42.15 (2019)
3. Battaglino, D., Frosini, A., Rinaldi, S.: A decomposition theorem for homogeneous sets with respect to diamond probes. Comput. Vis. Image Underst. **17**, 319–325 (2013)
4. Berge, C.: Hypergraphs. North-Holland, Amsterdam (1989)
5. Di Marco, N., Frosini, A., Kocay, W.L.: A study on the existence of null labelling for 3-hypergraphs. In: Flocchini, P., Moura, L. (eds.) IWOCA 2021. LNCS, vol. 12757, pp. 282–294. Springer, Cham (2021). https://doi.org/10.1007/978-3-030-79987-8_20
6. Frosini, A., Nivat, M.: Binary matrices under the microscope: a tomographical problem. Theoret. Comput. Sci. **370**(1–3), 201–217 (2007)
7. Herman, G.T., Kuba, A. (eds.): Discrete Tomography: Foundations Algorithms and Applications. Birkhauser, Boston (1999)
8. Herman, G.T., Kuba, A. (eds.): Advances in Discrete Tomography and Its Applications (Applied and Numerical Harmonic Analysis). Birkhauser, Boston (2007)
9. Janczewski, R., Obszarski, P., Turowski, K.: Weighted 2-sections and hypergraph reconstruction. Theoret. Comput. Sci. (2002, in press)
10. Nivat, M.: Sous-ensembles homogénes de Z2 et pavages du plan. C.R. Math. **335**(1), 83–86 (2002)

Author Index

Printed in the United States
by Baker & Taylor Publisher Services